Pitman Research Notes in Mathe[matics]
Series

Main Editors
H. Brezis, Université de Paris
R.G. Douglas, State University of New York at Stony Brook
A. Jeffrey, University of Newcastle upon Tyne *(Founding Editor)*

Editorial Board
R. Aris, University of Minnesota
A. Bensoussan, INRIA, France
S. Bloch, University of Chicago
B. Bollobás, University of Cambridge
W. Bürger, Universität Karlsruhe
S. Donaldson, University of Oxford
J. Douglas Jr, Purdue University
R.J. Elliott, University of Alberta
G. Fichera, Università di Roma
R.P. Gilbert, University of Delaware
R. Glowinski, Université de Paris
K.P. Hadeler, Universität Tübingen
K. Kirchgässner, Universität Stuttgart
B. Lawson, State University of New York at Stony Brook
W.F. Lucas, Claremont Graduate School
R.E. Meyer, University of Wisconsin-Madison
S. Mori, Nagoya University
L.E. Payne, Cornell University
G.F. Roach, University of Strathclyde
J.H. Seinfeld, California Institute of Technology
B. Simon, California Institute of Technology
S.J. Taylor, University of Virginia

Submission of proposals for consideration
Suggestions for publication, in the form of outlines and representative samples, are invited by the Editorial Board for assessment. Intending authors should approach one of the main editors or another member of the Editorial Board, citing the relevant AMS subject classifications. Alternatively, outlines may be sent directly to the publisher's offices. Refereeing is by members of the board and other mathematical authorities in the topic concerned, throughout the world.

Preparation of accepted manuscripts
On acceptance of a proposal, the publisher will supply full instructions for the preparation of manuscripts in a form suitable for direct photo-lithographic reproduction. Specially printed grid sheets can be provided and a contribution is offered by the publisher towards the cost of typing. Word processor output, subject to the publisher's approval, is also acceptable.

Illustrations should be prepared by the authors, ready for direct reproduction without further improvement. The use of hand-drawn symbols should be avoided wherever possible, in order to maintain maximum clarity of the text.

The publisher will be pleased to give any guidance necessary during the preparation of a typescript, and will be happy to answer any queries.

Important note
In order to avoid later retyping, intending authors are strongly urged not to begin final preparation of a typescript before receiving the publisher's guidelines. In this way it is hoped to preserve the uniform appearance of the series.

Longman Scientific & Technical
Longman House
Burnt Mill
Harlow, Essex, CM20 2JE
UK
(Telephone (0279) 426721)

Titles in this series. A full list is available on request from the publisher.

51 Subnormal operators
J B Conway
52 Wave propagation in viscoelastic media
F Mainardi
53 Nonlinear partial differential equations and their applications: Collège de France Seminar. Volume I
H Brezis and J L Lions
54 Geometry of Coxeter groups
H Hiller
55 Cusps of Gauss mappings
T Banchoff, T Gaffney and C McCrory
56 An approach to algebraic K-theory
A J Berrick
57 Convex analysis and optimization
J-P Aubin and R B Vintner
58 Convex analysis with applications in the differentiation of convex functions
J R Giles
59 Weak and variational methods for moving boundary problems
C M Elliott and J R Ockendon
60 Nonlinear partial differential equations and their applications: Collège de France Seminar. Volume II
H Brezis and J L Lions
61 Singular Systems of differential equations II
S L Campbell
62 Rates of convergence in the central limit theorem
Peter Hall
63 Solution of differential equations by means of one-parameter groups
J M Hill
64 Hankel operators on Hilbert Space
S C Power
65 Schrödinger-type operators with continuous spectra
M S P Eastham and H Kalf
66 Recent applications of generalized inverses
S L Campbell
67 Riesz and Fredholm theory in Banach algebra
B A Barnes, G J Murphy, M R F Smyth and T T West
68 Evolution equations and their applications
K Kappel and W Schappacher
69 Generalized solutions of Hamilton–Jacobi equations
P L Lions
70 Nonlinear partial differential equations and their applications: Collège de France Seminar. Volume III
H Brezis and J L Lions
71 Spectral theory and wave operators for the Schrödinger equation
A M Berthier
72 Approximation of Hilbert space operators I
D A Herrero
73 Vector valued Nevanlinna theory
H J W Ziegler
74 Instability, nonexistence and weighted energy methods in fluid dynamics and related theories
B Straughan
75 Local bifurcation and symmetry
A Vanderbauwhede
76 Clifford analysis
F Brackx, R Delanghe and F Sommen
77 Nonlinear equivalence, reduction of PDEs to ODEs and fast convergent numerical methods
E E Rosinger
78 Free boundary problems, theory and applications. Volume I
A Fasano and M Primicerio
79 Free boundary problems, theory and applications. Volume II
A Fasano and M Primicerio
80 Symplectic geometry
A Crumeyrolle and J Grifone
81 An algorithmic analysis of a communication model with retransmission of flawed messages
D M Lucantoni
82 Geometric games and their applications
W H Ruckle
83 Additive groups of rings
S Feigelstock
84 Nonlinear partial differential equations and their applications: Collège de France Seminar. Volume IV
H Brezis and J L Lions
85 Multiplicative functionals on topological algebras
T Husain
86 Hamilton–Jacobi equations in Hilbert spaces
V Barbu and G Da Prato
87 Harmonic maps with symmetry, harmonic morphisms and deformations of metric
P Baird
88 Similarity solutions of nonlinear partial differential equations
L Dresner
89 Contributions to nonlinear partial differential equations
C Bardos, A Damlamian, J I Díaz and J Hernández
90 Banach and Hilbert spaces of vector-valued functions
J Burbea and P Masani
91 Control and observation of neutral systems
D Salamon
92 Banach bundles, Banach modules and automorphisms of C^*-algebras
M J Dupré and R M Gillette
93 Nonlinear partial differential equations and their applications: Collège de France Seminar. Volume V
H Brezis and J L Lions
94 Computer algebra in applied mathematics: an introduction to MACSYMA
R H Rand
95 Advances in nonlinear waves. Volume I
L Debnath
96 FC-groups
M J Tomkinson
97 Topics in relaxation and ellipsoidal methods
M Akgül
98 Analogue of the group algebra for topological semigroups
H Dzinotyiweyi
99 Stochastic functional differential equations
S E A Mohammed

100 Optimal control of variational inequalities
V Barbu
101 Partial differential equations and dynamical systems
W E Fitzgibbon III
102 Approximation of Hilbert space operators Volume II
C Apostol, L A Fialkow, D A Herrero and D Voiculescu
103 Nondiscrete induction and iterative processes
V Ptak and F-A Potra
104 Analytic functions – growth aspects
O P Juneja and G P Kapoor
105 Theory of Tikhonov regularization for Fredholm equations of the first kind
C W Groetsch
106 Nonlinear partial differential equations and free boundaries. Volume I
J I Díaz
107 Tight and taut immersions of manifolds
T E Cecil and P J Ryan
108 A layering method for viscous, incompressible L_p flows occupying R^n
A Douglis and E B Fabes
109 Nonlinear partial differential equations and their applications: Collège de France Seminar. Volume VI
H Brezis and J L Lions
110 Finite generalized quadrangles
S E Payne and J A Thas
111 Advances in nonlinear waves. Volume II
L Debnath
112 Topics in several complex variables
E Ramírez de Arellano and D Sundararaman
113 Differential equations, flow invariance and applications
N H Pavel
114 Geometrical combinatorics
F C Holroyd and R J Wilson
115 Generators of strongly continuous semigroups
J A van Casteren
116 Growth of algebras and Gelfand–Kirillov dimension
G R Krause and T H Lenagan
117 Theory of bases and cones
P K Kamthan and M Gupta
118 Linear groups and permutations
A R Camina and E A Whelan
119 General Wiener–Hopf factorization methods
F-O Speck
120 Free boundary problems: applications and theory. Volume III
A Bossavit, A Damlamian and M Fremond
121 Free boundary problems: applications and theory. Volume IV
A Bossavit, A Damlamian and M Fremond
122 Nonlinear partial differential equations and their applications: Collège de France Seminar. Volume VII
H Brezis and J L Lions
123 Geometric methods in operator algebras
H Araki and E G Effros
124 Infinite dimensional analysis–stochastic processes
S Albeverio
125 Ennio de Giorgi Colloquium
P Krée
126 Almost-periodic functions in abstract spaces
S Zaidman
127 Nonlinear variational problems
A Marino, L Modica, S Spagnolo and M Degliovanni
128 Second-order systems of partial differential equations in the plane
L K Hua, W Lin and C-Q Wu
129 Asymptotics of high-order ordinary differential equations
R B Paris and A D Wood
130 Stochastic differential equations
R Wu
131 Differential geometry
L A Cordero
132 Nonlinear differential equations
J K Hale and P Martinez-Amores
133 Approximation theory and applications
S P Singh
134 Near-rings and their links with groups
J D P Meldrum
135 Estimating eigenvalues with *a posteriori/a priori* inequalities
J R Kuttler and V G Sigillito
136 Regular semigroups as extensions
F J Pastijn and M Petrich
137 Representations of rank one Lie groups
D H Collingwood
138 Fractional calculus
G F Roach and A C McBride
139 Hamilton's principle in continuum mechanics
A Bedford
140 Numerical analysis
D F Griffiths and G A Watson
141 Semigroups, theory and applications. Volume I
H Brezis, M G Crandall and F Kappel
142 Distribution theorems of L-functions
D Joyner
143 Recent developments in structured continua
D De Kee and P Kaloni
144 Functional analysis and two-point differential operators
J Locker
145 Numerical methods for partial differential equations
S I Hariharan and T H Moulden
146 Completely bounded maps and dilations
V I Paulsen
147 Harmonic analysis on the Heisenberg nilpotent Lie group
W Schempp
148 Contributions to modern calculus of variations
L Cesari
149 Nonlinear parabolic equations: qualitative properties of solutions
L Boccardo and A Tesei
150 From local times to global geometry, control and physics
K D Elworthy

151 A stochastic maximum principle for optimal control of diffusions
U G Haussmann
152 Semigroups, theory and applications. Volume II
H Brezis, M G Crandall and F Kappel
153 A general theory of integration in function spaces
P Muldowney
154 Oakland Conference on partial differential equations and applied mathematics
L R Bragg and J W Dettman
155 Contributions to nonlinear partial differential equations. Volume II
J I Díaz and P L Lions
156 Semigroups of linear operators: an introduction
A C McBride
157 Ordinary and partial differential equations
B D Sleeman and R J Jarvis
158 Hyperbolic equations
F Colombini and M K V Murthy
159 Linear topologies on a ring: an overview
J S Golan
160 Dynamical systems and bifurcation theory
M I Camacho, M J Pacifico and F Takens
161 Branched coverings and algebraic functions
M Namba
162 Perturbation bounds for matrix eigenvalues
R Bhatia
163 Defect minimization in operator equations: theory and applications
R Reemtsen
164 Multidimensional Brownian excursions and potential theory
K Burdzy
165 Viscosity solutions and optimal control
R J Elliott
166 Nonlinear partial differential equations and their applications: Collège de France Seminar. Volume VIII
H Brezis and J L Lions
167 Theory and applications of inverse problems
H Haario
168 Energy stability and convection
G P Galdi and B Straughan
169 Additive groups of rings. Volume II
S Feigelstock
170 Numerical analysis 1987
D F Griffiths and G A Watson
171 Surveys of some recent results in operator theory. Volume I
J B Conway and B B Morrel
172 Amenable Banach algebras
J-P Pier
173 Pseudo-orbits of contact forms
A Bahri
174 Poisson algebras and Poisson manifolds
K H Bhaskara and K Viswanath
175 Maximum principles and eigenvalue problems in partial differential equations
P W Schaefer
176 Mathematical analysis of nonlinear, dynamic processes
K U Grusa
177 Cordes' two-parameter spectral representation theory
D F McGhee and R H Picard
178 Equivariant K-theory for proper actions
N C Phillips
179 Elliptic operators, topology and asymptotic methods
J Roe
180 Nonlinear evolution equations
J K Engelbrecht, V E Fridman and E N Pelinovski
181 Nonlinear partial differential equations and their applications: Collège de France Seminar. Volume IX
H Brezis and J L Lions
182 Critical points at infinity in some variational problems
A Bahri
183 Recent developments in hyperbolic equations
L Cattabriga, F Colombini, M K V Murthy and S Spagnolo
184 Optimization and identification of systems governed by evolution equations on Banach space
N U Ahmed
185 Free boundary problems: theory and applications. Volume I
K H Hoffmann and J Sprekels
186 Free boundary problems: theory and applications. Volume II
K H Hoffmann and J Sprekels
187 An introduction to intersection homology theory
F Kirwan
188 Derivatives, nuclei and dimensions on the frame of torsion theories
J S Golan and H Simmons
189 Theory of reproducing kernels and its applications
S Saitoh
190 Volterra integrodifferential equations in Banach spaces and applications
G Da Prato and M Iannelli
191 Nest algebras
K R Davidson
192 Surveys of some recent results in operator theory. Volume II
J B Conway and B B Morrel
193 Nonlinear variational problems. Volume II
A Marino and M K V Murthy
194 Stochastic processes with multidimensional parameter
M E Dozzi
195 Prestressed bodies
D Iesan
196 Hilbert space approach to some classical transforms
R H Picard
197 Stochastic calculus in application
J R Norris
198 Radical theory
B J Gardner
199 The C^*-algebras of a class of solvable Lie groups
X Wang
200 Stochastic analysis, path integration and dynamics
K D Elworthy and J C Zambrini

201 Riemannian geometry and holonomy groups
S Salamon
202 Strong asymptotics for extremal errors and polynomials associated with Erdős type weights
D S Lubinsky
203 Optimal control of diffusion processes
V S Borkar
204 Rings, modules and radicals
B J Gardner
205 Two-parameter eigenvalue problems in ordinary differential equations
M Faierman
206 Distributions and analytic functions
R D Carmichael and D Mitrovic
207 Semicontinuity, relaxation and integral representation in the calculus of variations
G Buttazzo
208 Recent advances in nonlinear elliptic and parabolic problems
P Bénilan, M Chipot, L Evans and M Pierre
209 Model completions, ring representations and the topology of the Pierce sheaf
A Carson
210 Retarded dynamical systems
G Stepan
211 Function spaces, differential operators and nonlinear analysis
L Paivarinta
212 Analytic function theory of one complex variable
C C Yang, Y Komatu and K Niino
213 Elements of stability of visco-elastic fluids
J Dunwoody
214 Jordan decomposition of generalized vector measures
K D Schmidt
215 A mathematical analysis of bending of plates with transverse shear deformation
C Constanda
216 Ordinary and partial differential equations. Volume II
B D Sleeman and R J Jarvis
217 Hilbert modules over function algebras
R G Douglas and V I Paulsen
218 Graph colourings
R Wilson and R Nelson
219 Hardy-type inequalities
A Kufner and B Opic
220 Nonlinear partial differential equations and their applications: Collège de France Seminar. Volume X
H Brezis and J L Lions
221 Workshop on dynamical systems
E Shiels and Z Coelho
222 Geometry and analysis in nonlinear dynamics
H W Broer and F Takens
223 Fluid dynamical aspects of combustion theory
M Onofri and A Tesei
224 Approximation of Hilbert space operators. Volume I. 2nd edition
D Herrero
225 Operator theory: proceedings of the 1988 GPOTS–Wabash conference
J B Conway and B B Morrel
226 Local cohomology and localization
J L Bueso Montero, B Torrecillas Jover and A Verschoren
227 Nonlinear waves and dissipative effects
D Fusco and A Jeffrey
228 Numerical analysis 1989
D F Griffiths and G A Watson
229 Recent developments in structured continua. Volume III
D De Kee and P Kaloni
230 Boolean methods in interpolation and approximation
F J Delvos and W Schempp
231 Further advances in twistor theory. Volume I
L J Mason and L P Hughston
232 Further advances in twistor theory. Volume II
L J Mason and L P Hughston
233 Geometry in the neighborhood of invariant manifolds of maps and flows and linearization
U Kirchgraber and K Palmer
234 Quantales and their applications
K I Rosenthal
235 Integral equations and inverse problems
V Petkov and R Lazarov
236 Pseudo-differential operators
S R Simanca
237 A functional analytic approach to statistical experiments
I M Bomze
238 Quantum mechanics, algebras and distributions
D Dubin and M Hennings
239 Hamilton flows and evolution semigroups
J Gzyl
240 Topics in controlled Markov chains
V S Borkar
241 Invariant manifold theory for hydrodynamic transition
S Sritharan
242 Lectures on the spectrum of $L^2(\Gamma\backslash G)$
F L Williams
243 Progress in variational methods in Hamiltonian systems and elliptic equations
M Girardi, M Matzeu and F Pacella
244 Optimization and nonlinear analysis
A Ioffe, M Marcus and S Reich
245 Inverse problems and imaging
G F Roach
246 Semigroup theory with applications to systems and control
N U Ahmed
247 Periodic-parabolic boundary value problems and positivity
P Hess
248 Distributions and pseudo-differential operators
S Zaidman
249 Progress in partial differential equations: the Metz surveys
M Chipot and J Saint Jean Paulin
250 Differential equations and control theory
V Barbu

251 Stability of stochastic differential equations with respect to semimartingales
X Mao
252 Fixed point theory and applications
J Baillon and M Théra
253 Nonlinear hyperbolic equations and field theory
M K V Murthy and S Spagnolo
254 Ordinary and partial differential equations. Volume III
B D Sleeman and R J Jarvis
255 Harmonic maps into homogeneous spaces
M Black
256 Boundary value and initial value problems in complex analysis: studies in complex analysis and its applications to PDEs 1
R Kühnau and W Tutschke
257 Geometric function theory and applications of complex analysis in mechanics: studies in complex analysis and its applications to PDEs 2
R Kühnau and W Tutschke
258 The development of statistics: recent contributions from China
X R Chen, K T Fang and C C Yang
259 Multiplication of distributions and applications to partial differential equations
M Oberguggenberger
260 Numerical analysis 1991
D F Griffiths and G A Watson
261 Schur's algorithm and several applications
M Bakonyi and T Constantinescu
262 Partial differential equations with complex analysis
H Begehr and A Jeffrey
263 Partial differential equations with real analysis
H Begehr and A Jeffrey
264 Solvability and bifurcations of nonlinear equations
P Drábek
265 Orientational averaging in mechanics of solids
A Lagzdins, V Tamuzs, G Teters and A Kregers
266 Progress in partial differential equations: elliptic and parabolic problems
C Bandle, J Bemelmans, M Chipot, M Grüter and J Saint Jean Paulin
267 Progress in partial differential equations: calculus of variations, applications
C Bandle, J Bemelmans, M Chipot, M Grüter and J Saint Jean Paulin
268 Stochastic partial differential equations and applications
G Da Prato and L Tubaro
269 Partial differential equations and related subjects
M Miranda
270 Operator algebras and topology
W B Arveson, A S Mishchenko, M Putinar, M A Rieffel and S Stratila
271 Operator algebras and operator theory
W B Arveson, A S Mishchenko, M Putinar, M A Rieffel and S Stratila
272 Ordinary and delay differential equations
J Wiener and J K Hale
273 Partial differential equations
J Wiener and J K Hale

H Begehr

Freie Universität Berlin, Germany

and

A Jeffrey

University of Newcastle upon Tyne, England

(Editors)

Partial differential equations with real analysis

Longman Scientific & Technical

Copublished in the United States with
John Wiley & Sons, Inc., New York

Longman Scientific & Technical
Longman Group UK Limited
Longman House, Burnt Mill, Harlow
Essex CM20 2JE, England
and Associated companies throughout the world.

Copublished in the United States with
John Wiley & Sons Inc., 605 Third Avenue, New York, NY 10158

First published 1992

AMS Subject Classification: 35J15, 35J60, 35K22, 35Q53, 35K05, 35L10, 73C02, 35J05

ISSN 0269-3674

ISBN 0 582 09638 3

British Library Cataloguing in Publication Data

A catalogue record for this book is
available from the British Library

Library of Congress Cataloging-in-Publication Data

Partial differential equations with real analysis / H. Begehr (editor)
and A. Jeffrey (editor).
p. cm. -- (Pitman research notes in mathematics series, ISSN
0269-3674 ; 263)
1. Differential equations, Elliptic. 2. Heat equation.
3. Boundary value problems. I. Begehr, Heinrich G. W.
II. Jeffrey, A. (Alan) III. Series.
QA377.P313 1992
515'.353--dc20 92-4743
CIP

Printed and bound in Great Britain
by Biddles Ltd, Guildford and King's Lynn

Contents

Preface

The papers of this volume are dedicated to Professor Robert Pertsch Gilbert on the occasion of his 60th birthday. This volume contains contributions to partial differential equations based on real and functional analytic methods, while a companion volume contains papers using complex analytic methods.

Since the Korteweg–de Vries equation – the mathematical model for the unidirectional propagation of long waves and a principal subject in soliton theory – was considered for the first time about 100 years ago a large literature has grown on this and related equations. To an equation of this kind (A. Jeffrey and S. Xu) the Painlevé test is applied to find out if there are single–valued solutions near a movable singularity manifold. Existence and uniqueness results for a special second order elliptic–parabolic equation are developed in various function classes and used to discuss a more general degenerate evolution equation (G. Fichera). A noncharacteristic Cauchy problem is solved by variational methods for parabolic equations without any initial condition for $t = 0$ and a related inverse problem is handled (N.H. Dinh) determining one of the coefficients of the equation. Such a coefficient inverse problem is considered also for a second order elliptic equation (H. Li and Z. Hou). Identification of coefficients are studied for hyperbolic equations, too (G.C. Hsiao and J. Sprekels).

On the basis of an a priori estimate of solutions to linear second order elliptic equations weak solutions of nonlinear elliptic equations in the whole n–dimensional space, $2 \leq n$, are constructed via the Schauder continuation method. By prescribing polynomial behaviour at infinity uniqueness results are obtained (G.H. Hile and C.P. Mawata). Dirichlet, Neumann and mixed type boundary, respectively combined with transmission conditions each prescribed on half–planes in $\mathbb{R}^3$ for the Helmholtz equation with different wave numbers in each quadrant are explicitly solved (E. Meister, F. Penzel, F.O. Speck, and F.S. Teixeira). Using boundary integral methods existence and uniqueness is proved for solutions to scattering problems of time harmonic acoustic waves in a stratified medium expressed by Dirichlet, Neumann or mixed boundary conditions for the Helmholtz equation (S. Xu). Finally, some problems in elasticity are presented. Initial and boundary valuc problems for a partial differential–integral equation in radially symmetric thermoelasticity is contructively solved (L.R. Bragg, and P. Shi), using results for the radial heat equation. The influence of domain boundary perturbation on the decay of Saint–Venant end effects in linear isotropic elastostatics is investigated (C.O. Horgan and L.E. Payne). Real as well as complex analytic methods are used for a contact problem in orthotropic elasticity (H. Begehr and W. Lin). Using bianalytic functions it is reduced to some boundary value problem for two analytic functions.

The editors are grateful to Barbara Wengel, secretary at the I. Mathematical Institute of the Freie Universität Berlin, for retyping several of the manuscripts.

Biography of Robert Pertsch Gilbert

Robert Pertsch Gilbert was born on January 8, 1932 in New York City. He entered Brooklyn College of the City University of New York where he received his B.S. in 1952. In 1955 he earned an M.S. in mathematics as well as in physics at Carnegie-Mellon University and in 1958 he got his Ph.D. there. From September 1957 he worked for two years as an Instructor at the University of Pittsburgh, continuing there as an assistant professor until 1960. During this period he also had a contract as a consultant for the U.S. Bureau of Mines, Special Coal Research Division. After one year at Michigan State University he took a position as research assistant professor at the *Institute for Fluid Dynamics and Applied Mathematics* at the University of Maryland. Coinciding with this he consulted at the Naval Ordinance Laboratory, White Oak. In 1964 he was promoted to research associate professor at the University of Maryland. In 1966 he was appointed to the position of professor at Indiana University where he remainded until 1965. In 1975 Dr. Gilbert accepted the *Unidel Chair* of Mathematics at the University of Delaware.

In 1975 Dr. Gilbert received the *Senior Scientist Award* from the Alexander von Humboldt foundation spending 6 months at the Freie Universität Berlin followed by another 3 month period in 1986. In 1981 he got the *British Science Council Research Award* at Oxford University. In addition to these he was guest research professor at the universities of Glasgow (1972), Dortmund (1972), Hahn-Maltner-Institut Berlin (1974), Freie Universität Berlin (1974/75), Technical University of Kopenhagn (1979), Karlsruhe (1980), Oxford (1980/81), Guanzhou (1985), Mexico (1985) and Freie Universität Berlin (1985).

Dr. Gilbert's major achievements in his research areas are:

I. The generalization of Bergman's kernel method to harmonic functions of more than two variables as well as to general , linear elliptic equations.

II. The construction of integral operators in the sense of Bergman for differential equations in more than two variables based on his *method of ascent.* This approach for three variables leads to an operator known as the Whittaker-Bergman operator, whereas for four variables this extension is known as the Gilbert-Bergman operator.

III. Application of integral operator methods to axially symmetric potential scattering and quantum field theory.

IV. Extension of Nehari's results on singularities of Legendre series to series of eigenfunctions of second order Sturm-Liouville problems.

V. The creation, with Gerald Hile, of the theory of generalized hyperanalytic functions. This theory is the theory of solutions to first order elliptic systems of more than two equations and generalizes both Vekua's theory of generalized analytic functions and Douglis's theory of hyperanalytic functions.

VI. Application of complex analytic methods to wave propagation, elastic plate problems, and the development of a function theory for pseudoparabolic equations.

VII. The development, with Jim Buchanan, of a new theory of elastic shells. This theory is based on the idea of approximating the series expansion of the shell equations with respect to thickness, and using the Douglis hypercomplex algebra to represent the first n terms of the expansion. This approach leads to a new hyperanalytic function theory with which the shell equations may be solved to arbitrary order of accuracy.

His recent papers are devoted to Hele Shaw type flows and other free and moving boundary problems, underwater sound propagation, thermoelastic and elastoplastic shells and plates, transmutation theory and the use of symbolic algebras for solving differential equations.

Besides his fruitful research career, about 160 papers in refereed journals, Professor Gilbert has published three books on function theoretic and constructive methods for partial differential equations (1969,1974, 1983). Three more books are in preparation. He has also edited several proceedings of conferences which he had organized with colleagues.

Another important contribution to the mathematical community is the founding and editing by Professor Gilbert of two international journals *Applicable Analysis* (1971) and *Complex Variables, Theory and Application* (1982). Both journals have become well known and respected as high quality speciality journals in analysis. Moreover, Professor Gilbert is on the editorial boards of several journals and book series.

He is married to Nancy Gilbert, née (Page). They have a daughter Jennifer.

Robert Pertsch Gilbert

A JEFFREY AND S XU

On the integrability of $u_t + auu_x + bu_{3x} + cu_{5x} = 0$

1. Introduction

The partial differential equation

$$u_t + auu_x + bu_{3x} + cu_{5x} = 0 \quad , \tag{1.1}$$

where a, b and c are constants and $u_{rx} = \partial^r u/\partial x^r$ arises from an asymptotic argument in connection with nonlinear wave propagation. It is found that this equation governs the evolution of long weakly nonlinear waves in which stronger dispersive effects than those present in the KdV equation ($c = 0$) are necessary in order to balance the nonlinearity so that a travelling wave solution may exist. The question that then arises naturally is whether (1.1) is completely integrable, and so possesses soliton–type solutions.

Whereas for n–dimensional dynamical systems the system will be completely integrable if n constants of motion (integrals) can be found governing the evolution process, the corresponding situation for (1.1) is infinitely dimensional so it will only be completely integrable if an infinite number of such integrals can be shown to exist.

All the known completely integrable systems possess novel properties, like the demonstration of the existence of an infinite number of conservation laws, the existence of a Bäcklund transformation, Hirota's formalism arising as the integrability condition for a pair of linear systems of equations $V_x = PV$, $V_t = QV$ [1], and satisfying the Painlevé test [2].

When an equation is completely integrable it may be solved, subject to suitable initial conditions, by considering the scattering problem $V_x = PV$ and the time evolution problem $V_t = QV$.

Non–integrability can be inferred from the non–existence of an infinite number of conservation laws or by numerical calculations which show that travelling wave solutions do not interact like solitons [3]. The Painlevé test introduced by Ablowitz *et al.* [2] is based on the fact that every ordinary differential equation arising as result of a similarity reduction of a completely integrable partial differential equation is of the Painlevé type; and thus has no movable singularities except poles (perhaps after a transformation of variables). This test is awkward to apply because of the necessity of transforming a partial differential equation into an ordinary differential equation by finding a similarity reduction. An easier approach is due to Weiss [4–10] who introduced the Painlevé property for partial differential equations as a means of applying the Painlevé conjecture directly to a given partial differential equation without first reducing it to an ordinary differential equation. When using this approach, a partial differential equation is said to possess the Painlevé property if its solutions are single valued in the neighbourhood of a movable singularity manifold.

In Section 2 by way of example, and using the approach due to Weiss, we apply the Painlevé test to the KdV equation to show the connection between the Painlevé test, the Bäcklund transformation, Hirota's formalism and the inverse scattering problem.

Section 3 applies the Painlevé test to (1.1) and shows the equation does not possess the Painlevé property if $c \neq 0$. Finally, in Section 4, we show (1.1) only has three conservation laws which are of polynomial type.

2. An Application To the KdV Equation

Using the work of Weiss [4–10] we prevent a brief discussion of the connection between the Painlevé property and the Bäcklund transformation, the Lax pair and Hirota's formalism. By definition, a partial differential equation is said to have the Painlevé property, or to satisfy the Painlevé test, if its solutions are single valued about a movable singularity manifold. Now for a meromorphic function $f(x_1, x_2, ..., x_n)$ of n complex variables $x_1, x_2, ..., x_n$, the singularities will occur along an analytic manifold if dimension $n - 1$, and such manifolds are determined by conditions of the form

$$\phi(x_1, x_2, ..., x_n) = 0 \quad , \tag{2.1}$$

where ϕ is an analytic function in the neighbourhood of $\phi = 0$. The Painlevé test proceeds as follows. Assuming $u = u(x_1, x_2, ..., x_n)$ is a solution of the partial differential equation under investigation, we first expand u locally in a generalized Laurent series of the form

$$u(x_1, x_2, ..., x_n) = \frac{1}{\phi^m} \sum_{j=0}^{\infty} u_j(x_1, x_2, ..., x_n)\phi^j \quad , \tag{2.2}$$

where the Laurent coefficients u_j are analytic functions of their arguments.

Expansion (2.2) must have as many arbitrary functions as the order of the system, and these arbitrary functions are seen to be the functions u_j. We then determine all possible leading orders m and the function u_0, together with all of the corresponding resonances of ϕ^n, which are defined to be of the order of ϕ^n at which the corresponding u_n should be arbitrary. The arbitrariness must then be checked by developing recursion relations for u_n to ensure the validity of expansion (2.2). The recursion relation takes the form of a system of coupled partial differential equations involving u_j, $j < n$ and ϕ. The application of the Painlevé test for partial differential equations is then strictly analogous to the test for ordinary differential equations.

As an illustration we take the KdV equation which may be written in the form

$$u_t + 12uu_x + u_{3x} = 0 \quad . \tag{2.3}$$

All possible leading order terms are determined by substituting $u = \phi^{-m}u_0$ into (2.3). Any value m that causes two or more terms in (2.3) to balance and the rest to be ignored as $\phi \to \infty$ is called a **leading order**, while the balancing terms are called the **leading terms**.

The leading order for (2.3) is seen to be -2 and the substitution of (2.2) with $m = 2$ into (2.3) yields the recursion relation

$$\phi_x^{\,2}u_j(j+1)(j-4)(j-6) = F(\phi_x,\phi_t,...,u_0,u_1,...,u_{j-1}) \tag{2.4}$$

in which F is a nonlinear function of the derivatives of ϕ and u_i,

$1 \le i \le j-1$, with $j = 0,1,2,\ldots$. This recursion relation ceases to be defined when $j = -1$, 4 and 6, which are called **resonances**. The resonance at $j = -1$ corresponds to the arbitrariness of the singular manifold. The resonance at $j = 4$ induces an arbitrary function u_4 together with a compatibility condition on the functions $\{\phi, u_i, i = 1,2,3\}$ which requires the right–hand side of the recursion relation to vanish identically. A similar result applies when $j = 6$. Apart for $j = 4$ and $j = 6$, the u_j can be expressed in terms of the previous u_j. From the first few terms in the expansion (2.2) we find that

$$(j = 0) \quad u_0 = -\phi_x^{\,2}$$

$$(j = 1) \quad u_1 = \phi_{xx}$$

$$(j = 2) \quad \phi_x\phi_t + 12u_2\phi_x^{\,2} + 4\phi_x\phi_{3x} - 3\phi_{xx}^{\,2} = 0$$

$$(j = 3) \quad \phi_{xt} = u_2\phi_{xx} - u_3\phi_x^{\,2} + \phi_{3x} = 0 \tag{2.5}$$

$$(j = 4) \quad \frac{\partial}{\partial x}\left(\phi_{xt} + \phi_{3x} + \phi_{2x}u_2 - \phi_x^{\,2}u_3\right) = 0 \ .$$

On account of the relationship at $j = 3$, the compatibility condition at $j = 4$ is satisfied identically. After some calculation, it is possible to show that the compatibility condition at $j = 6$ is also satisfied identically.

Let us now consider the special case of (2.2) in which

$$u_3 = u_4 = u_6 = 0 \ . \tag{2.6}$$

Then it is easy to show that $u_j = 0$ for $j \ge 0$, provided u_2 satisfies the KdV equation (2.3). The truncated expansion (2.2) then takes the form

$$u = \frac{\partial^2}{\partial x^2}\log\phi + u_2 \tag{2.7a}$$

with

$$\phi_x\phi_t + 12u_2\phi_x^2 + 4\phi_x\phi_{3x} - 3\phi_{2x}^2 = 0 \quad , \tag{2.7b}$$

$$\phi_{xt} + 12u_2\phi_{2x} + \phi_{4x} = 0 \quad , \tag{2.7c}$$

and

$$u_{2,t} + 12u_2u_{2,x} + u_{2,3x} = 0 \quad , \tag{2.7d}$$

and so the KdV equation is seen to satisfy the Painlevé test.

We remark that (2.7) becomes an auto–Bäcklund transformation for the KdV equation provided (2.7b,c,d) are consistent. Thus in this case the Painlevé test is seen to have generated the Bäcklund transformation for the KdV equation in a very direct manner.

We remark also that (2.7b) and (2.7c) imply a Lax pair for the KdV equation. This can be seen by making the direct substitution $\phi_x = \psi^2$, which leads to the actual Lax pair

$$\psi_{2x} + (2u_2 + \xi)\psi = 0 \tag{2.8a}$$

$$\psi_t + (6u_2 + \xi)\psi_x + \psi_{3x} = 0 \quad . \tag{2.8b}$$

The connection between the Painlevé property and Hirota's method [11] can be seen as follows. Equation (2.3) can be transformed into a bilinear form, in terms of the Hirota bilinear operator, of the form

$$(D_x^4 + D_xD_t)f \circ f = 0 \quad , \tag{2.9}$$

by the transformation

$$u = \frac{\partial^2}{\partial x^2} \log f \quad , \tag{2.10}$$

where the Hirota bilinear operator D is defined by the expression $D_x^m(f \circ f) = (\partial_x - \partial_{x'})^m f(x)f(x')|_{x=x'}$.

If each solution of (2.3) is indexed by writing $u^{(n)}$, we can associate each $u^{(n)}$ with a corresponding $f^{(n)}$ through the expression

$$u^{(n)} = \frac{\partial^2}{\partial x^2} \log f^{(n)} \quad . \tag{2.11}$$

We may then consider the solutions u, u_2 in the Bäcklund transformation (2.7) as a pair of adjacent solutions in the set $\{u^{(n)}\}$, and then each interaction of the Bäcklund transformation generates a new singular manifold ϕ_{n-1}, where

$$u^{(n)} = \frac{\partial^2}{\partial x^2} \log \phi_{n-1} + u^{(n-1)} \quad . \tag{2.12}$$

It then follows from (2.11) and (2.12) that

$$f^{(n)} = \phi_{n-1} f^{(n-1)} \quad , \tag{2.13}$$

which is true for all $n > 0$ with $f^{(0)} = 1$.

The following theorem, which we merely state, was first given by Gibbon *et al.* [12]. It is easily proved and we refer to the original source reference for the details.

Theorem 1. If the functions $f^{(n)}$, $n = 1,2,3,...$, satisfy the Hirota equation

$$(D_x^{\ 4} + D_x D_t) f^{(n)} \circ f^{(n)} = 0 \quad ,$$

for every n, and if

$$f^{(n)} = \phi_{n-1} f^{(n-1)} \quad ,$$

then the resulting equation in ϕ_{n-1} and $u^{(n-1)}$ is satisfied by the Painlevé relations (2.7b) and (2.7c). Furthermore,

$$f^{(n)} = \prod_{i=0}^{n-1} \phi_i \quad . \tag{2.14}$$

This theorem, taken together with the Painlevé expansion, clarifies the reason for the choice of transformation (2.10) and explains why self–truncation occurs in such expansions. It is necessary to remark at this point that the results illustrated in this section hold for other known integrable equations.

3. Application of the Painlevé Test to $u_t + auu_x + bu_{3x} + cu_{5x} = 0$

We now apply the ideas illustrated in the case of the familiar KdV equation in Section 2 to (1.1), which can be written as

$$u_t + uu_x + u_{3x} + u_{5x} = 0 \tag{3.1a}$$

if $bc > 0$, and as

$$u_t + uu_x - u_{3x} + u_{5x} = 0 \quad , \tag{3.1b}$$

if $bc < 0$, by means of a suitable change of the variables x, t and u.

For simplicity, we now apply the Painlevé test to (3.1a), though it applies equally well to (3.1b) without additional complication.

Expansion (2.2) takes the form

$$u(x,t) = \frac{1}{\phi(x,t)^m} \sum_{j=0}^{\infty} u_j(x,\ t)\phi^j(x,\ t) \quad . \tag{3.2}$$

To determine the leading order term we substitute

$$u = u_0\phi^{-m} \tag{3.3}$$

into (3.1a) and, as a result, find that at $m = 4$ the fifth derivative term balances the nonlinear term. Thus the leading order occurs at $m = 4$, and there is no other leading order with $m < 4$.

Substitution of (3.2) with $m = 4$ into (3.1a) using machine symbolic manipulation then gives

$$\sum_{j=0}^{\infty} \left\{\phi^{j-4} u_{j,t} + (j - 4)\phi^{j-5} u_j \phi_t\right\} +$$

$$\sum_{j=0}^{\infty} u_j \phi^{j-4} \sum_{k=0}^{\infty} \left\{\phi^{j-4} u_{k,x} + (j - 4) u_k \phi_x \phi^{j-5}\right\} + \sum_{j=0}^{\infty} \left\{\phi^{j-4} u_{j,3x}\right.$$

$$+ (j - 4)\phi^{j-5}(3u_{j,2x}\phi_x + 3u_{j,x}\phi_{2x} + u_j\phi_{3x}) + (3j^2 - 9j$$

$$\left. + 20)\phi^{j-6}(u_{j,x}\phi_x^{\ 2} + u_j\phi_x\phi_{2x}) + (j^3 - 15j^2 + 74j - 120)\phi^{j-7} u_j \phi_x^{\ 3}\right\}$$

$$+ \sum_{j=0}^{\infty} \left\{\phi^{j-4} u_{j,5x} + (j - 4)\phi^{j-5}(5u_{j,4x}\phi_x + 10u_{j,3x} + \phi_{2x} + 10u_{j,2x}\phi_{3x}\right.$$

$$+ 5u_{j,x}\phi_{4x} + u_j\phi_{5x}) + (j^2 - 9j + 20)\phi^{j-6}(10u_{j,3x}\phi_x^{\ 2} + 30u_{j,2x}\phi_x\phi_{2x}$$

$$+ 20u_{j,x}\phi_x\phi_{3x} + 5u_j\phi_x\phi_{4x} + 15u_{j,x}\phi_{2x}^{\ 2}) + (j^3 - 15j^2 + 74j$$

$$- 120)(\phi^{j-7} 10u_{j,2x}\phi_x^{\ 3} + 30u_{j,x}\phi_{2x}\phi_x^{\ 2} + 10u_j\phi_{3x}\phi_x^{\ 2} + 15\phi_{2x}^{\ 2}\phi_x)$$

$$+ \phi^{j-8}(j^4 - 22j^3 + 179j^2 - 638j + 840)(5u_{j,x}\phi_x^{\ 4} + 10u_j\phi_{2x}\phi_x^{\ 3})$$

$$\left. + \phi^{j-9}(j^5 - 30j^4 + 355j^3 - 2020j^2 + 5944j - 6720)u_j\phi_x^{\ 5}\right\} = 0 \ . \qquad (3.4)$$

A recursion relation can be derived from (3.4), though it is first necessary to know u_0. From (3.4), a direct calculation with $u_j = 0$ for $j < 0$ yields

$$u_0 = - 1680\phi_x^{\ 4} \ . \qquad (3.5)$$

We observe that the time derivative and third order spatial derivative of (3.1a) make no contribution to u_0, and that if the coefficient of the nonlinear term is a, then in place of (3.5) we have the result

$$au_0 = -1690\phi_x^4 \quad . \tag{3.6}$$

The recursion relation obtained from (3.4) is

$$\begin{aligned}
&\{u_{j-5,t} - (j-8)u_{j-4}\phi_t\} + \sum_{m=0}^{\infty} u_{j-m}\left\{u_{m-1,x} + (m-4)u_m\phi_x\right\} \\
&+ \Big\{u_{j-5,3x} + (j-8)(3u_{j-4,2x}\phi_x + 3u_{j-4,x}\phi_{2x} + u_{j-4}\phi_{3x}) \\
&+ 3(j^2 - 15j + 56)(u_{j-3,x}\phi_x^2 + u_{j-3}\phi_x\phi_{2x} + (j^3 - 21j^2 + 146j \\
&- 366)u_{j-2}\phi_x^3\Big\} + \Big\{u_{j-5,5x} + (j-8)(5u_{j-4,4x}\phi_x + 10u_{j-4,3x}\phi_{2x} \\
&+ 10u_{j-4,2x}\phi_{3x} + 5u_{j-4,x}\phi_{4x} + u_{j-4}\phi_{5x}) + (j^2 - 15j \\
&+ 56)(10u_{j-3,3x}\phi_x^2 + 30u_{j-3,2x}\phi_x\phi_{2x} + 20u_{j-3,x}\phi_x\phi_{3x} + 5u_{j-3}\phi_x\phi_{4x} \\
&+ 15u_{j-3,x}\phi_{2x}^2) + (j^3 - 21j^2 + 146j - 336)(10u_{j-2,2x}\phi_{3x}^3 \\
&+ 30u_{j-2,x}\phi_{2x}\phi_x^2 + 10u_{j-2}\phi_{3x}\phi_x^2 + 15u_{j-2}\phi_{2x}^2\phi_x) + (j^4 - 26j^3 \\
&+ 251j^2 - 1066j + 1680)(5u_{j-1,x}\phi_x^4 + 10u_{j-1}\phi_{2x}\phi_x^3) + (j^5 - 30j^4 \\
&+ 355j^3 - 2020j^2 + 5944j - 6720)u_j\phi_x^5\Big\} = 0 \quad .
\end{aligned} \tag{3.7}$$

When (3.7) is re–arranged by collecting terms proportional to u_j it becomes

$$u_j\phi_x^{\,5}(j-8)(j+1)(j-12)(j^2-11j+70)$$
$$-F(\phi_x, \phi_{2x}, \ldots; u_0, u_1, \ldots, u_{j-1}) \tag{3.8}$$

where F is an analytic function of $\{\phi_x, \phi_{2x}, \ldots; u_0, u_1, \ldots, u_{j-1}\}$. This shows that the resonances occur at $j = -1$, 8 and 12 and we notice from (3.6) to (3.8) that the resonances do not depend on the coefficients a, b and c if we use (1.1) in place of (3.1). The compatibility conditions arrived at in this manner are

$$(j = 0) \quad u_0 = -1680\phi_x^{\,4},$$

$$(j = 1) \quad u_1 = 3360\phi_x^{\,2}\phi_{2x},$$

$$(j = 2) \quad u_2 = -\frac{1}{11}\Big\{43680\phi_x\phi_{3x} + 32760\phi_{2x}^{\,2} - 47040\phi_x^{\,4} - 840\phi_x^{\,2}\Big\},$$

$$(j = 3) \quad u_3 = \Big\{69720\phi_{4x}\phi_x^{\,2} + 6569360\phi_{3x}\phi_{2x}\phi_x + 7560\phi_{3x}\phi_x^{\,3}$$
$$45309600\phi_{2x}^{\,3} - 10442880\phi_{2x}\phi_x^{\,4} - 64680\phi_{2x}\phi_x^{\,2}\Big\}\,. \tag{3.9}$$

Proceeding in this manner, we find that when $j = 8$ the compatibility induced by the arbitrariness of u_8 cannot be satisfied. As a result, we conclude that (1.1) and, equivalently, (1.2) do not possess the Painlevé property and so are **not** completely integrable. As a result the equations do not possess soliton solutions.

4. Polynomial Type Conservation Laws

In this section we establish the existence of three polynomial type conservation laws for equations (1.1) and (1.2), and in doing so we make use of the terminology introduced by Kruskal *et al.* [13].

Any polynomial type conservation law of equations (1.1) and (1.2) can be written in the form

$$T_t + X_x = 0 \quad , \tag{4.1}$$

where T is the conserved density and X, the flux of T, are polynomials in u and its derivatives with respect to x which do not depend explicitly on x and t. All terms in the polynomials are of the form $ax_0^{a_0} u_1^{a_1} u_2^{a_2} \ldots u_p^{a_p}$, where $a_i \geq 0$ and, u_i denotes $\partial^i u/\partial x^i$ with $u_0 = u$. The degree m of a term is defined as

$$m = \sum_{i=1}^{p} ia_i \quad . \tag{4.2}$$

It is easily verified by direct substitution that the following T_i and X_i are, indeed, conserved densities and the corresponding fluxes for (1.1):

$$T_1 = u_0 \quad ,$$

$$X_1 = \tfrac{1}{2} u_0^2 + u_2 + u_4 \quad ,$$

$$T_2 = \tfrac{1}{2} u_0^2 \quad ,$$

$$X_2 = \tfrac{1}{3} u_0^3 + u_0 u_2 - \tfrac{1}{2} u_1^2 + u_0 u_4 - u_1 u_3 + \tfrac{1}{2} u_2^2 \quad ,$$

$$T_3 = \tfrac{1}{3} u_0^3 - u_1^2 + u_2^2 \quad , \tag{4.3}$$

$$\begin{aligned} X_3 = \tfrac{1}{4} u_0^4 &- 2u_1 u_3 + u_2^2 - 2u_1 u_5 + 4u_2 u_4 - 2u_3^2 + 2u_2 u_6 \\ &- 2u_3 u_5 + u_4^2 + u_0^2 u_2 - 2u_0 u_1^2 + u_0^2 u_4 - 2u_0 u_1 u_3 \\ &- 2u_1^2 u_2 + 2u_0 u_2^2 \quad . \end{aligned}$$

The corresponding expressions for (1.2) are

$$T_1 = u_0 \ ,$$

$$X_1 = \tfrac{1}{2} u_0^{\,2} + u_2 - u_4 \ ,$$

$$T_2 = \tfrac{1}{2} u_0^{\,2} \ ,$$

$$X_2 = \tfrac{1}{3} u_0^{\,3} + u_0 u_2 - \tfrac{1}{2} u_1^{\,2} - u_0 u_4 + u_1 u_3 - \tfrac{1}{2} u_2^{\,2} \ ,$$

$$T_3 = \tfrac{1}{3} u_0^{\,3} - u_1^{\,2} + u_2^{\,2} \tag{4.4}$$

$$\begin{aligned} X_3 = \tfrac{1}{4} u_0^{\,4} &- 2u_1 u_3 + u_2^{\,2} + 2u_1 u_5 - 4u_2 u_4 + 2u_3^{\,2} - 2u_2 u_6 \\ &+ 2u_3 u_5 - u_4^{\,2} + u_0^{\,2} u_2 - 2u_0 u_1^{\,2} - u_0^{\,2} u_4 + 2u_0 u_1 u_3 \\ &- 2u_1^{\,2} u_2 - 2u_0 u_2^{\,2} \ . \end{aligned}$$

We observe that inspection of (4.3) and (4.4) shows

$$\frac{\partial T_r}{\partial u_0} = (r - 1) T_{r-1} \ . \tag{4.5}$$

Kruskal [13] has shown that this relation holds for the KdV equation for all T_r, though further calculations show that for (1.1) and (1.2) it is only true for T_1, T_2 and T_3, so these equations only possess three polynomial type conservation laws.

For a discussion of the explicit form of the travelling wave solutions to (1.1) and (1.2) we refer to the paper by Jeffrey and Xu [14]. A related integrability problem for a variable coefficient KdV equation together with its Bäcklund transformation is considered in the paper by Jeffrey and Xu [15]. An account of a direct method by which to construct exact solutions to (1.1) and (1.2) is to be found in Jeffrey and Mohamad [16].

References

[1] Newell, A.C., Solitons in Mathematical Physics, Regional Conference Series in Applied Mathematics. SIAM, 1985.

[2] Ablowitz, M.J., Ramani, A. and Segur, H., A connection between evolution equations and ordinary differential equations of P–type: Part 1. J.Math.Phys. 21, (1979), 715–721; Part 2, J.Math.Phys. 21 (1979), 1006–1015.

[3] Benjamin, T.B., Bona, J.L. and Mahony, T., Model equations for long waves in dispersive systems. Phil.Trans.Roy.Soc.Lond., A 272 (1972), 47–78.

[4] Weiss, J., Tabor, M. and Carnevale, G., The Painlevé property for partial differential equations. J.Math.Phys. 24 (1983), 522–526.

[5] Weiss, J., The Painlevé property for partial differential equations, 2: Bäcklund transformation, Lax pairs and Schwarzian derivative. J.Math.Phys. 24 (1983), 1405–1413.

[6] Weiss, J., On a class of integrable systems and the Painlevé property. J.Math.Phys. 25 (1984), 13–24.

[7] Weiss, J., The Sine–Gordon equation: complete and partial integrability. J.Math. Phys. 25 (1984), 2226–2235.

[8] Weiss, J., The Painlevé property and Bäcklund transformations for the sequence of Boussinesq equations. J.Math.Phys. 26 (1985), 258–269.

[9] Weiss, J., Modified equations, rational solutions, and the Painlevé property and the Hirota–Satsuma equation. J.Math.Phys. 26 (1985), 2174–2180.

[10] Weiss, J., Bäcklund transformations and the Painlevé property. J.Math.Phys. 26 (1986), 1293–1305.

[11] Hirota, R., Direct method for finding exact solutions of nonlinear evolution equations. In: Bäcklund Transformations (R.M. Miura, ed), Lecture Notes in Mathematics 515, Springer, Berlin (1976), 40–68.

[12] Gibson, J.D., Radmore, P., Tabor, M. and Wood, D., The Painlevé property and Hirota's method. Studies in Appl.Math., 72 (1985), 39–65.

[13] Kruskal, M.D., Miura, R.M. and Gardner, C.S., Korteweg–de Vries equation and generalizations 5: Uniqueness and nonexistence of polynomial conservation laws. J.Math.Phys. 11 (1969), 952–960.

[14] Jeffrey, A., and Xu, S., Travelling wave solutions to certain nonlinear evolution equations. Int.J.Non–linear Mech., 24 (1989), 425–429.

[15] Jeffrey, A. and Xu, S., Bäcklund transformation for the variable coefficient Korteweg–de Vries equation. Appl.Anal. 34 (1989), 53–66.

[16] Jeffrey, A. and Mohamad, M.N.B., Travelling wave solutions to a higher order KdV equation. Chaos, Solitons and Fractals, 1 (1991), 187–194.

Department of Engineering Mathematics
The University
Newcastle upon Tyne, NE1 7RU
England, U.K.

G FICHERA

On a degenerate evolution problem

This paper is originated by a lecture that Prof. Francesco Altomare presented at the Symposium "Recenti sviluppi in Analisi Matematica e sue Applicazioni" held in Bari, 8-9 Nov. 1990, to honor Giovanni Aquaro on his 70th anniversary.(1)

Altomare considers a solution in the classical sense of the following problem

$$a(x)v_{xx}(x,t) - v_t(x,t) = 0 \qquad (0<x<1,\ t>0) \tag{1}$$

$$\lim_{x\to 0^+} a(x)v_{xx}(x,t) = \lim_{x\to 1^-} a(x)v_{xx}(x,t) = 0 \tag{2}$$

$$v(x,0) = v_o(x). \tag{3}$$

He supposes that $a(x)$ is a polynomial which has simple zeroes in 0 and 1 and is positive in $(0,1)$, $v_o(x)$ is a given function satisfying certain conditions. (See Sect.9 of the present paper).

The aim of Altomare is to represent the solution of the problem by special operators. He assumes as known the existence and the uniqueness of a classical solution for the above problem. Actually existence and uniqueness have been studied in the papers [8],[3] concerned with some evolution equations which include as particular case Eq.(1).

The scope of the present paper is to consider Eq.(1) under the general hypothesis $a(x) \equiv x(1-x)p(x)$, with $p(x)$ continuous and positive in $[0,1]$ and to prove existence and uniqueness theorems in various function classes for a boundary value problem which, in the general theory of 2nd order elliptic-parabolic equations, is considered as <u>well-posed</u>.(2) Thereafter this boundary value problem will be related to the boundary value problem (1),(2),(3) and it will be shown how existence theorem for the latter can be deduced from existence theorem for the former, while uniqueness theorem for (1),(2),(3) could fail to hold in some function class.

(1) See [2]. See also the paper [1].

(2) See [4], [5], [9], [10].

1. STATEMENT OF THE PROBLEM IN THE CLASS $\mathcal{X}$.

Let $\mathcal{H}$ be the function space of the real valued functions defined on the closed interval $[0,1]$ of the real axis satisfying the following conditions:

i) $u(x)$ is absolutely continuous in $[0,1]$;

ii) $u(0) = u(1) = 0$;

iii) $u'(x) \in L^2(0,1)$.

If we introduce in $\mathcal{H}$ the following norm

$$\|u\|^2 = \int_0^1 [u'(x)]^2 \, dx,$$

$\mathcal{H}$ can be viewed as a Hilbert space.

Let $p(x)$ be a function continuous and positive in the interval $[0,1]$. Set $a(x) = x(1-x)p(x)$.

Let us denote by $\mathcal{X}$ the class of the real valued functions defined by th conditions:

1) $u(x,t)$ is continuous in the half strip $x \in [0,1]$, $t>0$ with all of it partial derivatives $u_{t^h}(x,t)$ $(h=1,2,\dots)$.

2) For each h the partial derivatives $u_{xt^h}(x,t)$ and $u_{x^2t^h}(x,t)$ exist. Moreover $u_{xt^h}(x,t) \in C^\circ\{[0,1] \times (0,+\infty)\}$, $u_{x^2t^h}(x,t) \in C^\circ\{(0,1) \times (0,+\infty)\}$ (h=0,1,2,

3) If we denote by $C^\circ\{[0,+\infty); \mathcal{H}\}$ the space of functions defined in [0, valued in $\mathcal{H}$ and continuous in $[0,+\infty)$, the function $u(x,t)$ can be viewed as a funtion of t valued in $\mathcal{H}$ and belonging to $C^\circ\{[0,+\infty); \mathcal{H}\}$.

If $v \in \mathcal{H}$ we have for $0 \le x \le 1$

$$|v(x)| = \left|\int_0^x v'(\xi)d\xi\right| \le \int_0^1 |v'(\xi)| d\xi \le \|v\| .$$

Then for any $u(x,t) \in \mathcal{X}$

$$|u(x,t) - u(x,0)| \le \|u(x,t) - u(x,0)\| .$$

Since $u(x,t) \in C^\circ\{[0,+\infty); \mathcal{H}\}$ we deduce $u(x,t) \in C^\circ\{[0,1] \times [0,+\infty)\}$.

Our problem consists in proving existence and uniqueness of a function $u(x,t)$ such that

I) $u(x,t) \in \mathcal{H}$,

II) $$a(x) \frac{\partial^2 u}{\partial x^2} - \frac{\partial u}{\partial t} = 0 \quad (0<x<1,\ t>0) \tag{1.1}$$

III) $$u(0,t) = 0, \tag{1.2}$$

$$u(1,t) = 0, \tag{1.3}$$

$$u(x,0) = f(x), \tag{1.4}$$

where $f(x)$ is a given function of $\mathcal{H}$.

2. AN AUXILIARY EIGENVALUE PROBLEM.

Let us consider the following eigenvalue problem

$$a(x)v''(x) + \lambda v(x) = 0 \tag{2.1}$$

$$v(0) = 0 , \qquad v(1) = 0, \tag{2.2}$$

where $v(x) \in C^1[0,1] \cap C^2(0,1)$.

We have for $0<x<1$

$$v''(x) = \frac{-\lambda}{(1-x)p(x)} \frac{1}{x} \int_0^x v'(\xi)d\xi =$$

$$= \frac{\lambda}{xp(x)} \frac{1}{1-x} \int_x^1 v'(\xi)d\xi .$$

Hence

$$\lim_{x\to 0^+} v''(x) = \frac{-\lambda}{p(0)} v'(0) , \qquad \lim_{x\to 1^-} v''(x) = \frac{\lambda}{p(1)} v'(1),$$

i.e. $v(x) \in C^2[0,1]$.

Set $v''(x) = \varphi(x)$. We have for (2.2)

$$v(x) = \int_0^1 G(x,\xi)\,\varphi(\xi)\,d\xi$$

where

$$G(x,\xi)\begin{cases} = \xi(x-1) & , \quad 0 \le \xi < x \le 1, \\ = x(\xi-1) & , \quad 0 \le x < \xi \le 1. \end{cases}$$

From (2.1) we obtain

$$a(x)\,\varphi(x) + \lambda\int_0^1 G(x,\xi)\,\varphi(\xi)\,d\xi = 0. \tag{2.3}$$

Set

$$\psi(x) = \sqrt{a(x)}\;\varphi(x). \tag{2.4}$$

From (2.3) we have for $0<x<1$

$$\psi(x) + \lambda\int_0^1 K(x,\xi)\,\psi(\xi)\,d\xi = 0 \tag{2.5}$$

where

$$K(x,\xi)\begin{cases} = -\sqrt{\dfrac{\xi(1-x)}{x(1-\xi)p(x)p(\xi)}} & , \quad 0 \le \xi < x \le 1\ , \\ = -\sqrt{\dfrac{x(1-\xi)}{\xi(1-x)p(x)p(\xi)}} & , \quad 0 \le x < \xi \le 1. \end{cases}$$

Hence

$$|K(x,\xi)| \le \frac{1}{\sqrt{p(x)p(\xi)}}\ .$$

If we consider the eigenvalue problem (2.5) in the space $L^2(0,1)$ we know that this problem has like eigenvalues only a sequence of real eigenvalues,each of them with a finite (geometric) multiplicity.

It is easy to see that if $\psi(x) \in L^2(0,1)$ and $\psi(x)$ satisfies (2.5), then $\psi(x) \in C^\circ(0,1)$ and $\psi(x)$ is bounded in $(0,1)$. From this, through (2.4),(2.3), we see that the eigenvalue problem (2.1),(2.2) is perfectly equivalent to the eigenvalue problem (2.5). This means that the two problems have the same eigenvalues and if $\{\psi_k(x)\}$ is a complete sequence of eigenfunctions of (2.5),then the functions

$$v_k(x) = \int_0^1 G(x,\xi) \frac{\psi_k(\xi)}{\sqrt{a(\xi)}} d\xi$$

form a complete sequence of eigenfunctions for the problem (2.1),(2.2). From (2.1),(2.2) we obtain

$$-\int_0^1 [v'(x)]^2 dx + \lambda \int_0^1 \frac{[v(x)]^2}{a(x)} dx = 0.$$

It follows that each eigenvalue of problem (2.1), (2.2) is positive. Let λ_k be the eigenvalue corresponding to v_k. We may assume that

$$0 < \lambda_1 \leq \lambda_2 \leq \cdots \leq \lambda_k \leq \cdots$$

where each λ_k is repeated according to its multiplicity.

Let us consider the particular case $a(x) = x(1-x)$, i.e.

$$x(1-x)v'' + \lambda v = 0, \tag{2.6}$$

$$v(0) = v(1) = 0. \tag{2.7}$$

Assume $\lambda = k(k+1)$ $(k=1,2,\ldots)$ and

$$w_k(x) = x + \sum_{h=1}^{k} (-1)^h \binom{k}{h} \frac{(k+1)\ldots(k+h)}{(h+1)!} x^{h+1}.$$

Since

$$w_k(x) = x F(k+1, -k, 2; x),$$

where F is the hypergeometric function, we see that $w_k(x)$ satisfies (2.6) with $\lambda = k(k+1)$ (see [7] p.11) and moreover $w_k(0) = 0$. We have for $0 < \varepsilon < 1$ (see [12] p.281-282)

$$F(k+\varepsilon+1, -k, 2; 1) = \frac{\Gamma(2)\,\Gamma(1-\varepsilon)}{\Gamma(1-k-\varepsilon)\Gamma(2+k)}$$

and

$$w_k(1) = \lim_{\varepsilon\to 0} F(k+\varepsilon+1, -k, 2; 1) = \lim_{\varepsilon\to 0} \frac{\Gamma(2)\,\Gamma(1-\varepsilon)}{\Gamma(1-k-\varepsilon)\Gamma(2+k)} = 0.$$

The values $\tilde{\lambda}_k = k(k+1)$ $(k=1,2,\ldots)$ and the functions $w_k(x)$ are the only eigenvalues and the only eigenfunctions of the problem (2.6), (2.7). Let $\tilde{\lambda}$ be an eigenvalue of (2.6) (2.7) and w(x) a corresponding eigenfunction.

If $\tilde{\lambda} \neq k(k+1)$ for any k, we have for $k = 1,2,\ldots$

$$-\int_0^1 w'(x)w'_k(x)dx + \tilde{\lambda}\int_0^1 \frac{w(x)w_k(x)}{a(x)} dx = 0,$$

$$-\int_0^1 w'_k(x)w'(x)dx + k(k+1)\int_0^1 \frac{w_k(x)w(x)}{a(x)} dx = 0.$$

Hence

$$\int_0^1 \frac{w_k(x)w(x)}{a(x)} dx = 0 \qquad (k = 1,2,\ldots). \tag{2.8}$$

We have

$$\frac{w_k(x)}{a(x)} = \frac{a_{k\,k-1}x^{k-1} + a_{k\,k-2}x^{k-2} + \ldots + a_{k0}}{p(x)}$$

where

$$a_{k\,k-1} = (-1)^{k+1}\frac{(k+1)\ldots(2k)}{(k+1)!}, \quad a_{10} = 1.$$

Hence Eq.s (2.8) can be written

$$\sum_{h=0}^{k-1} a_{kh}\int_0^1 x^h \frac{w(x)}{p(x)} dx = 0 \qquad (k = 1,2,\ldots),$$

which imply

$$\int_0^1 \frac{w(x)}{p(x)} x^{k-1} dx = 0 \qquad (k = 1,2,\ldots).$$

For the classical Lerch theorem we deduce $w(x) \equiv 0$. If $\tilde{\lambda} = k_0(k_0+1)$ for some positive integer k_0 and if an eigenfunction $\tilde{w}(x)$ exists corresponding to $\tilde{\lambda}$ which is linearly independent from $w_{k_0}(x)$, then a constant c can be determined such that

$$\int_0^1 \frac{w_{k_0}(x)[cw_{k_0}(x) + \tilde{w}(x)]}{a(x)} dx = 0.$$

Assuming now $w(x) = cw_{k_0}(x) + \tilde{w}(x)$, we see that Eq.s (2.8) are satisfied and again $w(x) = 0$, which contradicts the linear independence of $w_{k_0}(x)$ and $\tilde{w}(x)$

To the eigenvalue problem (2.1) (2.2) we associate the Rayleigh-Ritz nctional

$$R(v) = \frac{\int_0^1 [v'(x)]^2 dx}{\int_0^1 \frac{[v(x)]^2}{a(x)} dx}$$

ere $v \in C^1[0,1]$, $v(0) = v(1) = 0$. Let p_0 and P_0 be real numbers such that r $x \in [0,1]$

$$0 < p_0 \leq p(x) \leq P_0 .$$

en

$$p_0 \frac{\int_0^1 [v'(x)]^2 dx}{\int_0^1 \frac{[v(x)]^2}{x(1-x)} dx} \leq R(v) \leq P_0 \frac{\int_0^1 [v'(x)]^2 dx}{\int_0^1 \frac{[v(x)]^2}{x(1-x)} dx} .$$

Hence from the maximum-minimum theory of the eigenvalues (see [11] p.4-13) deduce the following bounds for each eigenvalue λ_k of (2.1) (2.2):

$$p_0 k(k+1) \leq \lambda_k \leq P_0 k(k+1). \tag{2.9}$$

We may suppose that

$$\int_0^1 \frac{v_h(x) v_k(x)}{a(x)} dx = \delta_{hk} . \tag{2.10}$$

Let h(x) be a function of $\mathcal{H}$ such that

$$\int_0^1 v'_k(x) h'(x) dx = 0 \qquad (k = 1,2,\ldots)$$

Set $g(x) = [a(x)]^{-\frac{1}{2}} h(x)$. We have

$$\int_0^1 v'_k(x)h'(x)dx = -\int_0^1 v''_k(x)h(x)dx = -\int_0^1 v''_k(x)\sqrt{a(x)}\, g(x)dx = -\int_0^1 \psi_k(x)g(x)dx=0.$$

From $h(x) \in \mathcal{H}$, it follows that g(x) is bounded in (0,1). Since $\psi_k(x)$ is mplete in $L^2(0,1)$, we obtain $h(x) \equiv 0$. This means that $\{v_k(x)\}$ is a complete stem in the space $\mathcal{H}$. We have

$$\int_0^1 \frac{v'_h(x)}{\sqrt{\lambda_h}} \frac{v'_k(x)}{\sqrt{\lambda_k}} dx = -\int_0^1 \frac{v_h(x)}{\sqrt{\lambda_h}} \frac{v''_k(x)}{\sqrt{\lambda_k}} dx = \frac{\sqrt{\lambda_k}}{\sqrt{\lambda_h}} \int_0^1 \frac{v_h(x)v_k(x)}{a(x)} dx = \delta_{hk} . \tag{2.11}$$

en for any $u(x) \in \mathcal{H}$

$$u'(x) = \sum_{k=1}^{\infty} \frac{1}{\lambda_k} v'_k(x) \int_0^1 u'(\xi)v'_k(\xi)d\xi = \sum_{k=1}^{\infty} v'_k(x) \int_0^1 \frac{u(\xi)v_k(\xi)}{a(\xi)} d\xi$$

where the series converges in the space $L^2(0,1)$. It follows

$$u(x) = \sum_{k=1}^{\infty} v_k(x) \int_0^1 \frac{u(\xi)v_k(\xi)}{a(\xi)} d\xi \tag{2.12}$$

with uniform convergence of the last series in [0,1].

We have

$$v_k(x) = \int_0^1 G(x,\xi)v''_k(\xi)d\xi , \qquad v'_k(x) = \int_0^1 G_x(x,\xi)v''_k(\xi)d\xi .$$

Set

$$q_0 = \sup_{0<x<1} \left(\int_0^1 \frac{|G(x,\xi)|^2}{a(\xi)} d\xi \right)^{1/2}, \quad q_1 = \sup_{0<x<1} \left(\int_0^1 \frac{|G_x(x,\xi)|^2}{a(\xi)} d\xi \right)^{1/2}.$$

Then

$$|v_k(x)| = \lambda_k \left| \int_0^1 G(x,\xi) \frac{v_k(\xi)}{a(\xi)} d\xi \right| \leq \lambda_k q_0 \left(\int_0^1 \frac{[v_k(\xi)]^2}{a(\xi)} d\xi \right)^{1/2}$$

$$|v'_k(x)| = \lambda_k \left| \int_0^1 G_x(x,\xi) \frac{v_k(\xi)}{a(\xi)} d\xi \right| \leq \lambda_k q_1 \left(\int_0^1 \frac{[v_k(\xi)]^2}{a(\xi)} d\xi \right)^{1/2}$$

and for (2.10) we have for $0 \leq x \leq 1$

$$|v_k(x)| \leq q_0 \lambda_k , \tag{2.13}$$

$$|v'_k(x)| \leq q_1 \lambda_k . \tag{2.14}$$

We can write

$$v''_k(x) \begin{cases} = \dfrac{-\lambda_k}{(1-x)p(x)} \dfrac{1}{x} \displaystyle\int_0^x v'_k(\xi)d\xi , & 0 \leq x \leq \tfrac{1}{2}, \\[2ex] = \dfrac{\lambda_k}{xp(x)} \dfrac{1}{1-x} \displaystyle\int_x^1 v'_k(\xi)d\xi , & \tfrac{1}{2} \leq x \leq 1. \end{cases}$$

Hence for $0 \leq x \leq 1$

$$|v''_k(x)| \leq \frac{2}{p_0} q_1 \lambda_k^2 . \tag{2.15}$$

Remark. In the paper [8] the eigenvalue problem for the Eq.(2.6) is considered without imposing on v the classical boundary conditions (2.7), but by requiring that

$$\int_0^1 \frac{[v(x)]^2}{x(1-x)}\,dx < +\infty. \tag{2.16}$$

The Authors of [8] do not notice that for a solution of (2.6) belonging to $C^2(0,1)$ condition (2.16) is equivalent to $v \in C^1[0,1] \cap C^2(0,1)$, $v(0) = v(1) = 0$.

It is evident that these last conditions imply (2.16). Let us now suppose that v(x) is a $C^2(0,1)$ solution of (2.6) satisfying (2.16).

From (2. 6) we get

$$\int_{\frac{1}{2}}^{x} v\, v''\, d\xi = -\lambda \int_{\frac{1}{2}}^{x} \frac{[v(\xi)]^2}{\xi(1-\xi)}\, d\xi .$$

Hence

$$v(x)v'(x) = v(\tfrac{1}{2})v'(\tfrac{1}{2}) + \int_{\frac{1}{2}}^{x} [v'(\xi)]^2 d\xi - \lambda \int_{\frac{1}{2}}^{x} \frac{[v(\xi)]^2}{\xi(1-\xi)}\, d\xi ,$$

$$\int_{\frac{1}{2}}^{x} v(\xi)v'(\xi)d\xi = v(\tfrac{1}{2})v'(\tfrac{1}{2})(x-\tfrac{1}{2}) + \int_{\frac{1}{2}}^{x} d\xi \int_{\frac{1}{2}}^{\xi} [v'(t)]^2 dt$$

$$- \lambda \int_{\frac{1}{2}}^{x} d\xi \int_{\frac{1}{2}}^{\xi} \frac{[v(t)]^2}{t(1-t)}\, dt .$$

Consecutively applying integration by parts and using (2.6) we get

$$\tfrac{1}{2}\,[v(x)]^2 =$$

$$\tfrac{1}{2}\,[v(\tfrac{1}{2}) + v'(\tfrac{1}{2})(x-\tfrac{1}{2})]^2 + 2\lambda\,[v(\tfrac{1}{2})]^2 (x-\tfrac{1}{2})^2$$

$$+ \frac{\lambda}{2} \int_{\frac{1}{2}}^{x} \frac{(\xi - x)(\xi + x - 2x\xi)}{\xi^2(1-\xi)^2}\, [v(\xi)]^2 d\xi + \lambda \int_{\frac{1}{2}}^{x} \frac{\xi - x}{\xi(1-\xi)}\, [v(\xi)]^2 d\xi .$$

Then continuity of v(x) in [0,1] follows.

From (2.6) we deduce

$$v'(x) = v'(\tfrac{1}{2}) - \lambda \int_{\frac{1}{2}}^{x} \frac{v(\xi)}{\xi(1-\xi)} d\xi ,$$

which implies for $0 < x < \frac{1}{2}$

$$|v'(x)| \leq |v'(\tfrac{1}{2})| + |\lambda| \sup_{(0,\frac{1}{2})} \left\{ \frac{|v(\xi)|}{1-\xi} \right\} \left[\log \tfrac{1}{2} - \log x \right] . \tag{2.17}$$

Analogously for $\frac{1}{2} < x < 1$

$$|v'(x)| \leq |v'(\tfrac{1}{2})| + |\lambda| \sup_{(\frac{1}{2},1)} \left\{ \frac{|v(\xi)|}{\xi} \right\} \left[\log \tfrac{1}{2} - \log (1-x) \right] . \tag{2.18}$$

We have for $0 < x < 1$

$$\left| \int_{x}^{\frac{1}{2}} \frac{[v(\xi)]^2}{\xi} d\xi \right| \leq \int_{0}^{1} \frac{[v(\xi)]^2}{\xi(1-\xi)} d\xi .$$

Integration by parts on the integral on the left gives

$$[v(x)]^2 = \frac{1}{\log x} \left\{ [v(\tfrac{1}{2})]^2 \log \tfrac{1}{2} - \int_{x}^{\frac{1}{2}} \frac{[v(\xi)]^2}{\xi} d\xi - 2 \int_{x}^{\frac{1}{2}} v(\xi)v'(\xi)\log\xi \, d\xi \right\}$$

which implies $v(0) = 0$. Analogously we have $v(1) = 0$. Let us suppose $0 < x < \frac{1}{2}$, then

$$|v''(x)| \leq |\lambda| \frac{|v(x)|}{x(1-x)} \leq 2 \frac{|\lambda|}{x} \int_{0}^{x} |v'(\xi)| d\xi ,$$

hence, for (2.17), we have in $(0,\frac{1}{2})$

$$|v''(x)| \leq c \, |\log x| ;$$

c is a positive constant. For (2.18) we have in $(\frac{1}{2},1)$

$$|v''(x)| \leq c \, |\log(1-x)| .$$

Hence $v \in C^1 [0,1]$.

3. UNIQUENESS THEOREM.

We need now to prove that there exists at most one solution of the Problem I), II), III) stated in Sect.1.

For any fixed $t \in [0,+\infty)$ because of property 3 of the class $\mathcal{H}$ and of (2.12) we have in $[0,1]$

$$u(x,t) = \sum_{k=1}^{\infty} v_k(x) \int_0^1 \frac{u(\xi,t)v_k(\xi)}{a(\xi)}\, d\xi .$$

Set

$$U_k(t) = \int_0^1 \frac{u(x,t)v_k(x)}{a(x)}\, dx.$$

Since $u_t(x,t) \in C^\circ \{[0,1] \times (0,+\infty)\}$,we have for $t>0$

$$U_k'(t) = \int_0^1 \frac{u_t(x,t)v_k(x)}{a(x)}\, dx. \tag{3.1}$$

From (1.1) we deduce for $0<x<1$, $t>0$

$$u_{x^2}(x,t) = \frac{1}{a(x)}\, u_t(x,t),$$

which implies that, for any fixed $t>0$, $u_{x^2}(x,t) \in C^\circ[0,1]$. Hence

$$\int_0^1 u_{x^2}(x,t)v_k(x)dx = \int_0^1 u(x,t)v_k''(x)dx = = -\lambda_k U_k(t). \tag{3.2}$$

From (1.1),(3.1),(3.2) we deduce

$$U_k'(t) + \lambda_k U_k(t) = 0.$$

Then

$$U_k(t) = c_k e^{-\lambda_k t}.$$

Since $u(x,t) \in \mathcal{H}$, because of property 3) we have

$$\lim_{t\to 0^+} U_k(t) = \int_0^1 \frac{f(x)v_k(x)}{a(x)}\, dx = c_k .$$

Then for any $t>0$ we deduce for $x \in [0,1]$

$$\boxed{u(x,t) = \sum_{k=1}^{\infty} v_k(x)e^{-\lambda_k t} \int_0^1 \frac{f(\xi)v_k(\xi)}{a(\xi)}\, d\xi} \tag{3.3}$$

which for $f\equiv 0$ implies $u(x,t) \equiv 0$.

4. EXISTENCE THEOREM.

Let us now prove that the development on the right hand side of (3.3) gives the unique solution of Problem I),II),III) of Sect.1. We have

$$f(x) = \sum_{k=1}^{\infty} c_k v_k(x) \qquad , \qquad f'(x) = \sum_{k=1}^{\infty} c_k v'_k(x)$$

where the first series converges uniformly in $[0,1]$ and the second one in $L^2(0,1)$.

For (2.11) we have

$$\sum_{k=1}^{\infty} \lambda_k c_k^2 < +\infty .$$

Let c_0 be a positive constant such that for $k=1,2,...$

$$c_k \le \frac{c_0}{\sqrt{\lambda_k}} .$$

Assuming $0 \le x \le 1$, $t > 0$ we have from (3.3) differentiating formally

$$u_{t^h}(x,t) = \sum_{k=1}^{\infty} (-1)^h c_k \lambda_k^h v_k(x)e^{-\lambda_k t} ,$$

$$u_{xt^h}(x,t) = \sum_{k=1}^{\infty} (-1)^h c_k \lambda_k^h v'_k(x)e^{-\lambda_k t} ,$$

$$u_{x^2t^h}(x,t) = \sum_{k=1}^{\infty} (-1)^h c_k \lambda_k^h v''_k(x)e^{-\lambda_k t}$$

$$(h = 0,1,2,...) .$$

Because of (2.13), (2.14), (2.15) the h-th term of each series on the right hand sides is, for $t\ge t_0>0$, dominated, respectively, by

$$c_o q_o \lambda_k^{h+\frac{1}{2}} e^{-\lambda_k t_o},$$

$$c_o q_1 \lambda_k^{h+\frac{1}{2}} e^{-\lambda_k t_o},$$

$$2c_o \frac{q_1}{p_o} \lambda_k^{h+\frac{3}{2}} e^{-\lambda_k t_o}.$$

Then the above series, because of (2.9), are totally convergent in the half strip $0\leq x\leq 1$, $t_o\geq 0$. Because of the abitrariness of t_o, it is proved that u(x,t) enjoys properties 1) and 2) of functions of class $\mathcal{K}$ and moreover that Eq.s (1.1),(1.2),(1.3) are satisfied.

We have for t>0

$$\|u(x,t) - f(x)\|^2 = \int_0^1 [u_x(x,t) - f'(x)]^2 dx = \sum_{k=1}^{\infty} \lambda_k c_k^2 (1 - e^{-\lambda_k t})^2.$$

Given $\varepsilon>0$, let n_ε be such that

$$\sum_{k=n_\varepsilon+1}^{\infty} \lambda_k c_k^2 < \frac{\varepsilon}{2} .$$

Assume $t_\varepsilon>0$ such that for $0<t<t_\varepsilon$

$$\sum_{k=1}^{n_\varepsilon} \lambda_k c_k^2 (1 - e^{-\lambda_k t})^2 < \frac{\varepsilon}{2} .$$

Hence, for $0<t<t_\varepsilon$

$$\|u(x,t) - f(x)\| < \varepsilon .$$

This proves that property 3) of the functions of the class $\mathcal{K}$ and Eq. (1.4) are satisfied by u(x,t).

5. MAXIMUM PRINCIPLE.

5.1 Let u(x,t) be the solution of problem (1.1),(1.2),(1.3),(1.4) in the class $\mathcal{K}$. The following maximum principle holds:

$$|u(x,t)| \leq \max_{[0,1]} |f(x)| \tag{5.1}$$

$$(x,t) \in \bar{S} = [0,1] \times [0,+\infty).$$

Let c>0 be an arbitrarily given constant. Set

$$u(x,t) = e^{ct}\, v(x,t).$$

The function v(x,t) belongs to $\mathcal{B}$ and satisfies the following equations

$$a(x)v_{xx}(x,t) - v_t(x,t) - cv(x,t) = 0 \qquad (5.2)$$

$$v(0,t) = v(1,t) = 0 \qquad t \geq 0 \qquad (5.3)$$

$$v(x,0) = f(x) \qquad 0 \leq x \leq 1\ . \qquad (5.4)$$

Set

$$Lw = w_{xx} - \frac{1}{a(x)}\, w_t - \frac{cw}{a(x)}$$

$$L^* \varphi = \varphi_{xx} + \frac{1}{a(x)}\, \varphi_t - \frac{c\varphi}{a(x)}\ .$$

Assume w(x,t), φ (x,t) such that $w, \varphi, w_x, \varphi_x, w_t, \varphi_t, w_{xx}, \varphi_{xx} \in [0,1]\times(0,+\infty)$. Suppose that w(0,t) = w(1,t) = 0 for t > 0.

We have, denoting by $R_{\varepsilon,T}$ the rectangle $0\leq x\leq 1$, $\varepsilon \leq t\leq T$ $(0<\varepsilon<T)$,

$$\begin{aligned}\iint_{R_{\varepsilon,T}} \varphi\, Lw\, dxdt - \iint_{R_{\varepsilon,T}} wL^*\varphi\, dxdt \\ = \int_0^T \varphi(1,t)w_x(1,t)dt - \int_0^T \varphi(0,t)w_x(0,t)dt \qquad (5.5)\\ - \int_0^1 \frac{w(x,t)\,\varphi(x,t)}{a(x)}\, dx + \int_0^1 \frac{w(x,\varepsilon)\,\varphi(x,\varepsilon)}{a(x)} dx\ .\end{aligned}$$

If m is any positive integer, because of (5.3) we may assume $w=v^{2m}$. Take $\varphi \equiv 1$. From (5.5), (5,3) we get

$$2m(2m-1)\iint_{R_{\varepsilon,T}} v^{2m-2} v_x^2\, dx\, dt + 2m\iint_{R_{\varepsilon,T}} \frac{cv^{2m}}{a(x)}\, dxdt =$$

$$\int_0^1 \frac{[v(x,\varepsilon)]^{2m}}{a(x)} dx - \int_0^1 \frac{[v(x,T)]^{2m}}{a(x)} dx,$$

which implies

$$(2mc)^{\frac{1}{2m-2}} \left(\iint_{R_{\varepsilon,T}} v^{2m-2}\, \frac{v^2}{a(x)}\, dx \right)^{\frac{1}{2m-2}}$$

$$\leq \left(\int_0^1 [v(x,\varepsilon)]^{2m-2}\, \frac{[v(x,\varepsilon)]^2}{a(x)}\, dx \right)^{\frac{1}{2m-2}}$$

and for $m \to \infty$

$$\max_{R_{\varepsilon,T}} |v(x,t)| = \max_{[0,1]} |v(x,\varepsilon)| .$$

Hence for $0 \leq x \leq 1$, $0 < \varepsilon \leq t \leq T$

$$|u(x,t)| \leq e^{cT} |v(x,t)| \leq e^{cT} \max_{[0,1]} |v(x,\varepsilon)|$$

which for $c \to 0^+$, $\varepsilon \to 0^+$, and for the arbitrariness of T implies (5.1).

6. CONTINUOUS SOLUTIONS.

Let us denote by $\mathcal{H}_0$ the function class defined by conditions 1) and 2) which we have used for defining $\mathcal{H}$. Condition 3) is replaced by the following:

3_0) $u(x,t)$ is continuous in $\bar{S} = [0,1] \times [0, +\infty)$.

6.I <u>Given</u> $f(x)$ <u>belonging to</u> $C^0[0,1]$ <u>and vanishing at</u> 0 <u>and</u> 1, <u>there exists one and only one solution</u> $u(x,t)$ <u>in the class</u> $\mathcal{H}_0$ <u>of the problem (1,1), (1,2),(1,3),(1,4)</u>; $u(x,t)$, <u>for</u> $t>0$, <u>is given by (3.3)</u>.

By an argument similar to the one used in Sect.3 we see that if $u(x,t)$ exists it is given, for $t>0$, by (3.3). This proves uniqueness.

Let $\{f_n(x)\}$ be a sequence of functions of the space $\mathcal{H}$ which converges uniformly to $f(x)$ in the interval $[0,1]$ (3).

By the arguments used in Sect.4 and based on inequalities (2.13),(2.14), (2.15), we see that $u(x,t)$ defined by (3.3) is a function satisfying conditions 1),2) which define $\mathcal{H}_0$. Moreover $u(x,t)$ is a solution of (1.1) in $(0,1)\times(0,+\infty)$.

Set

$$u_n(x,t) = \sum_{k=1}^{\infty} v_k(x) e^{-\lambda_k t} \int_0^1 \frac{f_n(\xi) v_k(\xi)}{a(\xi)} d\xi ,$$

---

(3) Set

$$\varphi_m(x) = \sum_{k=1}^{m} 2\sin k\pi x \int_0^1 f(\xi) \sin k\pi\xi \, d\xi ,$$

we may assume

$$f_n(x) = \frac{1}{n} \sum_{m=1}^{n} \varphi_m(x).$$

we have in $[0,1]\times[t_0, +\infty)$ for any $t_0 > 0$

$$|u(x,t) - u_n(x,t)| \leq \sum_{k=1}^{\infty} |v_k(x)|\ e^{-\lambda_k t_0} \frac{1}{\lambda_k} \int_0^1 |f(x) - f_n(x)|\,|v''_k(\xi)|\,d\xi$$

$$\leq \frac{2q_0 q_1}{p_0} \max_{[0,1]} \left| f(x) - f_n(x) \right| \sum_{k=1}^{\infty} \lambda_k^2 e^{-\lambda_k t_0}.$$

This implies

$$\lim_{n\to\infty} u_n(x,t) = u(x,t)$$

uniformly in $[0,1] \times [t_0, +\infty)$.

On the other hand we have in $[0,1]\times[0, +\infty)$ for arbitrary n and s (maximum principle):

$$|u_{n+s}(x,t) - u_n(x,t)| \leq \max_{[0,1]} |f_{n+s}(x) - f_n(x)|.$$

Hence the function

$$u(x,t) \begin{cases} = \sum_{k=1}^{\infty} v_k(x)\, e^{-\lambda_k t} \int_0^1 \frac{f(\xi)v\ (\xi)}{a(\xi)}\, d\xi & \text{for } t>0 \\ = f(x) & \text{for } t=0 \end{cases}$$

belongs to $\mathcal{K}_0$ and is the solution of the problem.

Let us remark that the function u(x,t) satisfies in $\bar{S}$ the maximum principle (5.1).

7. DISCONTINUOUS SOLUTIONS.

It is well known that in heat diffusion, for the classical problem analogous to problem (1.1),(1.2),(1.3),(1.4), not only solutions continuous in $[0,1]\times[0,+\infty)$ exist but also solutions which are continuous in this closed half-strip $\bar{S}$ except in the points $\{0,0\}$ and $\{0,1\}$.

We intend to carry out a similar analysis for Eq.(1.1). To this end let us denote by $\mathcal{K}_1$ the function class of real valued functions u(x,t) which satisfy conditions 1) and 2) used for defining the class $\mathcal{K}$. Condition 3) is replaced by the following one:

3_1) A (necessarily unique) function f(x) exists belonging to $L^1(0,1)$ such that for any $w(x) \in \mathcal{H}$ one has

$$\lim_{t\to 0^+} \int_0^1 \frac{u(x,t)w(x)}{a(x)}\,dx = \int_0^1 \frac{f(x)w(x)}{a(x)}\,dx . \tag{7.1}$$

The following theorem holds:

7.I <u>Given</u> $f(x) \in L^q(0,1)$, <u>with</u> $q>2$, <u>there exists one and only one function</u> $u(x,t)$ <u>of the class</u> $\mathcal{K}_1$ <u>satisfying the conditions (1.1),(1.2),(1.3),(7.1)</u>.

Let us suppose that u(x,t) is a solution of the considered problem. Arguing like in Sect.3 we see that

$$U_k(t) = \int_0^1 \frac{u(x,t)v_k(x)}{a(x)}\,dx = c_k e^{-\lambda_k t} .$$

Hence, because of (7.1), we have

$$c_k = \int_0^1 \frac{f(x)v_k(x)}{a(x)}\,dx .$$

This implies that, for any $t>0$, (3.3) holds. Then uniqueness follows.

We have, because of (2.15),

$$|c_k| = \frac{1}{\lambda_k}\left|\int_0^1 f(x)v''_k(x)\,dx\right| \leq \frac{2}{p_0}\, q_1\, \lambda_k \int_0^1 |f(x)|\,dx .$$

An argument similar to the one used in Sect.4 shows that the series obtained from (3.3) by the formal differentiations

$$\frac{\partial}{\partial t^h} , \quad \frac{\partial}{\partial x\, \partial t^h} , \quad \frac{\partial}{\partial x^2\, \partial t^h} \qquad (h=0,1,2,\ldots)$$

are totally convergent for $0\leq x\leq 1$, $t\geq t_0>0$.

Hence u(x,t) given by (3.3) satisfies condition 1),2) and, moreover, $u_{x^2 t^h}(x,t) \in C^0\{[0,1]\times(0,+\infty)\}$.

Let w(x) be any function of $\mathcal{H}$ and let u(x,t) be given by (3.3). We have for any $t > 0$

$$\int_0^1 \frac{u(x,t)w(x)}{a(x)}\,dx = \sum_{k=1}^{\infty} e^{-\lambda_k t} \int_0^1 \frac{w(x)v_k(x)}{a(x)}\,dx \int_0^1 \frac{f(\xi)v_k(\xi)}{a(\xi)}\,d\xi$$

$$= \int_0^1 \frac{W(\xi,t)f(\xi)}{a(\xi)}\,d\xi ;$$

W(x,t) is the solution in the class $\mathcal{K}$ of problem (1.1),(1.2),(1.3),(1.4) where f(x) is replaced by w(x).

Set $\tilde{w}(\xi,t) = W(\xi,t) - w(\xi)$, we have

$$\left| \int_0^1 \frac{\tilde{w}(\xi,t)f(\xi)}{a(\xi)}\, d\xi \right| \le \left| \int_0^{\frac{1}{2}} \left(\frac{f(\xi)}{a(\xi)} \int_0^{\xi} \tilde{w}_x(x,t)dx \right) d\xi \right|$$

$$+ \left| \int_{\frac{1}{2}}^1 \left(\frac{f(\xi)}{a(\xi)} \int_1^{\xi} \tilde{w}_x(x,t)dx \right) d\xi \right| \le \left(\int_0^{\frac{1}{2}} \frac{|f(\xi)|\, d\xi}{\xi^{\frac{1}{2}}(1-\xi)p(\xi)} \right.$$

$$\left. + \int_{\frac{1}{2}}^1 \frac{|f(\xi)|\, d\xi}{\xi(1-\xi)^{\frac{1}{2}}\, p(\xi)} \right) \left(\int_0^1 |W_x(x,t) - w'(x)|^2\, dx \right)^{\frac{1}{2}}$$

which implies

$$\lim_{t\to 0^+} \int_0^1 \frac{u(x,t)w(x)}{a(x)}\, dx = \lim_{t\to 0^+} \int_0^1 \frac{W(\xi,t)f(\xi)}{w(\xi)}\, d\xi = \int_0^1 \frac{f(x)w(x)}{a(x)}\, dx$$

i.e. (7.1). Proof of existence is now complete.

It must be remarked that for uniqueness only $f(x) \in L^1(0,1)$ was needed. The stronger hypothesis $f(x) \in L^q(0,1)$, $q>2$ was used only for proving existence.

Let us now prove a regularization theorem for the solution u(x,t) of problem (1.1),(1.2),(1.3),(7.1).

To this end from now on in this and in the next Section we shall assume that the function p(x), besides being continuous and positive in [0,1], is of class $C^1(0,1)$ and p'(x) satisfies a uniform Hölder condition in any interval $[a,b] \subset (0,1)$. We set

$$S = [0,1] \times (0,+\infty).$$

7.II Suppose that $u(x,t) \in \mathcal{K}_1$, with $f(x) \in L^q(0,1)$ $(q>2)$ and assume that f(x) is continuous in the interval (α,β), $0 < \alpha < \beta < 1$, then u(x,t) is continuous in $S \cup I_{\alpha\beta}$ where $I_{\alpha\beta}$ is the interval (α,β) of the x-axis. Moreover

$$u(x,0) = f(x) \qquad \text{for } \alpha < x < \beta.$$

Let us consider the space $\mathcal{H}[\alpha,\beta]$ formed by the functions w(x) which are absolutely continuous in $[\alpha,\beta]$, vanish in α and in β and are such that

$w'(x) \in L^2(\alpha,\beta)$. It is evident that $\mathcal{H} = \mathcal{H}\,[0,1]$ and that $\mathcal{H}\,[\alpha,\beta]$ is the restriction to $[\alpha,\beta]$ of all the functions of $\mathcal{H}\,[0,1]$ which vanish outside of (α,β). Since $u(x,t) \in \mathcal{K}_1$ we have for any $w \in \mathcal{H}\,[\alpha,\beta]$

$$\lim_{t\to 0^+} \int_\alpha^\beta \frac{u(x,t)w(x)}{a(x)}\,dx = \int_\alpha^\beta \frac{f(x)w(x)}{a(x)}\,dx\,. \tag{7.2}$$

Let us consider $\mathcal{H}\,[\alpha,\beta]$ as an Hilbert space endowed with the scalar product

$$(w,z)_{I_{\alpha\beta}} = \int_\alpha^\beta w'(x)z'(x)dx.$$

Set for $t>0$

$$F_t(w) = \int_\alpha^\beta \frac{u(x,t)w(x)}{a(x)}dx\,.$$

F_t is a linear bounded functional in the space $\mathcal{H}\,[\alpha,\beta]$. Because of (5.2) and of the Banach-Steinhaus theorem we have for $t>0$

$$\|F_t\|_{I_{\alpha\beta}} \leq L\,, \tag{7.3}$$

where L is a positive constant independent of t.

Set

$$\Gamma(x,\xi) \begin{cases} = \dfrac{(\xi-\alpha)(x-\beta)}{\beta-\alpha} & \alpha \leq \xi \leq x\,, \\[2ex] = \dfrac{(x-\alpha)(\xi-\beta)}{\beta-\alpha} & x \leq \xi \leq \beta\,, \end{cases}$$

and

$$U(x,t) = \int_\alpha^\beta \Gamma(x,\xi)\,\frac{u(\xi,t)}{a(\xi)}\,d\xi$$

we have

$$F_t(w) = \int_\alpha^\beta U_{xx}(x,t)w(x)dx = -\int_\alpha^\beta U_x(x,t)w'(x)dx.$$

Hence for (7.3)

$$\int_\alpha^\beta |U_x(x,t)|^2\,dx = -\int_\alpha^\beta\int_\alpha^\beta \Gamma(x,\xi)\,\frac{u(x,t)u(\xi,t)}{a(x)a(\xi)}\,dx\,d\xi \leq L^2.$$

On the other hand we have

$$-\int_\alpha^\beta\int_\alpha^\beta \Gamma(x,\xi)\,\frac{[u(x,t)-a(x)]\,[u(\xi,t)-\alpha(\xi)]}{a(x)a(\xi)}\,dx\,d\xi \;\geq 0 .$$

Hence

$$-2\int_\alpha^\beta\int_\alpha^\beta \Gamma(x,\xi)\,\frac{u(x,t)}{a(x)}\,dx\,d\xi$$

$$\leq -\int_\alpha^\beta\int_\alpha^\beta \Gamma(x,\xi)\frac{u(x,t)u(\xi,t)}{a(x)a(\xi)}\,dx d\xi \;-\; \int_\alpha^\beta\int_\alpha^\beta \Gamma(x,\xi)dx\,d\xi .$$

We deduce for any t>0

$$-\int_\alpha^\beta \frac{u(x,t)}{a(x)}\,(x-\alpha)(x-\beta)dx \;\leq\; L^2 + \frac{(\beta-\alpha)^3}{12}\,. \tag{7.4}$$

Let σ be a positive number less than $\frac{1}{2}$. Consider the function

$$f_\sigma(x)\;\begin{cases} = 0 & 0\leq x<\sigma\;,\quad 1-\sigma<x\leq 1 \\ = f(x) & \sigma\leq x\leq 1-\sigma\,. \end{cases}$$

Let $u_\sigma(x,t)$ be the solution in the class $\mathcal{K}_1$ of the problem (1.1),(1.2), (1.3),(7.1), when we substitute f(x) by $f_\sigma(x)$. We have for $t\geq t_0>0$

$$|u(x,t)-u_\sigma(x,t)| = \left|\sum_{k=1}^{\infty} v_k(x)\,e^{-\lambda_k t}\int_0^1 \frac{[f(\xi)-f_\sigma(\xi)]\,v_k(\xi)}{a(\xi)}dx\right|$$

$$\leq \frac{2q_0q_1}{p_0}\left(\int_0^\sigma |f(\xi)|\,d\xi + \int_{1-\sigma}^1 |f(\xi)|\,d\xi\right)\sum_{k=1}^{\infty}\lambda_k^2\,e^{-\lambda_k t_0}.$$

Hence

$$\lim_{\sigma\to 0} u_\sigma(x,t) = u(x,t) \tag{7.5}$$

uniformly in $[0,1]\times[t_0,+\infty)$.

Set

$$\rho(\tau)\;\begin{cases} = \gamma\exp\,[(\tau^2-1)^{-1}] & |\tau|<1 \\ = 0 & |\tau|\geq 1 \end{cases}$$

where

$$\gamma = \left\{\int_0^1 \exp\,[(\tau^2-1)]^{-1}\,d\tau\right\}^{-1}.$$

Assume $0<\varepsilon<\sigma$, $\quad \rho_\varepsilon(x) = (2\varepsilon)^{-1}\rho\left(\frac{|x|}{\varepsilon}\right)$

and

$$f_{\sigma\varepsilon}(x) = \int_0^1 \rho_\varepsilon(x-y)f_\sigma(y)dy.$$

It is well known that

$$\lim_{\varepsilon\to 0}\int_0^1 |f_{\sigma\varepsilon}(x) - f_\sigma(x)|^q\, dx = 0.$$

Moreover $f_{\sigma\varepsilon}(x) \in C_0^\infty(0,1)$, hence $f_{\sigma\varepsilon}(x) \in \mathfrak{H}$.

Let us suppose that $f(x) \geq 0$ in $(0,1)$ which implies $f_\sigma(x) \geq 0$ and $f_{\sigma\varepsilon}(x) \geq 0$ in $[0,1]$. Denote by $u_{\sigma\varepsilon}(x,t)$ the solution in the class $\mathfrak{X}$ of the problem (1.1),(1.2),(1.3),(1.4) when $f(x)$ is replaced by $f_{\sigma\varepsilon}(x)$. Assume that $u_{\sigma\varepsilon}(x,t) < 0$ for some (x,t) with $0 < x < 1$, $0<t \leq T$ ($T>0$ arbitrarily chosen). Let (x_0,t_0) be the point of the rectangle $0\leq x\leq 1$, $0\leq t\leq T$, where the function $\zeta(x,t) = e^{-t} u_{\sigma\varepsilon}(x,t)$ assumes its minimum value. We have

$$a(x_0)\zeta_{xx}(x_0,t_0) - \zeta_t(x_0,t_0) - \zeta(x_0,t_0) = 0 \ , \ \zeta_{xx}(x_0,t_0) \geq 0 \ ,$$

$$\zeta_t(x_0,t_0) \leq 0 \quad , \quad \zeta(x_0, t_0) < 0 \, .$$

This is impossible. Hence $u_{\sigma\varepsilon}(x,t) \geq 0$ in $[0,1]\times[0,+\infty)$. We have for $t \geq t_0 > 0$

$$|u_\sigma(x,t) - u_{\sigma\varepsilon}(x,t)| \leq \frac{2q_0q_1}{p_0}\left(\int_0^1 |f_\sigma(\xi) - f_{\sigma\varepsilon}(\xi)|\, d\xi\right)\sum_{k=1}^\infty \lambda_k^2 e^{-\lambda_k t_0}$$

which implies

$$\lim_{\varepsilon\to 0} u_{\sigma\varepsilon}(x,t) = u_\sigma(x,t) \tag{7.6}$$

uniformly in $[0,1]\times[t_0, +\infty)$. From (7.6) and (7.5) we deduce that the function $u(x,t)$ is non-negative in $[0,1]\times(0, +\infty)$.

Let η be any positive number less than $\frac{1}{2}(\beta-\alpha)$. We have from (7.4) for any $t>0$

$$\int_{\alpha+\eta}^{\beta-\eta} u(x,t)dx \leq \frac{\max\limits_{\alpha\leq x\leq\beta} a(x)}{\eta(\beta-\alpha-\eta)}\left(L^2 + \frac{(\beta-\alpha)^3}{12}\right).$$

This means that for any interval (a,b) such that $[a,b] \subset (\alpha,\beta)$ the function of t ($t>0$)

$$\int_a^b u(x,t)\, dx$$

is bounded.

If $f(x)$ assumes in $(0,1)$ positive and negative values, we set

$$f_1(x) = \frac{|f(x)| + f(x)}{2} \quad , \qquad f_2(x) = \frac{|f(x)| - f(x)}{2} \,.$$

We have in (0,1): $f_1(x) \geq 0$, $f_2(x) \geq 0$, $f(x) = f_1(x) - f_2(x)$. If we denote by $u_h(x,t)$ the solution in the class $\mathcal{K}_1$ of the problem (1.1),(1.2),(1.3),(7.1) where the datum is $f_h(x)$, we have

$$u(x,t) = u_1(x,t) - u_2(x,t) \quad , \quad |u(x,t)| \leq u_1(x,t) + u_2(x,t).$$

Then the boundedness of

$$\int_a^b |u(x,t)| \, dx$$

for $t > 0$, follows.

Given $T > 0$, the integral

$$\int_0^T dt \int_a^b |u(x,t)| \, dx$$

is finite, hence, by Fubini's theorem we can choose a and b almost everywhere in (α, β) such that the integrals

$$\int_0^T |u(a,t)| \, dt \quad , \quad \int_0^T |u(b,t)| \, dt$$

be finite.

Because of the assumptions on a(x) a Green function $G(x,t;\xi,\tau)$ exists in the strip $a < x < b$ for the operators

$$L = \frac{\partial^2}{\partial x^2} - \frac{1}{a(x)} \frac{\partial}{\partial x} \quad , \quad L^* = \frac{\partial^2}{\partial x^2} + \frac{1}{a(x)} \frac{\partial}{\partial t}$$

(see [6] p.81-85).

Let $\varphi(x)$ be a function of $C_0^\circ(a,b)$. Assume $t>0$, the function

$$v(\xi,\tau) = \int_a^b \varphi(x) \, G(x,t;\xi,\tau) \, dx \tag{7.7}$$

is continuous in $[0,1]\times[0,t]$, the derivatives $v_\xi(\xi,\tau)$, $v_{\xi\xi}(\xi,\tau)$, $v_\tau(\xi,\tau)$ exist and are continuous in $[0,1]\times[0,t)$. Moreover we have

$$v_\xi(\xi,\tau) = \int_a^b \varphi(x) G_\xi(x,t;\xi,\tau) dx \quad , \quad L^* v = 0,$$

$$v(\xi,t) = \varphi(\xi) \quad ; \quad v(0,\tau) = v(1,\tau) = 0 \text{ for } 0 \leq \tau \leq t.$$

By using integration by parts we get for $0 < \varepsilon < t$

$$\int_0^t u(b,\tau)\, v_\xi(b,\tau)\, d\tau - \int_0^t u(a,\tau)\, v_\xi(a,\tau)\, d\tau =$$
$$\int_a^b \frac{u(\xi,\varepsilon)\, v(\xi,\varepsilon)}{a(\xi)}\, d\xi - \int_a^b \frac{u(\xi,t)\, v(\xi,t)}{a(\xi)}\, d\xi . \tag{7.8}$$

We have

$$\int_a^b \frac{u(\xi,\varepsilon)\, v(\xi,\varepsilon)}{a(\xi)}\, d\xi - \int_a^b \frac{f(\xi)\, v(\xi,0)}{a(\xi)}\, d\xi =$$

$$\int_a^b \frac{u(\xi,\varepsilon)[v(\xi,\varepsilon) - v(\xi,0)]}{a(\xi)}\, d\xi + \int_a^b \frac{u(\xi,\varepsilon)\, v(\xi,0)}{a(\xi)}\, d\xi - \int_a^b \frac{f(\xi)\, v(\xi,0)}{a(\xi)}\, d\xi .$$

Because of the boundedness of

$$\int_a^b |u(\xi,\varepsilon)|\, d\xi ,$$

of the continuity of $v(\xi,\varepsilon)$ in $[0,1]\times[0,t]$ and being $v(\xi,0) \in \mathcal{H}[a,b]$, we have

$$\lim_{\varepsilon\to 0^+} \int_a^b \frac{u(\xi,\varepsilon)\, v(\xi,\varepsilon)}{a(\xi)}\, d\xi = \int_a^b \frac{f(\xi)\, v(\xi,0)}{a(\xi)}\, d\xi$$

From (7.8) we deduce for $\varepsilon \to 0^+$

$$\int_0^t u(b,\tau)\, v_\xi(b,\tau)d\tau - \int_0^t u(a,\tau)\, v_\xi(a,\tau)d\tau =$$
$$\int_a^b \frac{f(\xi)\, v(\xi,0)}{a(\xi)}\, d\xi - \int_a^b \frac{u(\xi,t)\, \varphi(\xi)}{a(\xi)}\, d\xi . \tag{7.9}$$

Inserting (7.7) into (7.9) we have

$$\int_a^b \varphi(x)dx \int_0^t u(b,\tau)G_\xi(x,t;b,\tau)d\tau - \int_a^b \varphi(x)dx \int_0^t u(a,\tau)G_\xi(x,t;a,\tau)d\tau =$$
$$\int_a^b \varphi(x)dx \int_a^b \frac{f(\xi)}{a(\xi)}\, G(x,t;\xi,0)d\xi - \int_a^b \varphi(x)\, \frac{u(x,t)}{a(x)}\, dx .$$

For the arbitrariness of $\varphi(x)$ we deduce for $a<x<b$, $t>0$,

$$\frac{u(x,t)}{a(x)} = \int_a^b \frac{f(\xi)}{a(\xi)}\, G(x,t;\xi,0)d\xi + \int_0^t u(a,\tau)\, G_\xi(x,t;a,\tau)d\tau$$
$$- \int_0^t u(b,\tau)\, G_\xi(x,t;b,\tau)d\tau .$$

This representation of $u(x,t)$ shows that $u(x,t)$ is continuous in $(a,b)\times[0,+\infty)$ and $u(x,0) = f(x)$ in (a,b). This proves the theorem.

If $f(x) \in C^\circ(0,1) \cap L^q(0,1)$ $(q>2)$ as a corollary of this theorem we deduce that a (unique) solution of the problem (1.1),(1.2),(1.3) exists which belongs to $\mathcal{K}_1$, is continuous in $S \cup I_{01}$, $I_{01} \equiv \{x \in (0,1), t=0\}$, and is such that $u(x,0) = f(x)$, $0<x<1$.

8. <u>BOUNDED SOLUTIONS.</u>

Let $\mathcal{K}_2$ be the function class defined by conditions 1), 2) of Sect.1 and by

3_2) $u(x,t)$ <u>is bounded in</u> $S = [0,1]\times(0,+\infty)$ <u>and continuous in</u> $S \cup I_{01}$.

8.I <u>Given</u> $f(x) \in C^\circ(0,1)$, <u>bounded in</u> $(0,1)$, <u>there exists one and only one solution of the problem (1.1),(1.2),(1.3),(1.4) in the class</u> $\mathcal{K}_2$.

If we prove that $\mathcal{K}_2 \subset \mathcal{K}_1$ then uniqueness follows from the uniqueness in the class $\mathcal{K}_1$.

For proving existence we need only to prove that the solution in the class $\mathcal{K}_1$ is bounded in S if $f(x)$ satisfies conditions stated in theor.8.I.

Let $w(x)$ be a function of $\mathcal{H}$. We have

$$\frac{|w(x)|}{a(x)} \begin{cases} = \dfrac{1}{a(x)} \left|\displaystyle\int_0^x w'(\xi)d\xi\right| \leq \dfrac{1}{\sqrt{x}\,(1-x)p(x)} \left(\displaystyle\int_0^1 |w'|^2 d\xi\right)^{\frac{1}{2}} & 0 < x \leq \frac{1}{2} \\ = \dfrac{1}{a(x)} \left|\displaystyle\int_1^x w'(\xi)d\xi\right| \leq \dfrac{1}{x\sqrt{1-x}\;p(x)} \left(\displaystyle\int_0^1 |w'|^2 dx\right)^{\frac{1}{2}} & \frac{1}{2} \leq x < 1 . \end{cases}$$

Let $|u(x,t)| \leq L$ for $(x,t) \in S$. We have

$$\lim_{t\to 0^+} \frac{u(x,t)w(x)}{a(x)} = \frac{f(x)w(x)}{a(x)} \qquad 0 < x < 1$$

and

$$\left|\frac{u(x,t)w(x)}{a(x)}\right| \leq L\,\frac{|w(x)|}{a(x)} \leq \frac{2}{p(x)}\left(\frac{1}{\sqrt{x}} + \frac{1}{\sqrt{1-x}}\right)\left(\int_0^1 |w'|^2 d\xi\right)^{\frac{1}{2}}.$$

Hence

$$\lim_{t\to 0^+} \int_0^1 \frac{u(x,t)w(x)}{a(x)}\, dx = \int_0^1 \frac{f(x)w(x)}{a(x)}\, dx .$$

This proves $\mathcal{K}_2 \subset \mathcal{K}_1$[4].

Let f(x) be a function of $C^\circ(0,1)$ such that $\sup_{0<x<1} |f(x)| = L < +\infty$.
Let $0 < \varepsilon < \frac{1}{2}$. Set

$$f_\varepsilon(x) \begin{cases} = \dfrac{f(\varepsilon)}{\varepsilon}\, x & 0 \le x \le \varepsilon , \\ = f(x) & \varepsilon \le x \le 1-\varepsilon , \\ = \dfrac{f(1-\varepsilon)}{\varepsilon}\,(1-x) & 1-\varepsilon \le x \le 1. \end{cases}$$

We have $|f_\varepsilon(x)| \le L$ $(0 \le x \le 1)$, $f_\varepsilon(x) \in C^\circ(0,1)$, $f_\varepsilon(0)=f_\varepsilon(1)= 0$,

$$\int_0^1 |f(x) - f_\varepsilon(x)|\, dx \le 3\, L\varepsilon .$$

If u(x,t) is given, for t > 0, by (3.3) and

$$u_\varepsilon(x,t) = \sum_{k=1}^{\infty} v_k(x)\, e^{-\lambda_k t} \int_0^1 \frac{f_\varepsilon(\xi)v_k(\xi)}{a(\xi)}\, d\xi$$

we have for $(x,t) \in [0,1]\times[t_\circ, +\infty)$ $(t_\circ>0)$

$$|u(x,t) - u_\varepsilon(x,t)| \le \varepsilon \frac{6q_\circ q_1}{p_\circ} L \sum_{k=1}^{\infty} \lambda_k^2 e^{-\lambda_k t_\circ}$$

which implies

$$\lim_{\varepsilon\to 0} u_\varepsilon(x,t) = u(x,t)$$

uniformly in $[0,1]\times[t_\circ, +\infty)$. Since we have, for the maximum principle $|u_\varepsilon(x,t)| \le L$ in $\bar{S}$, we have in $[0,1]\times[t_\circ, +\infty)$

$$|u(x,t)| \le L$$

which, for the arbitrariness of $t_\circ$, implies in S

$$|u(x,t)| \le \sup_{0<x<1} |f(x)|. \tag{8.1}$$

---

(4) We have $\mathcal{K} \subset \mathcal{K}_\circ \subset \mathcal{K}_2 \subset \mathcal{K}_1$.

This proves boundedness of $u(x,t)$ in S and moreover extends to $u(x,t) \in \mathcal{B}_2$, through (8.1), the maximum principle (5.1).

9. THE BOUNDARY VALUE PROBLEM (1),(2),(3).

Prof. Altomare assumes on $v_o(x)$ the following hypotheses:

$$v_o(x) \in C^\circ[0,1] \cap C^2(0,1), \quad \lim_{x \to 0^+} a(x)v_o''(x) = \lim_{x \to 1^-} a(x)v_o''(x) = 0 \tag{9.1}$$

and he considers a classical solution for the problem (1),(2),(3) (see [1]p. 18 of preprint report).

Let us now examine the connections existing between problems (1),(2), (3) and (1.1),(1.2),(1.3),(1.4), assuming for a(x) the hypothesis considered in this paper, i.e. $a(x) = x(1-x)p(x)$, with $p(x)$ continuous and positive in [0,1].

From (9.1), denoting by c a suitable positive constant, one gets, for $0<x<1$,

$$|v_o'(x)| \le |v'(\tfrac{1}{2})| + \left|\int_{\frac{1}{2}}^{x} |v_o''(\xi)|\,d\xi\right| \le |v_o'(\tfrac{1}{2})| + c\int_{\frac{1}{2}}^{x} \frac{d\xi}{a(\xi)}$$

$$\le |v_o'(\tfrac{1}{2})| + \frac{c}{p_o}\Big(|\log x| + |\log(1-x)| + 2\log 2\Big).$$

Hence $v_o'(x) \in L^2(0,1)$. Set

$$f(x) = v_o(x) - v_o(0) - x\,[v_o(1) - v_o(0)] \; ; \tag{9.2}$$

We have $f(x) \in \mathcal{H}$.

If we look for a solution of problem (1),(2),(3) satisfying conditions 1),2) of Sect.1 and continuous in $\bar{S}$, we have that this solution is given by

$$v(x,t) = v_o(0) + x[v_o(1) - v_o(0)] + u(x,t),$$

where $u(x,t)$ is the solution of the problem (1.1),(1.2),(1.3),(1.4) in the class $\mathcal{B}$, when $f(x)$ is given by (9.2).

To prove this one has only to observe that

$$\lim_{x \to 0^+} a(x)v_{xx}(x,t) = \lim_{x \to 0^+} u_t(x,t) = 0,$$

$$\lim_{x \to 1^-} a(x)v_{xx}(x,t) = \lim_{x \to 1^-} u_t(x,t) = 0.$$

Obviously the solution $v(x,t)$ is unique in the above specified class.

Let us remark that for the existence in this class, because of theor. 6.I, conditions (9.1) are not needed but only $v_0 \in C^\circ[0,1]$.

Let us now suppose that the function class of classical solutions for the problem (1),(2),(3) is the one determined by conditions 1) and 2) of Sect. 1 and by the requirements: v bounded in S, $v \in C^\circ(S \cup I_{01})$. Suppose that $p(x)$ satisfies the conditions stated in Sect.7. Let $u(x,t)$ be the solution of the problem (1.1),(1.2),(1.3),(1.4) belonging to $\mathfrak{B}_2$, when we assume $f(x) \equiv 1$ (see theor.8.I). The function

$$v(x,t) = u(x,t) - 1$$

is a non-trivial solution of the problem (1),(2),(3) with $v_0(x)=0$, in the class of functions satisfying 1),2), bounded in S and continuous in $S \cup I_{01}$. Hence uniqueness fails to hold in this class.

REFERENCES

[1] F.ALTOMARE, Lototsky-Schnabl operators on the unit interval and degenerate diffusion equations, to appear on "Progress in Functional Analysis", Proc. of the Peñiscola Meeting 1990 on the occasion of the 60th birthday of M.Valdiviva, Series Math.Studies, Elsevier. Preprint report of Dip.to di Matem. Univ. Bari 20/90, 1990.

[2] F.ALTOMARE, Positive projections, approximation process and degenerate diffusion equations, to appear on Atti del Convegno internaz. "Recenti sviluppi in Analisi Matematica e sue Applicazioni", Bari 1990. Preprint report of Dip.to di Matem. Univ. Bari 21/90, 1990.

[3] PH.CLÉMENT-C.A.TIMMERMANS, On C_0-semigroups generated by differential operators satisfying Ventcel's boundary conditions, Indag. Mathem. Proceed. A 89, 1986, 379-387.

[4] G.FICHERA, Sulle equazioni differenziali lineari ellittico-paraboliche del secondo ordine, Mem.Acc.Naz.Lincei I,8,5,1956, 1-30.

[5] G.FICHERA, On a unified theory of boundary value problems for elliptic-parabolic equations of second order in "Boundary Problems.Differential Equations", edit.R.Langer,Univ.of Wisconsin Press,Madison (Wis) 1960, 97-120.

[6] A.FRIEDMAN, Partial differential equations of parabolic type, Prentice-Hall Inc., Englewood Cliffs N.J., 1964.

[7] W.MAGNUS-F.OBERHETTINGER, Formeln und Sätze für die speziellen Funktionen der Mathematischen Physik, 2 Auflage, Springer-Verlag, 1948.

[8] R.MARTINI-W.L.BOER, On the construction of semi-groups of operators, Indag Mathem., Proceed. A 36, 1974, 392-405.

[9] O.A.OLEINIK, A problem of Fichera, Dokl.Akad.Nauk SSSR 157, 1964, 1297-1300; Soviet Math. Dokl. 5, 1964, 1129-1133.

[10] O.A.OLEINIK-E.V.RADKEVIC, Second Order Equations with Nonnegative Characteristic Form, American Math. Society, Providence, Rhode Island, Plenu Press, New York-London, 1973 (Russian original edition, Moscow 1971; Italian translation, Veschi, Roma, 1973).

[11] A.WEINSTEIN-W.STENGER, Methods of Intermediate Problems for Eigenvalues, Theory and Ramifications, Mathematics in Science and Engineering, 89, Academic Press, New York and London, 1972.

[12] E.T.WHITTAKER-G.N.WATSON, A Course of Modern Analysis, Cambridge,Universi Press, 1962.

DINH NHO HAO

A noncharacteristic Cauchy problem for linear parabolic equations and related inverse problems II: a variational method[1]

1. Introduction

In evaluating new heat-shield materials, testing of rocket nozzles and developing transfer calorimaters, it is sometimes required to calculate the transfer surface heat flux, the surface temperature, and to determine the heat capacity, the heat transfer coefficient,..., from a temperature history measured at fixed locations inside the body. Problems of this kind arise also in casting and welding in steel or in polymer processing, quenching of solids in a fluid, infrared computerized axial tomography,.... These problems are well known under the name "inverse heat conduction problems". References on such problems can be found in [1, 2, 3, 11] and references therein.

The first kind of the above mentioned inverse problems belongs to the class of boundary inverse problems for parabolic equations that are well known to be severely ill-posed [20], and up to now there have been many approaches to them (see [11] for a short survey). These problems lead us mainly to consider the following noncharacteristic Cauchy problem for (linear and nonlinear) parabolic equations of the form

$$\begin{aligned} \mathcal{P}u(x,t) &= u_t(x,t),\ 0<x<1,\ 0<t\le T,\\ \text{"surface heat flux"}|_{x=0} &= \varphi(t),\ 0<t\le T,\\ \text{"surface temperature"}|_{x=0} &= \psi(t),\ 0<t\le T, \end{aligned}$$

where $\mathcal{P}$ is an elliptic operator, φ and ψ are given functions. We emphasize here that *no initial condition at $t=0$ is given!*

In [7] we have proved that a weak solution of a noncharacteristic Cauchy problem for linear parabolic equations in divergence form with coefficients depending on the time variable as functions of Holmgren class two exists if and only if the Cauchy data are functions of Holmgren class two! A function $g(t)$ is said to be of Holmgren class two, if it has the property that $|g^{(n)}| < cs^n(2n)!$ for some positive constants c, s and for all $n \geq 0$.

Many authors when they worked with noncharacteristic Cauchy problems for parabolic equations required the initial condition (see, e.g., [1, 2, 3] and references therein). But it is indeed too much. The result of [7], that shortly said above, shows that this condition is, indeed, not necessary. Furthermore, in the paper [8] we have suggested a variational method for solving such an inverse problem without the initial condition. We transform the inverse problem into an optimal control problem. This optimal control problem is still

[1]Supported by the German Academic Exchange Service (DAAD) and by the Research Group "Regularization", Free University of Berlin.

ill-posed. An attempt of using iterative regularization methods is discussed. We succeed in obtaing the gradient of the functional in the optimal control problem, and therefore we can use one of the iterative regularization methods. The discretization in time Galerkin method and the finite difference methods with convergence results are described in [9].

It is interesting that, when the initial condition in a noncharacteristic Cauchy problem for linear parabolic equations is known, then *we can not only determine the solution of the problem, but also a coefficient of these equations, provided that the other coefficients are known.* Subject of the paper [10] is the study of such an inverse problem. Namely, we considered the following inverse problems of finding one of the coefficients $a(x)$, $b(x)$, $c(x)$, $q(x)$ as well as the function $u(x,t)$ satisfying the following system

$$\begin{aligned}(a(x)u_x(x,t))_x &= u_t(x,t) + b(x)b_1(x,t)u_x + c(x)c_1(x,t)u + q(x)f(x,t), \qquad (1)\\ &\qquad 0 < x < 1, \quad 0 < t \leq T,\\ u(0,t) &= \varphi(t), \quad 0 < t \leq T, \qquad (2)\\ -a(0)u_x(0,t) &= \psi(t), \quad 0 < t \leq T. \qquad (3)\\ u(x,0) &= u_0(x), \quad 0 < x < 1. \qquad (4)\end{aligned}$$

Where $a \in L_\infty(0,1)$, b_1, $c_1 \in L_\infty((0,1)\times(0,T))$, $\Lambda \geq a \geq \mu > 0$, $f \in L_2(Q_T)$, $Q_T := (0,1)\times(0,T)$, and for every fixed $x \in (0,1)$ the functions $b_1(x,t)$, $c_1(x,t)$, $f(x,t)$ are functions of t that belong to a Holmgren class two (see [7]). The functions $a(x)$, $b(x)$, $c(x)$, $q(x)$ are sought in $L_\infty(0,1)$.

In [10] various existence and uniqueness theorems for this problem are proved. Our inverse problem is close to those in [15, 18, 17] and [5, 6, 21]. But in contrast to these works, we are not working with semi-infinite axis x of the space variable, like [15, 18, 17], but with a finite interval (our method is still valid for the case of semi-infinite axis). It is interesting to note here that we do not need the heat flux at $x = 1$ like [5, 6, 21]. The heat flux on the other side of a body is not always available. Thus, we *omit an extra condition given in the other boundary*, and furthermore, we use *minimum requirements on the smoothness of the coefficients with respect to the space variable and we are able to deal with the case when coefficients depend on the time variable!* This is the case when the Gel'fand-Levitan theory does not work [21], [22]-[25].

2. Variational Method

We suggest the following variational method for the inverse problems (1)-(4): We look for a control that consists of the unknown coefficient and of *the unknown "heat flux" at* $x = 1$, and minimizes the defect

$$J_0(v) = \|u(0,\cdot) - \varphi(\cdot)\|^2_{L_2(0,T)}$$

in an appropriate domain. As the inverse problems are ill-posed, these optimal control problems are also ill-posed. Therefore a regularization should be suggested. Namely, we consider the following general regularized optimal control problem

Minimize the following functional:

$$J_\alpha(v) = \|u(0,\cdot) - \varphi(\cdot)\|^2_{L_2(0,T)} + \alpha\Omega(v_0, v_1, v_2, v_3) \qquad (5)$$

on the set $V = (V_0, V_1, V_2, V_3)$, where $V_0 \subset L_2(0,T)$, V_1, $V_2 \subset L_\infty(Q_T)$, $V_3 \in L_2(Q_T)$, $v = (v_0, v_1, v_2, v_3)$ is the control, α is a nonnegative regularization parameter, the regularization functional Ω will be defined bellow, and u satisfies the following problem

$$u_t(x,t) = (a(x,t,v_1)u_x)_x + c(x,t,v_2)u + f(x,t,v_3), \tag{6}$$
$$0 < x < 1, \quad 0 < t < T,$$
$$-a(x,t,v_1(x,t))u_x(x,t)|_{x=0} = g(t), \quad 0 < t \le T, \tag{7}$$
$$a(x,t,v_1(x,t))u_x(x,t)|_{x=1} = v_0(t), \quad 0 < t \le T, \tag{8}$$
$$u(x,0) = u_0(x), \quad 0 < x < 1. \tag{9}$$

Here,

$$0 < \nu \le a(x,t,v_1) \le \mu, \ \forall\, v_1 \in V_1, \text{ a.e } (x,t) \in Q_T,$$
$$|c(x,t,v_2)| \le \mu_1 \text{ a.e. in } Q_T, \forall\, v_2 \in V_2,$$
$$\text{and for all } v_3 \in V_3 \text{ the function } f(x,t,v_3) \in L_2(Q_T),$$

where ν, μ, μ_1 are given positive numbers.

The solution of the problem (6)-(9) is understood in the sense of $V^{1,0}(Q_T) := C([0,T]; L_2(\Omega)) \cap L_2((0,T); H^1(\Omega))$: *A function $u(x,t)$ is said to be a weak solution in $V^{1,0}(Q_T)$ of the problem (6)-(9), if it is a function of $V^{1,0}(Q_T)$ and satisfies the integral identity*

$$\int_{Q_T} (-u\eta_t + a(x,t,v_1)u_x\eta_x - c(x,t,v_2)u\eta - f(x,t,v_3)\eta)\, dxdt$$
$$= \int_0^1 u_0(x)\eta(x,0)dx + \int_0^T g(t)\eta(0,t)dt + \int_0^T v_0(t)\eta(1,t)dt,$$

$\forall\, \eta \in H^{1,1}(Q_T)$, $\eta(x,T) = 0$.

For convenience we summary here some results of [8].

Theorem A: ([8, Theorem 1], and [19, Plotnikov, 1965]) *There exists a unique solution from $V^{1,0}(Q_T)$ of Problem (6)-(9) and the following estimate holds*

$$\sup_{0\le t\le T} \|u(\cdot,t)\|^2_{L_2(0,1)} + \|u_x\|^2_{L_2(Q_T)} + \|u(0,\cdot)\|^2_{L_2(0,T)} + \|u(1,\cdot)\|^2_{L_2(0,T)}$$
$$\le c\left[\|u_0\|^2_{L_2(0,1)} + \|f\|^2_{L_2(Q_T)} + \|g\|^2_{L_2(0,T)} + \|v_0\|^2_{L_2(0,T)}\right]$$

where $c \ge 0$ is a constant independent of u.

The energy-balance equation for the problem (6)-(9): If u is a weak solution from $V^{1,0}(Q_T)$ of the problem (6)-(9), then

$$\frac{1}{2}\|u(\cdot,t)\|^2_{L_2(0,1)} + \int_0^1\int_0^t (cu_x^2)dxd\tau \tag{10}$$
$$= \frac{1}{2}\|u(\cdot,0)\|^2_{L_2(0,1)} + \int_0^1\int_0^t fudxd\tau + \int_0^t g(\tau)u(0,\tau)d\tau + \int_0^t v_0(\tau)u(1,\tau)d\tau. \tag{11}$$

Theorem B: ([8, Theorem 2]) *Let all conditions of Theorem A and the following ones be satisfied:*

$$\operatorname{vrai}\max_{x\in(0,1)}\left|\frac{\partial a}{\partial t}\right|,\quad \operatorname{vrai}\max_{x\in(0,1)}\left|\frac{\partial c}{\partial t}\right| \le \mu(t),\quad \int_0^T \mu(t)dt < \infty, \tag{12}$$

$$u_0 \in H^1(0,1),\quad g \in H^1(0,T),\quad v_0 \in H^1(0,T). \tag{13}$$

Then there exists a unique weak solution in $V^{1,0}(Q_T)$ of the problem (6)-(9) that belongs to $H^{1,1}(Q_T)$ and satisfies the following inequality

$$\max_{0\le t\le T}\|u(\cdot,t)\|_{L_2(0,1)} + \|u_x\|_{L_2(Q_T)} + \|u_t\|_{L_2(Q_T)} + \|u\|_{L_2(Q_T)} + \|u(0,\cdot)\|_{L_2(0,T)} +$$
$$+\|u(1,\cdot)\|_{L_2(0,T)} \le c\left(\|u_0\|_{H^1(0,1)} + \|f\|_{L_2(Q_T)} + \|g\|_{H^1(0,T)} + \|v_0\|_{H^1(0,T)}\right)$$

where c is a constant independent of u.

The control v_i may be appear only in one coefficient or in some coefficients. Since these problems are different in their nature, we treat them separately. For related problems, see [8, 9, 16, 13, 14, 26].

3. Control is the Right Hand Side Coefficient and the Right Boundary Condition: $v := (v_0, v_3)$.

Suppose that the functional $J_\alpha(v)$ has the form

$$J_\alpha(v) = \|u(0,\cdot) - \varphi(\cdot)\|^2_{L_2(0,T)} + \alpha\left(\|v_0\|^2_{L_2(0,1)} + \|v_3\|^2_{L_2(Q_T)}\right) \tag{14}$$

and the equation (6) has the spesific form:

$$u_t(x,t) = (a(x,t)u_x)_x + c(x,t)u + v_3(x,t). \tag{15}$$

Let V_0 and V_1 be convex and bounded sets in $L_2(Q_T)$ and $L_2(0,T)$, respectively. The problem is linear, and by the same method of [8] we can obtain the following results without any difficulties.

3a. Existence.

Theorem 1: *There exists a solution of Problem (5),(15),(7)-(9) and every minimizing consequence converges weakly to the set of all minimum point of the problem.*

3b. Gradient. Let ψ be a solution in $H^{1,0}(Q_T)$ of the following adjoint to (15),(7)-(9) problem

$$\begin{aligned}
\psi_t &= -(a\psi_x)_x - c\psi,\quad 0 < x < 1,\\
\psi(x,T) &= 0,\quad 0 < x < 1,\\
a\psi_x|_{x=0} &= -2[u(0,t) - \varphi(t)],\quad 0 < t \le T,\\
-a\psi_x|_{x=1} &= 0,\quad 0 < t \le T,
\end{aligned}$$

where $u = u(x,t)$ be a solution of Problem (15),(7)-(9).

Theorem 2: *The functional J_α is Fréchet differentiable and its gradient can be found as follows*

$$J'_\alpha(v) = \left\{ \begin{array}{c} \psi(x,t;v) + 2\alpha v_3 \\ \psi(0,t;v) + 2\alpha v_0 \end{array} \right\}. \tag{16}$$

Theorem 3: *The gradient $J'_\alpha(v)$ satisfies a Lipschitz condition.*

With these results we can write a necessary and sufficient condition for the optimal control and use a gradient method for solving the optimal control problem numerically as in [8, 14, 26]. Since no new idea appears, we do not write all of that here.

4. Control is $v_2(x,t) := c(x,t)$ and v_0: $v = (v_0, v_2)$.

The functional $J_\alpha(v)$ is

$$J_\alpha(v) = \|u(0,\cdot) - \varphi(\cdot)\|^2_{L_2(0,T)} + \alpha\left(\|v_0\|^2_{L_2(0,T)} + \|v_2\|^2_{L_2(Q_T)}\right), \tag{17}$$

and the equation (6) has the form

$$u_t(x,t) = (a(x,t)u_x)_x + v_2(x,t)u + f(x,t). \tag{18}$$

Theorem 4: *If $u_0 \in H^1(0,1)$, $g(t) \in H^1(0,T)$, $v_0 \in V_0 \subset H^1(0,T)$, and all conditions of Theorem B are satisfied. Then there exists an optimal control in V of Problem (17),(18),(7)-(9).*

Proof: With the condition of the theorem, $u(x,t)$ belongs to $H^{1,1}(Q_T)$ (Theorem B). Let $v^n = (v_0^n, v_2^n)$ be a minimizing sequence: $J_\alpha(v^n) \to \inf J_\alpha$, $v^n \in V$. Set $u^n := u(x,t;v^n)$. Since $\alpha > 0$, v^n is bounded in $H^1(0,T) \times L_\infty(Q_T)$. Then, by Theorem B

$$u^n \text{ is bounded in } H^1(Q_T).$$

It follows that we can extract a subsequence, again denoted by u^n, such that

$$\begin{aligned} u^n &\to u \text{ weakly in } L_2(0,T;H^1(0,1)), \\ u_t^n &\to u_t \text{ weakly in } L_2(Q_T), \\ u^n &\to u \text{ strongly in } L_2(Q_T). \end{aligned}$$

Since v^n is bounded in $H^1(0,T) \times L_\infty(Q_T)$, we can assume that

$$\begin{aligned} v_0^n &\to v_0^n \text{ strongly in } L_2(0,T), \\ v_2^n &\to v_2 \text{ weak star in } L_\infty(Q_T). \end{aligned}$$

We have

$$\begin{aligned} &\int_{Q_T} (-u^n\eta_t + au_x^n\eta_x - (v_2^n u^n + f)\eta)\,dxdt \\ &\qquad = \int_0^T g(t)\eta(0,t)dt + \int_0^T v_0^n(t)\eta(1,t)dt, \end{aligned}$$

$\forall\, \eta \in H^{1,0}(Q_T)$, $\eta(\cdot,T) = 0$.

Since $v_2^n \rightarrow v_2$ weak star in $L_\infty(Q_T)$, we have

$$\left|\int_{Q_T} v_2^n(u^n - u)\eta dxdt\right| \leq constant\|u^n - u\|_{L_2(Q_T)} \rightarrow 0.$$

It follows,

$$\begin{aligned}\int_{Q_T} v_2^n u^n \eta dxdt &= \int_{Q_T} v_2^n u\eta dxdt + \int_{Q_T} v_2^n(u^n - u)dxdt \\ &\rightarrow \int_{Q_T} v_2 u\eta dxdt.\end{aligned}$$

Hence, $u = u(x,t;v)$ and $u^n(0,t) \rightarrow u(0,t;v)$ weakly in $L_2(0,T)$. Consequently, $\liminf J_\alpha(v^n) \geq J_\alpha(v)$ and therefore v is an optimal control. The theorem is proved.

Remark 1: If $\alpha = 0$, then with the requirement that V_0 is closed and bounded in $H^1(0,T)$, the set V_2 is supposed to be in $L_\infty(Q_T)$ and convex, closed in the weak topology of the dual of $L_1(Q_T)$, the statement of Theorem 4 remains valid.

5. Control is in All Coefficients: $v = (v_0, v_1, v_2, v_3)$.

Suppose that the control appears in all coeficients:

$$u_t(x,t) = (a(x,t,v_1)u_x)_x + c(x,t,v_2)u + f(x,t,v_3), \tag{19}$$

and the functional J_α has the form (5), where

$$\Omega(v) = \|v_0 - w_0\|^2_{L_2(0,T)} + \|v_1 - w_1\|^2_{L_2(Q_T)} + \|v_2 - w_2\|^2_{L_2(Q_T)} + \|v_3 - w_3\|^2_{L_2(Q_T)}.$$

Here $w = (w_0, w_1, w_2, w_3) \in L_2(0,T) \times L_2(Q_T) \times L_2(Q_T) \times L_2(Q_T)$.

Let $a(x,t,v_1)$, $c(x,t,v_2)$, $f(x,t,v_3)$ for every fixed $v_1 \in V_1$, $v_2 \in V_2$, $v_3 \in V_3$ be measurable functions of (x,t), and for almost every $(x,t) \in Q_T$ be continuously differentiable functions of v_1, v_2, v_3, respectively.

The sets V_i are defined as follows

$$\begin{aligned} &V_0 \text{ is closed bounded set in} L_2(0,T), \\ V_1 &= \{v_1 | v_1 \in L_\infty(Q_T), 0 < a_1 \leq v_1 \leq a_2, \text{a.e. in } Q_T\}, \\ V_2 &= \{v_2 | v_2 \in L_\infty(Q_T), c_1 \leq v_2 \leq c_2, \text{a.e. in } Q_T\}, \\ V_3 &= \{v_3 | v_3 \in L_2(Q_T), \|v_3\|_{L_2(Q_T)} \leq F\}, \end{aligned}$$

here a_i, c_i, F are given and finite numbers. Suppose also that $\alpha > 0$ and we call this optimal problem by Problem $\mathcal{G}$.

5a. Existence.

Theorem 5: *There exists an everywhere dense G_δ-subset G of $L_2(0,T) \times L_\infty(Q_T)^3$ such that, for every $w = (w_0, w_1, w_2, w_3) \in G$, Problem $\mathcal{G}$ has a unique optimal control.*

Proof: At first, we prove that J_0 is continuous. Let $\Delta v \in V$ be an increment of the control at $v \in V$ such that $v + \Delta v \in V$. Let $\Delta u(x,t) \equiv u(x,t;v+\Delta v) - u(x,t;v)$, and put $u \equiv u(x,t;v)$. It is clear that $\Delta u(x,t)$ satisfies the integral identity

$$\int_{Q_T} [-\Delta u \eta_t + a(x,t,v_1+\Delta v_1)\Delta u_x \eta_x - \\ -(c(x,t,v_2+\Delta v_2)\Delta u + \Delta a u_x \eta_x - \Delta c u \eta + \Delta f \eta)\,]\,dxdt \\ = \int_0^T \Delta v_0(t)\eta(1,t)dt,$$

$\forall\, \eta \in H^{1,1}(Q_T)$, $\eta(\cdot,T) = 0$. Here

$$c\Delta a = a(x,t,v_1+\Delta v_1) - a(x,t,v_1),$$
$$\Delta c = c(x,t,v_2+\Delta v_2) - c(x,t,v_2),$$
$$\Delta f = f(x,t,v_3+\Delta v_3) - f(x,t,v_3).$$

By the method receiving the energy-balance equation (11) (see [8]) we have

$$\frac{1}{2}\int_0^1 \Delta u^2(x,t)dx + \int_{Q_T} \left[a(x,t,v_1+\Delta v_1)(\Delta u_x)^2 - c(x,t,v_2+\Delta v_2)(\Delta u)^2\right] dxdt \\ + \int_{Q_T} (\Delta a \Delta u_x u_x - (\Delta c u + \Delta f)\eta)\,dxdt = \int_0^T \Delta v_0(t)\eta(1,t)dt.$$

Again, by the method receiving the stability estimate in Theorem A, we get

$$\|\Delta u\|_{V^{1,0}(Q_T)} \leq c_1 \left[\|\Delta f - \Delta c u\|_{L_2(Q_T)} + \|\Delta a u_x\|_{L_2(Q_T)} + \|\Delta v_0\|_{L_2(0,T)}\right],$$

where c_1 is a non-negative constant. Hence, $\|\Delta u\|_{V^{1,0}(Q_T)} \to 0$ as $\|\Delta v\|_V \to 0$. Consequently, $\|\Delta u(0,\cdot)\|_{L_2(0,T)} \to 0$, when $\|\Delta v\|_V \to 0$. (Here and henceforth we use the notation

$$\|v\|_V := \|v_0\|_{L_2}(0,T) + \|v_1\|_{L_2(Q_T)} + \|v_2\|_{L_2(Q_T)} + \|v_3\|_{L_2(Q_T)},$$

and

$$(v^1,v^2)_V := (v_0^1,v_0^2)_{L_2(0,T)} + (v_1^1,v_1^2)_{L_2(Q_T)} + (v_2^1,v_2^2)_{L_2} + (v_3^1,v_3^2)_{L_2(Q_T)}).$$

Observing that

$$J_0(v+\Delta v) - J_0(v) = 2\int_0^T (u(0,t) - \varphi(t))\Delta u(0,t)dt + \|\Delta u(0,\cdot)\|^2_{L_2(0,T)},$$

we conclude that the functional J_0 is continuous. On the other hand it is bounded bellow, and the spaces $L_2(\cdot)$ are uniformly convex. Therefore, the theorem is proved by the following lemma.

Lemma: (M. F. Bidaut [4], M. Goebel [12]) *Let X be a uniformly convex Banach space, V be a closed, bounded set in X, the functional $I(v)$ on V be lower semi-continuous and bounded from below, $\alpha > 0$, $\beta \geq 1$ be given numbers. Then there exists an everywhere dense G_δ-subset G of X such that for every $\omega \in G$, the functional*

$$I_\alpha(v) \equiv I(v) + \alpha\|v-\omega\|_X^\beta$$

attains its infimum on V. If $\beta > 1$ then the element at which $I_\alpha(v)$ attains its infimum is unique.

5b. A Necessary and Sufficient Condition for the Optimal Control. Let $\psi(x,t) := \psi(x,t;v)$ be a solution from $V^{1,0}(Q_T)$ of the following adjoint to (19), (7)-(9) problem

$$\psi_t = -(a(x,t,v_1)\psi_x)_x - c(x,t,v_2)\psi, \quad 0 < x < 1, \quad 0 < t \le T, \tag{20}$$
$$\psi(x,T) = 0, \quad 0 < x < 1, \tag{21}$$
$$a(x,t,v_1(x,t))\psi_x(x,t)|_{x=0} = -2[u(0,t) - \varphi(t)], \quad 0 < t \le T, \tag{22}$$
$$-a(x,t,v_1(x,t))\psi_x(x,t)|_{x=1} = 0, \quad 0 < t \le T, \tag{23}$$

where $u = u(x,t)$ be a solution of Problem (19),(7)-(9). The function $\psi = \psi(x,t)$ satisfies the integral identity

$$\int_{Q_T} (\psi\eta_t + a(x,t,v_1)\psi_x\eta_x - c(x,t,v_2)\psi\eta)\,dxdt$$
$$= -\int_0^1 \psi(x,0)\eta(x,0)dx + 2\int_0^T [u(0,t) - \varphi(t)]\eta(0,t)dt \tag{24}$$

$\forall\, \eta \in H^{1,1}(Q_T)$.

If we replace t in (20)-(24) by the new variable $\tau := T-t$, then we obtain a boundary value problem of the same type as (19),(7)-(9). Since $u \in V^{1,0}(Q_T)$, $u(0,t) \in L_2(0,T)$. From Theorem A we see that there exists only a solution in $V^{1,0}(Q_T)$ of Problem (20)-(23).

We introduce now so called the Hamilton-Pontryagin function of Problem (19),(7)-(9) as follows

$$H(x,t;u,\psi,v) = -\left[v_0\psi(1,t;v) - a(x,t,v_1)u_x(x,t;v)\psi_x(x,t;v)\right.$$
$$\left. -c(x,t,v_2)u(x,t;v)\psi(x,t;v) + f(x,t,v_3)\psi(x,t;v) + \alpha\|v-w\|_V^2\right].$$

Theorem 6: *Let all above mentioned conditions be satisfied and*

i) the functions $a(x,t,p)$, $c(x,t,p)$, $f(x,t,p)$ satisfy a Lipschitz condition as a function for p for almost $(x,t) \in Q_T$.

ii) the partial derivatives $\partial a(x,t,v_1)/\partial v_1$, $\partial c(x,t,v_2)/\partial v_2$ are bounded in $L_\infty(Q_T)$, and $\partial f(x,t,v_3)/\partial v_3$ is bounded in $L_2(Q_T)$.

Then the functional J_α is Fréchet differentiable and its gradient can be found as follows

$$J'_\alpha(v) = -\frac{\partial H}{\partial v}. \tag{25}$$

If $a(x,t,v_1(x,t)) \equiv v_1(x,t)$, $c(x,t,v_2(x,t)) \equiv v_2(x,t)$ and $f(x,t,v_3) \equiv v_3(x,t)$ then

$$J'_\alpha(v) = -\frac{\partial H}{\partial v} = \left\{\begin{array}{c} \psi(x,1;v) + 2\alpha(v_0 - w_0) \\ -u_x(x,t;v)\psi_x(x,t;v) + 2\alpha(v_1 - w_1) \\ -u(x,t;v)\psi(x,t;v) + 2\alpha(v_2 - w_2) \\ \psi(x,t;v) + 2\alpha(v_3 - w_3) \end{array}\right\}. \tag{26}$$

Proof: We have observed that $J_\alpha(v)$ is continuous. Now we prove that it is Fréchet differentiable. To do this we prove that

$$\begin{aligned}
&2\int_0^T (u(0,t)-\varphi(t))\,\Delta u(0,t)dt = \\
&= \int_0^T \Delta a\psi(1,t)dt - \int_{Q_T} (u_x\psi_x\Delta v_1(t) + u\psi\Delta c - \psi\Delta f)\,dxdt \\
&+ \int_{Q_T} (-\Delta a\Delta u_x\psi + \Delta c\Delta u\psi)\,dxdt
\end{aligned} \tag{27}$$

Let $v_0^k \in V_0 \cap H^1(0,T)$ be a sequence such that $v_0^k \to v_0 \in V_0$ in $L_2(0,T)$, and $g^k(t)$ be a sequence in $H^1(0,T)$ such that $g^k \to g$ in $L_2(0,T)$, u_0^k be a sequence in $H^1(0,1)$ that converges to u_0 in $L_2(0,1)$. Further, let $v_1^k(x,t)$ be a sequence in $L_\infty(Q_T)$ such that $v_1^k \in V_1$, $\partial v_1^k(x,t)/\partial t \in L_\infty(Q_T)$, for $k \to \infty$ this sequence converges to v_1. Furthermore, let $a^k(x,t,p)$ be continuous in Q_T and be continuously differentiable function of p, $a^k \geq \nu$ and

$$\int_0^T vrai \max_{x\in(0,1)} \left|\frac{\partial a^k(x,t,p)}{\partial t}\right| dt < \infty$$

and $a^k(x,t,p) \to a(x,t,p)$ a.e. in Q_T as $k \to \infty$. It is clear that such sequences always exist. Let $u^k = u^k(x,t;v^k)$ and $\psi^k = \psi^k(x,t;v^k)$ be solutions of Problems (19)-(7)-(9) and (20)-(23) with u_0, v_0, a, g replaced by u_0^k, v_0^k, a^k, g^k. It follows from Theorem B that there exists a unique solution in $H^{1,1}(Q_T)$ of the problem of finding a function u^k under the foregoing conditions. By the same method in [8, 14] it can be proved that $u^k(x,t) \to u(x,t)$ in $V^{1,0}(Q_T)$.

On the other hand, the fact $u^k(x,t) \in H^{1,1}(Q_T)$ follows that $u^k(0,t)$, $u^k(1,t) \in H^{1/2}(Q_T)$. Consequently $u^k(0,t)-\varphi(t) \in L_2(0,T)$ and the function $\psi^k = \psi^k(x,t)$ satisfies the integral identity

$$\begin{aligned}
&\int_{Q_T} \left(\psi^k\eta_{1t} + a^k(x,t,v_1^k)\psi_x^k\eta_{1x} - c(x,t,v_2)\psi^k\eta_1\right) dxdt \\
&= -\int_0^1 \psi^k(x,0)\eta_1(x,0)dx + 2\int_0^T [u^k(0,t)-\varphi(t)]\eta_1(0,t)dt
\end{aligned} \tag{28}$$

$\forall\ \eta_1 \in H^{1,1}(Q_T)$. Using Theorem A and reasoning as in the proof that $\{u^k(x,t)\}$ converges to $u(x,t)$, we can establish the convergence of the sequence $\{\psi^k(x,t)\}$ to $\psi(x,t)$ in the norm $V^{1,0}(Q_T)$.

Let $\Delta v^k = (\Delta v_0^k(t),\ \Delta v_1^k(x,t),\ \Delta v_2(x,t),\ \Delta v_3(x,t))$ be such that $\Delta v_0^k \in H^1(0,T)$, $\Delta v_1^k \in L_\infty(Q_T)$, $\dfrac{\partial \Delta v_1^k}{\partial t} \in L_\infty(Q_T)$, $v^k + \Delta v^k \in V$, and for $k \to \infty$ let $v_0^k \to v_0$ in $L_2(0,T)$, $v_1^k \to v_1$ in $L_\infty(Q_T)$. Let $\Delta u^k = u^k(x,t;v^k+\Delta v^k) - u^k(x,t;v^k)$. Theorem B implies that Δu^k belongs to $H^{1,1}(Q_T)$ and satisfies

$$\begin{aligned}
&\int_{Q_T} \Big[\Delta u_t^k\eta + a^k(x,t,v_1^k+\Delta v_1^k)\Delta u_x^k\eta_x \\
&- \left(c(x,t,v_2+\Delta v_2)\Delta u^k + \Delta f\right)\eta\Big]\,dxdt
\end{aligned} \tag{29}$$

$$+\int_{Q_T} \left(\Delta a^k u_x^k\eta_x - \Delta c u^k\eta\right) dxdt = \int_0^T \Delta v_0^k(t)\eta(1,t)dt, \tag{30}$$

$\forall\, \eta \in H^{1,0}(Q_T)$, where $\Delta a^k = a^k(x,t,v_1^k+\Delta v_1^k) - a^k(x,t,v_1^k)$. Putting $\eta = \psi^k(x,t)$ in (30) and $\eta_1 = \Delta u^k(x,t)$ in (28) and substracting one result from the other, we find that

$$\int_{Q_T} \left(\Delta a^k \Delta u_x^k \psi_x^k - \Delta c \Delta u^k \psi^k - \Delta f \psi^k + \Delta a^k u_x^k \psi_x^k - \Delta c u^k \psi^k\right)$$
$$= \int_0^T \Delta v_0^k(t)\psi^k(1,t)dt - 2\int_0^T [u^k(0,t)-\varphi(t)]\Delta u^k(0,t)dt.$$

Now letting $k \to \infty$ in this identity, we obtain (27).

On the other hand, we have proved in the proof of Theorem 5 that $\|\Delta u(0,\cdot)\|_{L_2(0,T)}$ tends to 0 as $\|\Delta v\|_V$ tends to 0. Consequently, from (27), and conditions of the theorem we obtain

$$\begin{aligned} &J_\alpha(v+\Delta v) - J_\alpha(v) = \\ &\quad = \int_0^T \Delta v_0(t)\psi(1,t)dt - \int_{Q_T}\left(u_x\psi_x\frac{\partial a}{\partial v_1} + u\psi\frac{\partial c}{\partial v_2} - \psi\frac{\partial f}{v_3}\right)dxdt \\ &\quad +2\alpha(v-w,\Delta v)_V + o(\|\Delta v\|_V) \\ &\quad = -\left(\frac{\partial H}{\partial v},\Delta v\right) + 2\alpha(v-w,\Delta v) + o(\|\Delta v\|_V). \end{aligned}$$

Hence, the Fréchet differensibility of the functional $J_\alpha(v)$ and the formula (25) are proved. The formula (26) is now clear.

It is not hard to prove the following theorem.

Theorem 7: *If the condition s of Theorem 6 are satisfied, then for v^* to be optimum, it is necessary that*

$$\begin{aligned} &(\psi^*(1,t,v^*), v_0 - v_0^*)_{L_2(0,T)} - \left(\frac{\partial a(x,t,v_1^*(x,t))}{\partial v_1}u_x^*\psi_x^*, v_1 - v_1^*\right)_{L_\infty(Q_T)} \\ &\quad - \left(\frac{\partial c(x,t,v_2)}{\partial v_2}u^*\psi^*, v_2 - v_2^*\right)_{L_2(Q_T)} \\ &\quad + \left(\frac{\partial f(x,t,v_3)}{\partial v_3}\psi^*, v_3 - v_3^*\right)_{L_2(Q_T)} + 2\alpha(v^*-w, v-v^*)_V \ge 0, \end{aligned}$$

$\forall\, v \in V$. Here $u^* = u^*(x,t)$, $\psi^* = \psi^*(x,t)$ are the solution of Problems (19),(7)-(9) and (20)-(23) respectively, for $v = v^*$.

6. Approximating by the Finite Fifference Method. Let all conditions in §5 be satisfied. We shall approximate the following optimal problem by the finite difference method:

Minimize the functional:

$$J_\alpha(v) = \|u(0,\cdot) - \varphi(\cdot)\|^2_{L_2(0,T)} + \alpha\|v\|^2_V. \tag{31}$$

Subject to

$$v \in V, \qquad \text{with } V_0 \in H^1(0,T), \tag{32}$$

$$
\begin{aligned}
u_t(x,t) &= (v_1(x,t)u_x)_x + v_2(x,t)u + v_3(x,t), && (33)\\
&\quad 0<x<1,\quad 0<t<T, &&\\
-v_1(x,t)u_x(x,t)|_{x=0} &= g(t) \in H^1(0,T),\quad 0<t\le T, && (34)\\
v_1(x,t)u_x(x,t)|_{x=1} &= v_0(t) \in H^1(0,T),\quad 0<t\le T, && (35)\\
u(x,0) &= u_0(x) \in H^1(0,1). && (36)
\end{aligned}
$$

The solution of the problem (33)-(35) is understood in the weak sense. We suppose that the problem (31)-(36) has a unique solution.

Let $h = 1/N$, $\tau = T/M$, where N and M are natural numbers. We discretize Problem (31)-(36) as follows:

Minimize the functional

$$
\begin{aligned}
I_\alpha^{h,\tau}(v^{h,\tau}) = \tau \sum_{l=0}^{M-1} \left| y^{h,\tau}(0,l) - \varphi(l)\right|^2 + \\
+\alpha\left(\tau \sum_{l=0}^{M-1}\left(v_1^\tau(l))^2 + h\tau\sum_{j=1}^{N}\sum_{l=0}^{M-1}(v_1(j,l)+v_2(j,l)+v_3(j,l)\right)\right) \qquad (37)
\end{aligned}
$$

upon the set $V^{h,\tau} := V_0^\tau \times V_1^{h\tau} \times V_2^{h\tau} \times V_3^{h\tau}$, where

$$
V_0^\tau = \{[v_0^\tau] = (v_0^\tau(1),\dots,v_0^\tau(M-1)|v_0^\tau = v_0(l\tau), l=1,2,\dots,M-1, v_0 \in V_0\},
$$

and for $i = 1,\ 2,\ 3$

$$
V_i^{h\tau} = \left\{[v_i^{h\tau}] = \left(v_i^{h\tau}(j,l)\right), j=0,1,\dots,N-1, l=1,2,\dots,M-1, v_i \in V_i\right\}.
$$

Here $y^{h,\tau} := y$ is a solution of the difference scheme

$$
\begin{aligned}
y_{\bar t} &= (v_1(i,j)y_x)_{\bar x} + v_2(i,j)y + v_3(i,j), && (38)\\
&\quad i=0,1,\dots,N-1,\quad j=1,2,\dots,M-1, &&\\
y(i,0) &= u_0(i), i=0,1,\dots,N-1, && (39)\\
v_1(N,j)y_{\bar x}(N,j) &= v_1(j), j=1,2,\dots,M, && (40)\\
-v_1(0,j)y_x(0,j) &= -h\left(y_j(0,t) + y(0,j) - v_2(0,j)y(0,j)\right. &&\\
&\quad \left.+v_3(0,j)\right) + \tau g(0,j),\quad j=1,2,\dots,M. && (41)
\end{aligned}
$$

(see [8, 9]).

Set

$$
J_\alpha^* = \inf_V J_\alpha(v),\quad I_\alpha^{h,\tau\,*} = \inf_{V^{h,\tau}} I_\alpha^{h,\tau}(v^{h,\tau}).
$$

Suppose that for every N and M with the aid of a minimization method we can find an approximate value $I_\alpha^{h,\tau\,*} + \epsilon_{h,\tau}$ of $I_\alpha^{h,\tau\,*}$ and $v^{h,\tau} \in V^{h,\tau}$ such that

$$
I_\alpha^{h,\tau\,*} \le I_\alpha^{h,\tau}(v^{h,\tau}) \le I_\alpha^{h,\tau\,*} + \epsilon_{h,\tau}, \qquad (42)
$$

where $\epsilon_{h,\tau} \ge 0$ and $\epsilon_{h,\tau} \to 0$ as $n,\ m \to \infty$.

By the method in the proof of [9, 26] we can prove the following theorem

Theorem 8: *Let N, $M \to \infty$. Then*

$$\lim_{N,M\to\infty} I_\alpha^{h,\tau\,*} = J_\alpha^*.$$

Furthurmore, if $v^{h,\tau}$ satisfies the condition (42), then $[v^{h,\tau}] = (\widehat{[v_0^\tau]}, \widetilde{[v_1^{h\tau}]}, \widetilde{[v_2^{h\tau}]}, \widetilde{[v_3^{h\tau}]})$ is a minimizing sequence of Problem (31)-(36):

$$\lim_{N,M\to\infty} J_\alpha([v^{h,\tau}]) = J_\alpha^*.$$

Here

$$\begin{aligned}
\widehat{v_0^\tau}(t) &= v_0^\tau(l\tau) + v_0^\tau(l\tau)(t - l\tau), \text{ for } t_l < t < t_{l+1}, l = 0, 1, \ldots, M-1. \\
\widetilde{v_k^{h\tau}}(x) &= v_k^{h\tau}(j,l), \text{ for } x_j < x < x_{j+1}, t_l < t < t_{l+1}, \\
& \qquad j = 0, 1, \ldots, N-1, l = 0, 1, \ldots, M-1.
\end{aligned}$$

7. About Gradient Methods.

As already mentioned, our variational method is aimed to minimize the defect $J_0(v) = \|u(0,\cdot;v) - \varphi(\cdot)\|^2_{L_2(0,T)}$ over the set V. Since this problem is ill-posed, we added a regularization term $\alpha\Omega(v)$. Such a method was used in many works for similar problems (see [1, 2, 3, 8, 14, 16, 26] and others). There are two strategies to deal with Problem (1)-(4) with the aid of the variational problem (5)-(9):

- To minimize directly the defect $J_0(v)$ over V. This problem is unstable. For this problem we have already found the gradient $J_0'(v)$, therefore we can use one of iterative regularization methods [1, 2, 27, 29] to solve it.
- To minimize the regularized functional $J_\alpha(v)$ over V and use also gradient methods.

In both ways we have to use the gradient $J_\alpha'(v)$. It is clear that we can find the gradient numerically: We solve Problem (6)-(9) and its adjoint system (20)-(24). We remember that (6)-(9) and its adjoint problem are the same if we change the direction of the time variable, therefore in computations we use the same algorithm for them. Using the defference scheme (38)-(41) we can find discrete gradients of our problem. This method has been suggested in [28] for other problems but there have been no convergence theorems.

References

[1] O. M. Alifanov: Inverse Heat Conduction Problems. "Mashinostroienie", Moscow, 1988 (in Russian).

[2] O. M. Alifanov, E. A. Artiukhin, S. V. Rumianziev: Extremal Methods for Solving Incorrect Problems and their Applications to Inverse Heat Mass Transfer Problems. Nauka, Moscow, 1988 (in Russian).

[3] J. V. Beck, B. Blackwell, S. R. St. Clair, Jr.: Inverse Heat Conduction Problems. John Wiley and Sons, New York 1985.

[4] M. F. Bidaut: Thèse, Université de Paris, 1973.

[5] J. R. Cannon: Determination of an unknown coefficient in a parabolic differential equation. *Duke Math. J.* 30(1963), 313-323.

[6] A. M. Denisov: Approximate solution of Volterra equation of the first kind associated with an inverse problem for the heat equation. *Moscow Univ. Comput. Math. Cyber.* 15(3)(1980), 57-60.

[7] Dinh Nho Hào: A noncharacteristic Cauchy problem for linear parabolic equations I: Solvability. *FB Mathematik, FU Berlin*, Preprint A-91-36.

[8] Dinh Nho Hào: A noncharacteristic Cauchy problem for linear parabolic equations II: A variational method. *FB Mathematik, FU Berlin*, Preprint A-91-37.

[9] Dinh Nho Hào: A noncharacteristic Cauchy problem for linear parabolic equations III: A variational method and its approximation schemes. *FB Mathematik, FU Berlin*, Preprint A-91-38.

[10] Dinh Nho Hào: A noncharacteristic Cauchy problem for linear parabolic equations and related inverse problems I: Solvability. *FB Mathematik, FU Berlin*, Preprint A-91-39.

[11] Dinh Nho Hào and R. Gorenflo: A noncharacteristic Cauchy problem for the heat equation. *Acta. Appl. Math.* 24(1991), 1-27.

[12] M. Goebel: On existence of optimal control. *Math. Nachr.* 93(1979), 67-73.

[13] A. D. Iskenderov: On variational formulations of multidimensional inverse problems of mathematical physics. *Soviet Math. Dokl.* 29:1(1984), 52-55.

[14] A. D. Iskenderov and R. K. Tagiev: Problems of optimization with controls in the coefficients of a parabolic equation. *Diff. Equations* 19(1983), 990-999.

[15] B. F. Jones: The determination of a coefficient in a parabolic equation. Part I, Existence and uniqueness. *J. Math. Mech.* 11(1962), 907-918.

[16] J.-L. Lions: Optimal Control of Systems Governed by Partial Differential Equations. Springer-Verlag, Berlin-Heidelberg-New York 1971.

[17] I. G. Malyshev: Inverse problems for the heat equation in a domain with moving boundary. *Ukrain. Math. J.* 27(1975), 687-691.

[18] C. D. Pagani: Determining a coefficient in a parabolic equation. *Appl. Anal.* 14(1982), 99-116.

[19] V. I. Plotnikov: Uniqueness and existence theorems and apriori properties of generalized solutions. *Doklady Akad. Nauk SSSR* 165(1965), 1405-1407.

[20] C. Pucci: Alcune limitazioni per le soluzioni di equazioni paraboliche. *Ann. Mat. Pura Appl.*, XLVIII(IV) (1959), 161-172.

[21] W. Rundell: An inverse problem for a parabolic partial differential equation. *Rocky Mountain J. Math.* 13:4(1983), 679-688.

[22] T. Suzuki: Remarks on the uniqueness in an inverse problem for the heat equation, I, *Proc. Japan Acad.* 58(1982), 93-96.

[23] T. Suzuki: Remarks on the uniqueness in an inverse problem for the heat equation, I, *Proc. Japan Acad.* 58(1982), 175-177.

[24] T. Suzuki: Uniqueness and nonuniqueness in an inverse problem for the parabolic equation. *J. Differential Equations* 47(1983), 296-316.

[25] T. Suzuki: Gel'fand-Levitan's theory, deformation formulas and inverse problems, *J. Fac. Sci. Univ. Tokyo. IA Math.* 32(1985), 223-271.

[26] R. K. Tagiev: Optimal Control Problems for Parabolic and Hyperbolic Equations and Finite Difference Methods for Solving. Thesis. Baku 1982 (in Russian).

[27] G. M. Vainikko, A. Iu. Veretennikov: Iterative Processes in Ill-Posed Problems. Nauka, Moscow, 1986 (in Russian).

[28] F. P. Vasil'ev: On gradient methods for solving optimal control of systems governed by parabolic equations. In "Optimal Control", Znanie, Moscow, 1978, 118-143 (in Russian).

[29] F. P. Vasil'ev: Methods for Solving Extremal Problems. Nauka, Moscow, 1981 (in Russian).

Dinh Nho Hào
Free University of Berlin
Institute of Mathematics I
Arnimallee 2-6
W-1000 Berlin 33, FRG

G N HILE AND C P MAWATA

Liouville theorems for nonlinear elliptic equations of second order

1. Introduction

We consider solutions in $\mathbf{R}^n$ - "entire" solutions - of the nonlinear second order elliptic partial differential equation

$$a(x) \cdot D^2u(x) + H(x,u(x),Du(x)) = 0 , \qquad (1.1)$$

The function $H = H(x,u,p)$, with $x,p \in \mathbf{R}^n$, $u \in \mathbf{R}$, maps $\mathbf{R}^{2n+1}$ into $\mathbf{R}$, and is assumed measurable in all variables, Lipschitz continuous in u and p. Decay conditions on H and associated Lipschitz constants at infinity also are assumed. The real symmetric $n \times n$ matrix a is strictly positive definite in $\mathbf{R}^n$, approaches the Laplace operator at infinity, and is Lipschitz continuous with Lipschitz constant again decaying appropriately at infinity. We discuss "strong" solutions of (1.1), in any space $W^{2,p}_{loc}(\mathbf{R}^n)$ with $p > n$. By well known embedding theorems, such solutions are continuous with continuous first order derivatives; hence pointwise behaviour of such solutions near infinity may be discussed. We show that, under certain hypotheses, there exists a unique entire solution of (1.1) vanishing at infinity. More generally we show that, given a real constant c, there exists a unique entire solution approaching c at infinity, and, given a harmonic polynomial P, there exista a unique entire solution u such that u - P vanishes at infinity.

These results extend those of Begehr and Hile [3], who studied entire classical solutions of the linear version of (1.1),

$$a(x) \cdot D^2u(x) + b(x) \cdot Du(x) + c(x) \cdot u(x) + f(x) = 0 , \qquad (1.2)$$

with coefficients a, b, c and the nonhomogeneous term f assumed Hölder continuous. When specialized to the linear equation (1.1), our results for the nonlinear equation reproduce the results of [3], but for strong, rather than classical, solutions, and with the weaker restriction that b, c, and f be only measurable and decay appropriately at infinity.

Several authors, for example [4,6,7,10], have investigated existence and/or uniqueness questions for entire solutions of linear elliptic equations similar to (1.2), but, except for [3], not with the inclusion of the zero order term $c \cdot u$. There has also been much work on entire solutions of certain nonlinear and quasilinear eliptic equations and systems, as for example in [9,11,13,14,15,16]. The usual result obtained for a nonlinear equation is the traditional statement of Liouville's theorem, namely, a bounded and entire solution is necessarily constant. Consequently, whatever hypotheses are placed on the nonlinear equation to obtain this strict version of Liouville's theorem, when specialized to the linear equation, require that $c \equiv 0$, $f \equiv 0$. In the present work, however, it is shown that, under appropriate decay conditions on the coefficients at infinity, entire solutions of the linear equation (1.2) exist which approach any specified constant at infinity; such solutions are bounded but not constant. Another common Liouville type result, found for example in [17], is the statement that any bounded solution of a given equation is identically 0. Such statements likewise do not apply to the equations we study, as we obtain nonconstant bounded solutions.

Liouville theorems for higher order linear elliptic equations have been obtained by Weck [17]; he shows that bounded solutions of a wide class of elliptic equations with constant coefficients must be constant, and that bounded solutions of some strongly coercive equations with variable coefficients must be identically zero.

Weak entire solutions of quasilinear equations in divergence form,

$$- D_{\beta}[a^{\alpha\beta}(x,u,\nabla u)\ D_{\alpha}u] = f(x,u,\nabla u)$$

were studied by Hildebrandt and Wildman [9], under a condition of the form

$$|f(x,u,p)| \le (\text{constant})\ |p|^2 \quad ;$$

bounded solutions again are necessarily constant.

Our results and those of [3] are closest to those of Friedman [6], who studied the linear equation

$$a \cdot D^2u + b \cdot Du = f \quad ,$$

with suitable decay conditions on a, b, and f at infinity. Under appropriate hypotheses, there exists a unique entire solution of this equation vanishing at infinity, and, for any given real constant, there exists a unique entire solution approaching that constant at infinity. These statements elaborate upon a result of Gilbarg and Serrin [7], who showed that under appropriate conditions any bounded solution near infinity of the homogeneous equation

$$a \cdot D^2 u + b \cdot Du = 0$$

has a limit at infinity; the maximum principle then implies that any entire and bounded solution is constant.

Entire classical solutions of linear parabolic equations with variable coefficients,

$$a \cdot D^2 u + b \cdot Du + c \cdot u - \frac{\partial u}{\partial t} = f \quad ,$$

were studied by Mawata [12]; he established a one-to-one correspondence between entire solutions having polynomial growth at infinity and polynomial solutions of the heat equation.

The authors dedicate this paper to Robert P. Gilbert, the mathematical father and grandfather, respectively, of G. N. Hile and C. P. Mawata, on the occasion of his sixtieth birthday.

2. L^p Estimates for the Linear Equation

In this section we establish estimates for entire solutions of linear elliptic equations; we bound L^p norms of the derivatives of a solution in terms of analogous norms on the solution itself and the nonhomogeneous term.

For Ω an open set in $\mathbf{R}^n$ and F a measurable function from Ω into $\mathbf{R}$, $\mathbf{R}^n$, or $\mathbf{R}^{n\times n}$, we define the usual L^p norm of F over Ω,

$$\| F \|_{p;\Omega} := \left\{ \int_\Omega |F|^p \, dx \right\}^{1/p} \text{ for } 1 \le p < \infty , \quad \| F \|_{\infty;\Omega} := \operatorname*{ess\,sup}_{\Omega} |F| .$$

(When F is a scalar or vector, $|F|$ is the Euclidean norm of F, and when F is a matrix, $|F|^2 := \sum F_{ij}{}^2$.) Whenever $\Omega = \mathbf{R}^n$, we write simply

$$\| F \|_{p;\mathbf{R}^n} =: \| F \|_p .$$

We let r refer to the function $r(x) := |x|$; then, for $\sigma \in \mathbf{R}$, we have the weighted norms

$$\| (1+r)^\sigma F \|_p := \begin{cases} \left[\int_{\mathbf{R}^n} (1+|x|)^{\sigma p} \, |F(x)|^p \, dx \right]^{1/p} & , \text{ if } 1 \le p < \infty , \\ \operatorname*{ess\,sup}_{x \in \mathbf{R}^n} \left\{ (1+|x|)^\sigma \, |F(x)| \right\} & , \text{ if } p = \infty . \end{cases}$$

Note that these norms increase with σ when p is fixed.

Let **L** be the differential operator defined on functions $u = u(x)$, $x \in \mathbf{R}^n$, $n \geq 2$, according to

$$Lu = a \cdot D^2u + b \cdot Du + c \cdot u = \sum_{i,j=1}^{n} a_{ij} D^{ij}u + \sum_{i=1}^{n} b_i D^i u + cu \quad .$$

Here $a = a(x)$ is an $n \times n$ symmetric real matrix with measurable entries, $b = b(x)$ an $n \times 1$ real vector with measurable entries, and $c = c(x)$ a measurable real scalar, all defined for $x \in \mathbf{R}^n$. The function u is a real valued function in $\mathbf{R}^n$ with Sobolev derivatives to order 2, Du is the $n \times 1$ vector, $(D^i u)$, of first order partial derivatives of u, and D^2u is the $n \times n$ matrix, $(D^{ij}u)$, of second order partial derivatives of u. We assume the following conditions on the coefficients of **L** (I is the $n \times n$ identity matrix) :

(L1) There exists a positive constant λ such that

$$a(x)\, \xi \cdot \xi \geq \lambda\, |\xi|^2 \quad , \text{ for all } x,\ \xi \in \mathbf{R}^n \quad .$$

(L2) There exist constants α, K, with $0 < \alpha \leq 1$, $K \geq 0$, such that

$$|a(x) - a(y)| \leq K \frac{|x - y|^\alpha}{(1 + |x| + |y|)^\alpha} \quad , \text{ for all } x,y \in \mathbf{R}^n \ .$$

(L3) There exist nonnegative constants Λ and δ such that, for all $x \in \mathbf{R}^n$,

$$\| (1+r)^\delta\ (a - I) \|_\infty \ , \ \| (1+r)^{1+\delta}\ b \|_\infty \ , \ \| (1 + r)^{2+\delta}\ c \|_\infty \leq \Lambda \quad .$$

Lemma 1A Assume conditions (L1), (L2), (L3) on the coefficients of **L**, let $u \in W^{2,p}_{loc}(\mathbf{R}^n)$, with $1 < p < \infty$, and let $\sigma \in \mathbf{R}$. Then

$$\begin{aligned} &\| (1+r)^{\sigma+1}\ Du \|_p + \| (1+r)^{\sigma+2}\ D^2u \|_p \\ &\leq C(n,p,\alpha,\lambda,\Lambda,K,\sigma) \left[\| (1+r)^\sigma\ u \|_p + \| (1+r)^{\sigma+2}\ \mathbf{L}u \|_p \right] \quad . \end{aligned} \tag{2.1}$$

Proof We may assume the right side is finite, as otherwise there is nothing to prove.

First we consider the region $|x| < 4$. Then (L1) holds in this region, and (L2), (L3) imply that

$$|a(x)-a(y)| \leq K\ |x-y|^\alpha \ , \ |a(x)| \leq \Lambda + \sqrt{n} \ , \ |b(x)| \leq \Lambda \ , \ |c(x)| \leq \Lambda \ .$$

Therefore, by a well known apriori bound for bounded domains ([8], Thm. 9.11, pg. 235),

$$\int_{|x|<2} \left[|Du|^p + |D^2u|^p \right] dx \le C(n,p,\alpha,\lambda,\Lambda,K) \int_{|x|<4} \left[|u|^p + |\mathbf{L}u|^p \right] dx \ .$$

But $|x| < 4$ implies $1 \le 1 + |x| < 5$; hence, letting C depend also on σ, we may write this inequality as

$$\int_{|x|<2} \left[\left| (1+r)^{\sigma+1} Du \right|^p + \left| (1+r)^{\sigma+2} D^2u \right|^p \right] dx$$
$$\le C(n,p,\alpha,\lambda,\Lambda,K,\sigma) \int_{|x|<4} \left[\left| (1+r)^{\sigma} u \right|^p + \left| (1+r)^{\sigma+2} \mathbf{L}u \right|^p \right] dx \ . \tag{2.2}$$

We next consider a region in $\mathbf{R}^n$ of the form $\rho < |x| < 8\rho$, where ρ is fixed and constant, $\rho \ge 1$. We define a function v in the region $1 < |x| < 8$ according to

$$v(x) := u(\rho x) \quad , \quad Dv(x) = \rho\, Du(\rho x) \quad , \quad D^2v(x) = \rho^2\, D^2u(\rho x) \quad ,$$

and set $f := \mathbf{L}u$; then v solves the equation

$$\tilde{a} \cdot D^2v + \tilde{b} \cdot Dv + \tilde{c}\, v = \tilde{f} \quad ,$$

where we define also

$$\tilde{a}(x) := a(\rho x) \ , \ \tilde{b}(x) := \rho\, b(\rho x) \ , \ \tilde{c}(x) := \rho^2\, c(\rho x) \ , \ \tilde{f}(x) := \rho^2\, f(\rho x) \ .$$

From (L1), (L2), (L3) we observe that, for $1 < |x|,|y| < 8$,

$$\tilde{a}(x)\, \xi \cdot \xi \ge \lambda\, |\xi|^2 \quad , \text{ for all } \xi \in \mathbf{R}^n \quad ,$$

$$|\tilde{a}(x) - \tilde{a}(y)| \le K \frac{\rho^\alpha\, |x - y|^\alpha}{(1 + |\rho x| + |\rho y|)^\alpha} \le K\, |x - y|^\alpha \quad .$$

$$|\tilde{a}(x) - I| \ , \ |\tilde{b}(x)| \ , \ |\tilde{c}(x)| \le \Lambda \quad .$$

Again the well known apriori bound for bounded domains ([8], pg. 235) gives

$$\int_{2<|x|<4} \left[|Dv|^p + |D^2v|^p \right] dx \le C(n,p,\alpha,\lambda,\Lambda,K) \int_{1<|x|<8} \left[|v|^p + |\tilde{f}|^p \right] dx,$$

or, in terms of u,

$$\int_{2\rho<|x|<4\rho} \left[\left| \rho\, Du \right|^p + \left| \rho^2\, D^2u \right|^p \right] dx$$
$$\le C(n,p,\alpha,\lambda,\Lambda,K) \int_{\rho<|x|<8\rho} \left[\left| u \right|^p + \left| \rho^2\, \mathbf{L}u \right|^p \right] dx \quad .$$

But for $\rho < |x| < 8\rho$, we have $\rho < 1 + |x| < 9\rho$; therefore, after multiplying the above integral inequality by $\rho^{\sigma p}$, we are led to an estimate of the form

$$\int_{2\rho<|x|<4\rho} \left[\; | \; (1+r)^{\sigma+1} \; Du \; |^p + | \; (1+r)^{\sigma+2} \; D^2u \; |^p \right] dx$$

$$\leq C(n,p,\alpha,\lambda,\Lambda,K,\sigma) \int_{\rho<|x|<8\rho} \left[\; | \; (1+r)^{\sigma} \; u \; |^p + | \; (1+r)^{\sigma+2} \; Lu \; |^p \right] dx \quad .$$

Now we choose in particular $\rho = 2^m$, and obtain for m = 0, 1, 2, ...,

$$\int_{2^{m+1}<|x|<2^{m+2}} \left[|(1+r)^{\sigma+1} \; Du|^p + |(1+r)^{\sigma+2} \; D^2u|^p \right] dx$$

$$\leq C(n,p,\alpha,\lambda,\Lambda,K,\sigma) \int_{2^m<|x|<2^{m+3}} \left[|(1+r)^{\sigma}u|^p + |(1+r)^{\sigma+2} \; Lu|^p \right] dx \; . \tag{2.3}$$

Finally, we add inequality (2.2) with inequalities (2.3), $m \geq 0$, and obtain

$$\int_{\mathbf{R}^n} \left[\; | \; (1+r)^{\sigma+1} \; Du \; |^p + | \; (1+r)^{\sigma+2} \; D^2u \; |^p \right] dx$$

$$\leq C(n,p,\alpha,\lambda,\Lambda,K,\sigma) \int_{\mathbf{R}^n} \left[\; | \; (1+r)^{\sigma} \; u \; |^p + | \; (1+r)^{\sigma+2} \; Lu \; |^p \right] dx \quad .$$

This inequality gives (2.1). ∎

Lemma 1B Let $u \in W^{2,p}_{loc}(\mathbf{R}^n)$, $n < p \leq \infty$, and let $\sigma \in \mathbf{R}$. Then

$$\| \; (1+r)^{\sigma + n/p} \; u \; \|_\infty + \| \; (1+r)^{\sigma + 1 + n/p} \; Du \; \|_\infty$$

$$\leq C(n,p,\sigma) \left[\| \; (1+r)^{\sigma} \; u \; \|_p + \| \; (1+r)^{\sigma+1} \; Du \; \|_p + \| \; (1+r)^{\sigma+2} \; D^2u \; \|_p \right] \; . \tag{2.4}$$

Proof Again, we may assume the right side is finite.

First assume $|x| \geq 1$. We set $\rho = |x|$, and

$$v(y) := u(\rho y) \quad , \quad Dv(y) = \rho \; Du(\rho y) \quad , \quad D^2v(y) = \rho^2 \; D^2u(\rho y) \quad .$$

By the Sovolev embedding theorem ([1], Thm. 5.4, pg. 97), for $p > n$,

$$\sup_{1 \leq |y| \leq 2} |v(y)| + \sup_{1 \leq |y| \leq 2} |Dv(y)|$$

$$\leq C(n,p) \left[\int_{1\leq|y|\leq2} |v(y)|^p + |Dv(y)|^p + |D^2v(y)|^p \; dy \right]^{1/p} \quad ,$$

which in terms of u gives

$$|u(x)| + \rho\, |Du(x)|$$
$$\le C(n,p)\rho^{-n/p}\left[\int\!\!\int_{\rho\le|y|\le 2\rho}\left\{|u(y)|^p + |\rho Du(y)|^p + |\rho^2 D^2u(y)|^p\right\} dy\right]^{1/p} . \tag{2.5}$$

Now in the integral we have the estimate

$$(1 + \rho)/2 \le \rho = |x| \le |y| \le 1 + |y| \le 1 + 2\rho \le 3\rho \; ;$$

thus we may multiply (2.5) by $\rho^{\sigma + n/p}$ to obtain

$$(1 + |x|)^{\sigma + n/p}\, |u(x)| + (1 + |x|)^{\sigma + 1 + n/p}\, |Du(x)|$$

$$\le C(n,p,\sigma)\left[\int\!\!\int_{\rho\le|y|\le 2\rho}\left\{|(1+r)^{\sigma}u|^p + |(1+r)^{\sigma+1}Du|^p + |(1+r)^{\sigma+2}D^2u|^p\right\} dy\right]^{1/p}$$

$$\le C(n,p,\sigma)\ \| (1+r)^{\sigma} u \|_p + \| (1+r)^{\sigma+1} Du \|_p + \| (1+r)^{\sigma+2} D^2u \|_p \quad .$$

Next assume that $|x| \le 1$. Applying the Sobolev embedding theorem ([1], pg. 97) now to the unit ball, we obtain

$$|u(x)| + |Du(x)| \le \sup_{|y|\le 1} |u(y)| + \sup_{|y|\le 1} |Du(y)|$$
$$\le C(n,p)\left[\int_{|y|\le 1} |u(y)|^p + |Du(y)|^p + |D^2u(y)|^p\, dx\right]^{1/p} . \tag{2.6}$$

But for $|y| \le 1$, we have $1 \le 1 + |y| \le 2$ and $c_1(\sigma) \le (1+r)^{\sigma} \le c_2(\sigma)$; thus (2.6) gives

$$(1+|x|)^{\sigma + n/p}\, |u(x)| + (1+|x|)^{\sigma + 1 + n/p}\, |Du(x)|$$

$$\le C(n,p,\sigma)\left[\, \| (1+r)^{\sigma} u \|_p + \| (1+r)^{\sigma+1} Du \|_p + \| (1+r)^{\sigma+2} D^2u \|_p \right. .$$

Our inequalities for $|x| \le 1$ and for $|x| \ge 1$ now give (2.4). ∎

Combining the two lemmas, we have

Theorem 1 Assume conditions (L1), (L2), (L3) on the coefficients of **L**, let $u \in W^{2,p}_{loc}(\mathbf{R}^n)$ with $n < p < \infty$, and let $\sigma \in \mathbf{R}$. Then

$$\| (1+r)^{\sigma + n/p}\, u \|_{\infty} + \| (1+r)^{\sigma + 1 + n/p}\, Du \|_{\infty}$$

$$+ \| (1+r)^{\sigma+1} Du \|_p + \| (1+r)^{\sigma+2} D^2u \|_p \tag{2.7}$$

$$\le C(n,p,\alpha,\lambda,\Lambda,K,\sigma) \left[\| (1+r)^{\sigma} u \|_p + \| (1+r)^{\sigma+2} Lu \|_p \right] ,$$

and

$$\| (1+r)^{\sigma+2} \Delta u \|_p \tag{2.8}$$

$$\le C(n,p,\alpha,\lambda,\Lambda,K,\sigma-\delta) \left[\| (1+r)^{\sigma-\delta} u \|_p + \| (1+r)^{\sigma+2} Lu \|_p \right] .$$

Proof To obtain (2.7), we combine Lemmas 1A and 1B. For (2.8), we write

$$\Delta u = Lu - (a - I) \cdot D^2 u - b \cdot Du - c\, u ,$$

and estimate

$$\|(1+r)^{\sigma+2} \Delta u\|_p \le \|(1+r)^{\sigma+2} Lu\|_p + \|(1+r)^{\delta} (a - I)\|_\infty \|(1+r)^{\sigma+2-\delta} D^2 u\|_p$$

$$+ \|(1+r)^{1+\delta} b\|_\infty \|(1+r)^{\sigma+1-\delta} Du\|_p + \|(1+r)^{2+\delta} c\|_\infty \|(1+r)^{\sigma-\delta} u\|_p .$$

Then by (L3) and Lemma 1A,

$$\| (1+r)^{\sigma+2} \Delta u \|_p \le \| (1+r)^{\sigma+2} Lu \|_p$$

$$+ \Lambda\, C(n,p,\alpha,\lambda,\Lambda,K,\sigma-\delta) \left(\| (1+r)^{\sigma-\delta} u \|_p + \| (1+r)^{\sigma+2-\delta} Lu \|_p \right) ,$$

which gives (2.8). ∎

3. Bounds for the Poisson Equation

We discuss in some detail entire solutions of the *Poisson equation*,

$$\Delta u = f .$$

We use the fundamental solution of the Laplace equation,

$$\Gamma(x) := \begin{cases} \dfrac{1}{(2-n)\,\omega_n} |x|^{2-n} & , \text{ if } n \ge 3 \\[2ex] \dfrac{1}{2\pi} \log |x| & , \text{ if } n = 2 \end{cases}$$

(where ω_n is the surface area of the unit ball in $\mathbf{R}^n$), and the *Newtonian potential* of the function f,

$$Sf(x) := \int_{\mathbf{R}^n} \Gamma(x-y)\, f(y)\, dy .$$

We also define, whenever f is Lebesgue integrable over $\mathbf{R}^n$,

$$f_o := \int_{\mathbf{R}^n} f(y)\, dy . \tag{3.1}$$

We shall require the following estimates on the Newtonian potential :

Theorem 2 Suppose $n \geq 2$, $\sigma \in \mathbf{R}$, $p \in [1,\infty]$, and let f be a real valued measurable function on $\mathbf{R}^n$ such that

$$\| (1+r)^{\sigma+2} f \|_p < \infty .$$

(a) If $n \geq 3$ and $- n/p < \sigma < n - 2 - n/p$, then

$$\| (1+r)^{\sigma} \mathbf{S}f \|_p \leq C(n,p,\sigma) \| (1+r)^{\sigma+2} f \|_p \quad , \text{ for } 1 \leq p \leq \infty , \tag{3.2}$$

$$\|(1+r)^{\sigma + n/p} \mathbf{S}f\|_\infty \leq C(n,p,\sigma) \|(1+r)^{\sigma+2} f\|_p , \text{ for } n/2 < p \leq \infty . \tag{3.3}$$

(b) If $n = 2$, $1 < p \leq \infty$, $- \infty < \tau < - 2/p < \sigma$, then

$$\| (1+r)^{\tau} \mathbf{S}f \|_p \leq C(p,\sigma,\tau) \| (1+r)^{\sigma+2} f \|_p , \tag{3.4}$$

and if moreover $\sigma < 1 - 2/p$, then for $|x| \geq 1$,

$$|\mathbf{S}f(x) - f_o \Gamma(x)| \leq C(p,\sigma) \| (1+r)^{\sigma+2} f \|_p |x|^{- \sigma - 2/p} . \tag{3.5}$$

The proof of Theorem 2, being quite computational in nature, is postponed until the last section of the paper.

Theorem 3 Suppose $n/2 < p < \infty$, $- n/p < \sigma$, and let f be a real valued measurable function on $\mathbf{R}^n$ such that

$$\| (1+r)^{\sigma+2} f \|_p < \infty .$$

(a) If $n \geq 3$, then there exists exactly one strong solution u in $W^{2,p}_{loc}(\mathbf{R}^n)$ of $\Delta u = f$ which vanishes at infinity; this solution is $u = \mathbf{S}f$, and, whenever $- n/p < \sigma < n - 2 - n/p$, we have for this solution the bound

$$\| (1+r)^{\sigma} \mathbf{S}f \|_p + \| (1+r)^{\sigma+1} D(\mathbf{S}f) \|_p + \| (1+r)^{\sigma+2} D^2(\mathbf{S}f) \|_p \leq C(n,p,\sigma) \| (1+r)^{\sigma+2} f \|_p . \tag{3.6}$$

(b) If $n = 2$, then there exists exactly one strong solution u in $W^{2,p}_{loc}(\mathbf{R}^2)$ of $\Delta u = f$ such that $u - \gamma \Gamma$ vanishes at infinity for some constant γ; this solution is $u = \mathbf{S}f$, and the uniquely determined constant is $\gamma = f_o$. For γ we have the bound

$$|\gamma| \le C(p,\sigma) \; \| (1+r)^{\sigma+2} f \|_p \; , \tag{3.7}$$

and for the solution **S**f and any τ such that $-\infty < \tau < -2/p$,

$$\begin{aligned} &\| (1+r)^{\tau} \mathbf{S}f \|_p + \| (1+r)^{\tau+1} D(\mathbf{S}f) \|_p + \| (1+r)^{\tau+2} D^2(\mathbf{S}f) \|_p \\ &\le C(p,\sigma,\tau) \; \| (1+r)^{\sigma+2} f \|_p \; . \end{aligned} \tag{3.8}$$

Moreover, if also $\sigma < 1 - 2/p$, then for $|x| \ge 1$,

$$|\mathbf{S}f(x) - \gamma\, \Gamma(x)| \le C(p,\sigma) \; \| (1+r)^{\sigma+2} f \|_p \; |x|^{-\sigma - 2/p} \; . \tag{3.9}$$

Proof Uniqueness of the solution in both cases is clear; the difference of two solutions is an entire harmonic function which (a) vanishes at infinity when $n \ge 3$ (b) behaves like a multiple of $\log |x|$ near infinity when $n = 2$. In either case, this difference must be identically zero. When $n = 2$, it is clear that $u - \gamma\,\Gamma$ can vanish at infinity for at most one constant γ.

We may choose σ smaller, if necessary, so that $-n/p < \sigma < n - 2 - n/p$ when $n \ge 3$, $-2/p < \sigma < 1 - 2/p$ when $n = 2$. Then (3.3) shows that **S**f vanishes at infinity when $n \ge 3$, and (3.5) shows that $\mathbf{S}f - f_o\,\Gamma$ vanishes at infinity when $n = 2$.

Given any large radius R, we break **S**f into two integrals as

$$\mathbf{S}f(x) := \int_{|y|<R} \Gamma(x-y)\, f(y)\, dy + \int_{|y|\ge R} \Gamma(x-y)\, f(y)\, dy \quad .$$

By well known results on the Newtonian potential for bounded domains ([8], Thm. 9.9, pg. 230), the first integral is a strong solution of $\Delta u = f$ in the region $\mathbf{B}_R := \{ x : |x| < R \}$, and in fact belongs to the space $W^{2,p}(\mathbf{B}_R)$. Obviously, the second integral is harmonic in $\mathbf{B}_R$. Since R is arbitrary, $u = \mathbf{S}f$ is in the space $W^{2,p}_{loc}(\mathbf{R}^n)$ and solves $\Delta u = f$ there.

It remains only to verify the bounds (3.6) - (3.9). In the case $n \ge 3$, we combine (3.2) of Theorem 2 with (2.1) of Lemma 1A and obtain (3.6). When $n = 2$ and $-\infty < \tau < -2/p$, Theorem 2 yields (3.4); then Lemma 1A, with $\mathbf{L} = \Delta$, shows that the left side of (3.8) is bounded by

$$C(n,p,\sigma,\tau) \left(\| (1+r)^{\sigma+2} f \|_p + \| (1+r)^{\tau+2} f \|_p \right),$$

which, since $\tau < -2/p < \sigma$, is smaller than the right side of (3.8).

To obtain (3.7), we let q be determined by $1/p + 1/q = 1$, and find that

$$|\gamma| = \left| \int_{\mathbf{R}^n} (1+r)^{\sigma+2} f\,(1+r)^{-\sigma-2}\,dy \right| \le \| (1+r)^{\sigma+2} f \|_p \, \| (1+r)^{-\sigma-2} \|_q ,$$

where $\| (1+r)^{-\sigma-2} \|_q = C(p,\sigma) < \infty$ because $\sigma > -2/p$. (This estimate does not require $\sigma < 1 - 2/p$.) Finally, (3.9) comes from (3.5) of Theorem 2. ∎

4. Entire Solutions of the Linear Equation with Conditions at Infinity

We next establish apriori bounds on entire solutions of linear equations with prescribed behaviour at infinity. We continue to assume conditions (L1), (L2), and (L3) on the coefficients of **L**, but in (L3) we require that $\delta > 0$; consequently, **L** approaches the Laplace operator at infinity.

Theorem 4 Assume conditions (L1), (L2), (L3) on the coefficients of **L**, with also (a) $c \le 0$ in $\mathbf{R}^n$ when $n \ge 3$, (b) $c \equiv 0$ in $\mathbf{R}^2$ when $n = 2$. Also assume that $0 < \delta \le 1$, that $p, \sigma \in \mathbf{R}$ with $n < p < \infty$, and that $u \in W^{2,p}_{loc}(\mathbf{R}^n)$.

(a) If $n \ge 3$, $-\frac{n}{p} < \sigma < n - 2 - \frac{n}{p}$, and if $u(\infty) = 0$, then

$$\| (1+r)^{\sigma} u \|_p + \| (1+r)^{\sigma+1} Du \|_p + \| (1+r)^{\sigma+2} D^2u \|_p \le C(n,p,\alpha,\lambda,\Lambda,K,\sigma) \left[\| (1+r)^{\sigma+2} \mathbf{L}u \|_p \right] . \tag{4.1}$$

(b) If $n = 2$, $-2/p < \sigma$, and if $u - \gamma\,\Gamma$ vanishes at ∞ for some real constant γ, then for $\tau < -2/p$,

$$\| (1+r)^{\tau} u \|_p + \| (1+r)^{\tau+1} Du \|_p + \| (1+r)^{\tau+2} D^2u \|_p \le C(p,\alpha,\lambda,\Lambda,K,\sigma,\tau) \, \| (1+r)^{\sigma+2} \mathbf{L}u \|_p . \tag{4.2}$$

Moreover, if $\sigma < -2/p + \delta$, then

$$\| (1+r)^{\sigma+2} \Delta u \|_p \le C(p,\alpha,\lambda,\Lambda,K,\sigma,\delta) \, \| (1+r)^{\sigma+2} \mathbf{L}u \|_p , \tag{4.3}$$

$$|\gamma| \le C(p,\alpha,\lambda,\Lambda,K,\sigma,\delta) \, \| (1+r)^{\sigma+2} \mathbf{L}u \|_p , \tag{4.4}$$

and for $|x| \ge 1$,

$$|u(x) - \gamma\,\Gamma(x)| \le C(p,\alpha,\lambda,\Lambda,K,\sigma,\delta) \, \|(1+r)^{\sigma+2} \mathbf{L}u\|_p \, |x|^{-\sigma - 2/p} . \tag{4.5}$$

Proof (a) Again, we may assume the norm on the right of (4.1) is finite. Then we show first that $\| (1+r)^{\sigma} u \|_p$ is finite. Since $u(\infty) = 0$, u is bounded; therefore, as is easily checked, $\| (1+r)^{\tau} u \|_p < \infty$ whenever $\tau < - n/p$. Now let τ be any real number such that $\| (1+r)^{\tau} u \|_p$ is finite and $- n/p - \delta < \tau \leq \sigma - \delta$. Applying (2.8) of Theorem 1, with σ replaced by $\tau + \delta$, we find that

$$\|(1+r)^{\tau+2+\delta} \Delta u\|_p \leq C(n,p,\alpha,\lambda,\Lambda,K,\tau) \left(\|(1+r)^{\tau} u\|_p + \|(1+r)^{\tau+2+\delta} Lu\|_p \right),$$

with the norm on the right involving Lu finite since $\tau+2+\delta \leq \sigma+2$. Observing that also $- n/p < \tau + \delta \leq \sigma < n - 2 - n/p$, we may apply Theorem 3 with σ replaced by $\tau + \delta$ and conclude that

$$\| (1+r)^{\tau+\delta} u \|_p \leq C(n,p,\tau+\delta) \| (1+r)^{\tau+2+\delta} \Delta u \|_p ,$$

and hence, since $\tau+2+\delta \leq \sigma+2$,

$$\|(1+r)^{\tau+\delta}u\|_p \leq C(n,p,\alpha,\lambda,\Lambda,K,\tau,\delta) \left(\|(1+r)^{\tau}u\|_p + \|(1+r)^{\sigma+2}Lu\|_p\right) , \tag{4.6}$$

with the right side, and hence the left, finite. We have shown that, for any τ in the range $- n/p - \delta < \tau \leq \sigma - \delta$, the finiteness of $\| (1+r)^{\tau} u \|_p$ implies also the finiteness of $\| (1+r)^{\tau+\delta} u \|_p$; hence, $\| (1+r)^{\sigma} u \|_p$ is finite.

Now we establish (4.1). By Theorem 1, it is sufficient to show that

$$\| (1+r)^{\sigma} u \|_p \leq C(n,p,\alpha,\lambda,\Lambda,K,\sigma) \| (1+r)^{\sigma+2} Lu \|_p ,$$

again for all u in $W^{2,p}_{loc}(\mathbf{R}^n)$ such that $u(\infty) = 0$.

If the assertion is false, then there exist sequences $\{a_m\}$, $\{b_m\}$, $\{c_m\}$, $\{u_m\}$ such that $\{a_m\}$, $\{b_m\}$, $\{c_m\}$ satisfy conditions (L1), (L2), (L3) with each $c_m \leq 0$, each u_m is in the space $W^{2,p}_{loc}(\mathbf{R}^n)$ with $u_m(\infty) = 0$, and

$$\| (1+r)^{\sigma} u_m \|_p = 1 \quad , \quad \| (1+r)^{\sigma+2} Lu_m \|_p \leq \frac{1}{m} . \tag{4.7}$$

By Theorem 1 and (4.7), for each m we have the bound

$$\begin{aligned} &\| (1+r)^{\sigma + n/p} u_m \|_\infty + \| (1+r)^{\sigma + 1 + n/p} Du_m \|_\infty \\ &+ \| (1+r)^{\sigma+1} Du_m \|_p + \| (1+r)^{\sigma+2} D^2u_m \|_p \\ &\leq C(n,p,\alpha,\lambda,\Lambda,K,\sigma) \left[\| (1+r)^{\sigma} u_m \|_p + \| (1+r)^{\sigma+2} Lu_m \|_p \right] \\ &\leq 2\, C(n,p,\alpha,\lambda,\Lambda,K,\sigma) . \end{aligned} \tag{4.8}$$

Conditions (L1), (L2), (L3), applied to each a_m, imply that we may pass to subsequences so that, for some $n \times n$ symmetric matrix function a in $\mathbf{R}^n$,

$$a_m \longrightarrow a \quad \text{uniformly on compact subsets of } \mathbf{R}^n \quad ,$$

with (L1), (L2), (L3) holding also for a.

Note that (L3), applied to each b_m, shows that

$$\|b_m\|_p \le \left[\int_{\mathbf{R}^n} \Lambda^p \, (1 + |x|)^{-p} \, dx \right]^{1/p} \le \Lambda \, C(n,p) \quad ;$$

hence we may again pass to a subsequence so that there exists a function b in $L^p(\mathbf{R}^n)$ with

$$b_m \longrightarrow b \quad \text{weakly in } L^p(\mathbf{R}^n) \quad , \quad \|b\|_p \le \Lambda \, C(n,p) \quad .$$

We show that b satisfies (L3) almost everywhere, and hence on all of $\mathbf{R}^n$ after modification on a set of measure zero. If not, there exists a measurable subset E of $\mathbf{R}^n$, with $0 < m(E) < \infty$, such that

$$(1 + |x|)^{1+\delta} \, |b(x)| > \Lambda \quad \text{on } E \; .$$

We set $v := \chi_E \operatorname{sign} b$; then $v \in L^q(\mathbf{R}^n)$, and

$$\lim \left| \int_{\mathbf{R}^n} b_m \, v \, dx \right| \le \lim \int_E |b_m| \, dx \le \int_E \Lambda \, (1 + |x|)^{-1-\delta} \, dx$$

$$< \int_E |b| \, dx = \int_{\mathbf{R}^n} b \, v \, dx = \left| \int_{\mathbf{R}^n} b \, v \, dx \right| \quad ,$$

contradicting the weak convergence of b_m to b in $L^p(\mathbf{R}^n)$. By a similar argument, we may pass to a subsequence so that, for some c in $L^p(\mathbf{R}^n)$,

$$c_m \longrightarrow c \quad \text{weakly in } L^p(\mathbf{R}^n) \quad , \quad \|c\|_p \le \Lambda \, C(n,p) \quad ,$$

and with c satisfying (L3) almost everywhere. An argument analogous to the preceding one for b shows that also $c \le 0$ in $\mathbf{R}^n$.

Inequality (4.7) implies that we may pass to a subsequence so that, for some function f in $\mathbf{R}^n$,

$$(1 + r)^{\sigma+2} \, \mathbf{L}u_m \longrightarrow (1 + r)^{\sigma+2} \, f \quad \text{weakly in } L^p(\mathbf{R}^n) \; ;$$

since also $\| \, (1+r)^{\sigma+2} \, f \, \|_p \le 1/m$ for each m, we have in fact that $f \equiv 0$, and

$$(1 + r)^{\sigma+2} \, \mathbf{L}u_m \longrightarrow 0 \text{ weakly in } L^p(\mathbf{R}^n) \quad .$$

The bounds (4.7) and (4.8) show that, on any ball $\mathbf{B}_R$ centered at 0 with radius R, we have an estimate

$$\| u_m \|_{W^{2,p}(\mathbf{B}_R)} \le C(n,p,\alpha,\lambda,\Lambda,K,\sigma,R) \quad .$$

The Rellich Kondrachov Theorem ([1], pg. 144) asserts that the space $W^{2,p}(\mathbf{B}_R)$ is compactly imbedded in the space $C^1(\overline{\mathbf{B}_{R/2}})$ whenever $p > n$; applying this result to successive balls of radius 1, 2, 3, ..., we conclude that we may once again pass to a subsequence so that, for some function u in $C^1(\mathbf{R}^n)$, $u_m \longrightarrow u$ and $Du_m \longrightarrow Du$ uniformly on compact subsets of $\mathbf{R}^n$.

The bound (4.8) also implies that, after passing to a subsequence, there are functions $\{v_\alpha\}$, $|\alpha| = 2$, such that

$$(1 + r)^{\sigma+2} D^\alpha u_m \longrightarrow (1 + r)^{\sigma+2} v_\alpha \quad \text{weakly in } L^p(\mathbf{R}^n) \quad ,$$

$$\| (1+r)^{\sigma+2} v_\alpha \|_p \le 2\, C(n,p,\alpha,\lambda,\Lambda,K,\sigma) \quad .$$

Note that for $\phi \in C_o^\infty(\mathbf{R}^n)$,

$$\int_{\mathbf{R}^n} u\, D^\alpha\phi\, dx = \lim \int_{\mathbf{R}^n} u_m\, D^\alpha\phi\, dx = \lim \int_{\mathbf{R}^n} D^\alpha u_m\, \phi\, dx$$

$$= \lim \int_{\mathbf{R}^n} (1+r)^{\sigma+2} D^\alpha u_m\, (1+r)^{-\sigma-2}\, \phi\, dx = \int_{\mathbf{R}^n} (1+r)^{\sigma+2}\, v_\alpha\, (1+r)^{-\sigma-2}\, \phi\, dx$$

$$= \int_{\mathbf{R}^n} v_\alpha\, \phi\, dx \quad ;$$

hence $v_\alpha = D^\alpha u$ for each α, and we have $u \in W^{2,p}_{loc}(\mathbf{R}^n)$, with

$$(1 + r)^{\sigma+2} D^\alpha u_m \longrightarrow (1 + r)^{\sigma+2} D^\alpha u \quad \text{weakly in } L^p(\mathbf{R}^n) \quad .$$

From these considerations, we conclude that all subsequences may be chosen so that, on any bounded open subset Ω of $\mathbf{R}^n$,

$$a_m \cdot D^2u_m \longrightarrow a \cdot D^2u \quad , \quad b_m \cdot Du_m \longrightarrow b \cdot Du \quad , \quad c_m u_m \longrightarrow cu \quad , \quad \mathbf{L}u_m \longrightarrow 0$$

weakly in $L^p(\Omega)$. Since weak limits are unique (a.e.), it follows that

$$a \cdot D^2u + b \cdot Du + c\, u = 0 \quad \text{in } \mathbf{R}^n \quad . \tag{4.9}$$

We let $m \to \infty$ in (4.8), recalling that $u_m \longrightarrow u$ uniformly on compact subsets of $\mathbf{R}^n$, to conclude that $\| (1+r)^{\sigma + n/p}\, u \|_\infty$ is finite; hence $u(\infty) = 0$. It is known that, under our assumptions (L1) - (L3) and with $c \le 0$, the maximum principle applies to strong solutions of (4.9) in the space $W^{2,p}_{loc}(\mathbf{R}^n)$, $p > n$. (See [2], [5], or Thm. 9.11, pg. 220 of [8].) In particular, in any large ball centered at the origin, $|u|$ is bounded by its maximum on the boundary of the ball. Letting the radius tend to infinity, we conclude that $u \equiv 0$.

We now have $u_m \longrightarrow 0$ uniformly on compact subsets of $\mathbf{R}^n$; then by (4.8),

$$\| (1+r)^{\sigma + n/p - \delta/2} u_m \|_\infty \longrightarrow 0 \quad ,$$

and hence

$$\| (1+r)^{\sigma-\delta} u_m \|_p = \| (1+r)^{\sigma + n/p - \delta/2} u_m (1+r)^{- n/p - \delta/2} \|_p$$

$$\leq \| (1+r)^{\sigma + n/p - \delta/2} u_m \|_\infty \left[\int_{\mathbf{R}^n} (1+r)^{- n - \delta p/2} dx \right]^{1/p} \longrightarrow 0 \quad .$$

Recall that (4.6) was established for τ such that $- n/p - \delta < \tau \leq \sigma - \delta$ and for u with $\| (1+r)^\tau u \|_p < \infty$; applying (4.6) to u_m with $\tau = \sigma - \delta$ yields

$$\|(1+r)^\sigma u_m\|_p \leq C(n,p,\alpha,\lambda,\Lambda,K,\sigma) \left(\|(1+r)^{\sigma+2} \mathbf{L}u_m\|_p + \|(1+r)^{\sigma-\delta} u_m\|_p \right) .$$

Since the right side of this inequality tends to 0 as $m \longrightarrow \infty$, we have contradicted (4.7).

(b) First assume that (4.2) holds for all $\tau < - 2/p$ and for all u in $W^{2,p}_{loc}(\mathbf{R}^2)$ such that $u - \gamma\,\Gamma$ vanishes at infinity for some real constant γ. The bound (2.8) of Theorem 1 gives

$$\|(1+r)^{\sigma+2} \Delta u\|_p \leq C(n,p,\alpha,\lambda,\Lambda,K,\sigma-\delta) \left[\|(1+r)^{\sigma-\delta} u\|_p + \|(1+r)^{\sigma+2} \mathbf{L}u\|_p \right] ,$$

and then, when $\sigma < - 2/p + \delta$, we may replace τ by $\sigma - \delta$ in (4.2) to bound $\| (1+r)^{\sigma-\delta} u \|_p$ and obtain (4.3). Theorem 3 then forces $u = \mathbf{S}(\Delta u)$, and (3.7) and (4.3) give (4.4). Since $\sigma < - 2/p + \delta \leq - 2/p + 1$, we may combine (3.9) of Theorem 3 with (4.3) to obtain (4.5).

It remains only to verify (4.2); we may assume that $\| (1+r)^{\sigma+2} \mathbf{L}u \|_p$ is finite. Since we assume $u - \gamma\,\Gamma$ vanishes at infinity for some real γ, u behaves like $\gamma\,\Gamma$ at infinity and the norm $\| (1+r)^\tau u \|_p$ must be finite whenever $\tau < - 2/p$. Since the left side of (4.2) increases as τ increases, while the right side decreases as σ decreases, we may take τ larger and σ smaller, if necessary, so that

$$\tau < - 2/p < \sigma < \tau + \delta < - 2/p + \delta \ . \tag{4.10}$$

Applying Lemma 1A with σ replaced by τ, we see that it is sufficient to establish instead of (4.2) the simpler estimate

$$\| (1+r)^{\tau} u \|_p \le C(p,\alpha,\lambda,\Lambda,K,\sigma,\tau) \| (1+r)^{\sigma+2} Lu \|_p . \qquad (4.11)$$

We assume the estimate (4.11) is false. Then there exist sequences $\{a_m\}$, $\{b_m\}$, $\{u_m\}$, $\{\gamma_m\}$ such that $\{a_m\}$, $\{b_m\}$ satisfy conditions (L1) - (L3), each u_m is in the space $W^{2,p}_{loc}(\mathbf{R}^n)$, each $u_m - \gamma_m \Gamma$ vanishes at infinity, and

$$\| (1+r)^{\tau} u_m \|_p = 1 \quad , \quad \| (1+r)^{\sigma+2} Lu_m \|_p \le 1/m . \qquad (4.12)$$

Combining Theorem 1 with (4.12) we derive the estimate

$$\begin{aligned} &\| (1+r)^{\tau + 2/p} u_m \|_\infty + \| (1+r)^{\tau + 1 + 2/p} Du_m \|_\infty \\ &+ \| (1+r)^{\tau+1} Du_m \|_p + \| (1+r)^{\tau+2} D^2 u_m \|_p \\ &\le C(p,\alpha,\lambda,\Lambda,K,\tau) \Big[\| (1+r)^{\tau} u_m \|_p + \| (1+r)^{\tau+2} Lu_m \|_p \Big] \\ &\le 2\, C(p,\alpha,\lambda,\Lambda,K,\tau) . \end{aligned} \qquad (4.13)$$

As in the proof of (a), we may pass to a subsequence so that $a_m \longrightarrow a$ uniformly on compact subsets of $\mathbf{R}^2$ with (L1) - (L3) holding for a, so that $b_m \longrightarrow b$ weakly in $L^p(\mathbf{R}^2)$, $\|b\|_p \le \Lambda\, C(p)$, and (L3) holds for b, and so that moreover $(1+r)^{\sigma+2} Lu_m \longrightarrow 0$ weakly in $L^p(\mathbf{R}^2)$; we may also assume that for some u in $W^{2,p}_{loc}(\mathbf{R}^2)$, $u_m \longrightarrow u$ and $Du_m \longrightarrow Du$ uniformly on compact subsets of $\mathbf{R}^2$, with the convergence $(1+r)^{\sigma+2} D^\alpha u_m \longrightarrow (1+r)^{\sigma+2} D^\alpha u$ weakly in $L^p(\mathbf{R}^2)$. Then in $\mathbf{R}^2$,

$$a \cdot D^2 u + b \cdot Du = 0 .$$

We let $m \longrightarrow \infty$ in (4.13), obtaining

$$\| (1+r)^{\tau + 2/p} u \|_\infty \le 2\, C(p,\alpha,\lambda,\Lambda,K,\tau) .$$

Combining (2.8) of Theorem 1 with (4.12), and recalling that $\sigma - \delta < \tau$, we have

$$\| (1+r)^{\sigma+2} \Delta u_m \|_p \le 2\, C(p,\alpha,\lambda,\Lambda,K,\sigma,\delta) ;$$

then the estimates (3.7) and (3.9) of Theorem 3 yield

$$|\gamma_m| \le C(p,\alpha,\lambda,\Lambda,K,\sigma,\delta) ,$$

and for $|x| \ge 1$,

$$|u_m(x) - \gamma_m \Gamma(x)| \le C(p,\alpha,\lambda,\Lambda,K,\sigma,\delta)\, |x|^{-\sigma - 2/p} .$$

Since $\{\gamma_m\}$ is a bounded sequence, we may again pass to a subsequence so that

$\gamma_m \longrightarrow \gamma$ for some real constant γ; then we may let $m \longrightarrow \infty$ and conclude that, for $|x| \geq 1$,

$$|u(x) - \gamma\,\Gamma(x)| \leq C(p,\alpha,\lambda,\Lambda,K,\sigma,\delta)\ |x|^{-\sigma - 2/p} \quad . \tag{4.14}$$

The maximum principle for strong solutions of $a \cdot D^2u + b \cdot Du = 0$ ([8], Theorem 9.6) asserts that on any ball of radius R about the origin, the maximum and minimum of u occurs on the boundary; letting $R \longrightarrow \infty$, we conclude from (4.14) that $\gamma = 0$, $u \equiv 0$.

Since now $u_m \longrightarrow 0$ uniformly on compact subsets of $\mathbf{R}^2$ and, by (4.13), the norms $\| (1+r)^{\tau + 2/p} u_m \|_\infty$ are uniformly bounded, elementary estimates show that, for any $\varepsilon > 0$, $\| (1+r)^{\tau + 2/p - \varepsilon} u_m \|_\infty \longrightarrow 0$ as $m \longrightarrow \infty$. Then we obtain

$$\| (1+r)^{\tau - 2\varepsilon} u_m \|_p \leq \| (1+r)^{\tau + 2/p - \varepsilon} u_m \|_\infty \, \| (1+r)^{-2/p - \varepsilon} \|_p \longrightarrow 0 \ .$$

By (4.10), we may choose ε sufficiently small that $\sigma - \delta \leq \tau - 2\varepsilon$; then (2.8) of Theorem 3, the latest inequality, and (4.12) yield the estimate

$$\| (1+r)^{\sigma+2} \Delta u_m \|_p \leq C(p,\alpha,\lambda,\Lambda,K,\sigma,\delta) \left(\| (1+r)^{\tau-2\varepsilon} u_m \|_p + 1/m \right) \longrightarrow 0 \ .$$

Finally, (3.8) of Theorem 3 gives

$$\| (1+r)^{\tau} u \|_p \leq C(p,\sigma,\tau) \, \| (1+r)^{\sigma+2} \Delta u_m \|_p \longrightarrow 0 \ ,$$

contradicting (4.12). ∎

5. The Nonlinear Equation

We consider the nonlinear partial differential equation

$$a(x) \cdot D^2u(x) + H(x,u(x),Du(x)) = 0 \ , \tag{5.1}$$

or more briefly,

$$a \cdot D^2u + H(\cdot,u,Du) = 0 \ .$$

The function $H : \mathbf{R}^{2n+1} \longrightarrow \mathbf{R}$ is assumed measurable, and we write $H = H(x,u,p)$, where $x \in \mathbf{R}^n$, $u \in \mathbf{R}$, $p \in \mathbf{R}^n$. We assume the following conditions on the $n \times n$ real and symmetric matrix a and the function H :

(N1) There exists a positive constant λ such that

$$a(x)\,\xi \cdot \xi \geq \lambda\, |\xi|^2 \quad , \text{ for all } x,\ \xi \in \mathbf{R}^n \ .$$

(N2) There exist constants α, K, with $0 < \alpha \leq 1$, $K \geq 0$ such that

$$|a(x) - a(y)| \leq K \frac{|x - y|^\alpha}{(1 + |x| + |y|)^\alpha} \quad , \text{ for all } x,y \in \mathbf{R}^n,\ x \neq y \ .$$

(N3) There exist constants Λ, δ, with $\Lambda \geq 0$, $\delta > 0$, such that for $x \in \mathbf{R}^n$,

$$|a(x) - I| \leq \Lambda\ (1 + |x|)^{-\delta} \quad ,$$

and for all x, u, v, p and q,

$$|H(x,u,p) - H(x,v,p)| \leq \Lambda\ |u - v|\ (1 + |x|)^{-2-\delta}$$

$$|H(x,u,p) - H(x,u,q)| \leq \Lambda\ |p - q|\ (1 + |x|)^{-1-\delta} \ .$$

(N4) For all x, u, v and p, with $u \neq v$,

$$\frac{H(x,u,p) - H(x,v,p)}{u - v} \begin{cases} \leq 0 & , \text{ if } n \geq 3 \\ = 0 & , \text{ if } n = 2 \ . \end{cases}$$

(N5) For some p and σ, with $n < p < \infty$ and $-n/p < \sigma$,

$$\| (1+r)^{\sigma+2}\ H(\cdot,0,0) \|_p < \infty \ .$$

We seek to "linearize" the equation (5.1). For this purpose, we introduce a function $C = C(x,u,v,p)$, where $x,p \in \mathbf{R}^n$ and $u,v \in \mathbf{R}$, with the formula

$$C(x,u,v,p) := \begin{cases} \dfrac{H(x,u,p) - H(x,v,p)}{u - v} & , \text{ if } u \neq v \\ 0 & , \text{ if } u = v \ . \end{cases} \tag{5.2}$$

For $p \in \mathbf{R}^n$ we also introduce the notation

$$p_{\bullet i} := \begin{cases} (p_1,\ldots,p_i,0,\ldots,0) & , \text{ if } 1 \leq i \leq n \\ (0,\ldots,0) & , \text{ if } i = 0 \ , \end{cases}$$

and we define functions $B_i = B_i(x,u,p,q)$, $i = 1, \ldots, n$, according to

$$B_i(x,u,p,q) = \begin{cases} \dfrac{H(x,u,q+(p-q)_{\bullet i}) - H(x,u,q+(p-q)_{\bullet i-1})}{p_i - q_i} & , \text{ if } p_i \neq q_i \\ 0 & , \text{ if } p_i = q_i . \end{cases} \tag{5.3}$$

Then conditions (N3) and (N4) yield the inequalities

$$|B_i(x,u,p,q)| \leq \Lambda\ (1 + |x|)^{-1-\delta} \quad , \ i = 1, \ldots, n \ , \tag{5.4}$$

$$|C(x,u,v,p)| \leq \Lambda\ (1 + |x|)^{-2-\delta} \quad , \tag{5.5}$$

$$C(x,u,v,p) \begin{cases} \leq 0 & , \text{ if } n \geq 3 \\ = 0 & , \text{ if } n = 2 \ . \end{cases} \tag{5.6}$$

Applying definitions (5.2) and (5.3), we may write

$$\begin{aligned} H(x,u,p) - H(x,v,q) &= \sum_{i=1}^{n} B_i(x,v,p,q)(p_i - q_i) + C(x,u,v,p)(u - v) \\ &= B(x,v,p,q) \cdot (p-q) + C(x,u,v,p)(u-v) \ , \end{aligned} \tag{5.7}$$

where we define B as the vector

$$B = (B_1, B_2, \ldots, B_n) \quad .$$

Theorem 5 Assume conditions (N1) - (N5) on H and the matrix a. Then there exists a unique entire solution u of (5.1) in $W^{2,p}_{loc}(\mathbf{R}^n)$ such that

(a) when $n \geq 3$, $u(\infty) = 0$

(b) when $n = 2$, $u - \gamma\,\Gamma$ vanishes at infinity for some real constant γ .

Moreover, for this solution u we have the bounds

(a) when $n \geq 3$ and $-n/p < \sigma < n - 2 - n/p$, $0 < \delta \leq 1$,

$$\begin{aligned} &\| (1+r)^{\sigma} u \|_p + \| (1+r)^{\sigma+1} Du \|_p + \| (1+r)^{\sigma+2} D^2u \|_p \\ &\leq C(n,p,\alpha,\lambda,\Lambda,K,\sigma) \left[\| (1+r)^{\sigma+2} H(\cdot,0,0) \|_p \right] \ . \end{aligned} \tag{5.8}$$

(b) when $n = 2$ and $\tau < -2/p < \sigma$, $0 < \delta \leq 1$,

$$\begin{aligned} &\| (1+r)^{\tau} u \|_p + \| (1+r)^{\tau+1} Du \|_p + \| (1+r)^{\tau+2} D^2u \|_p \\ &\leq C(p,\alpha,\lambda,\Lambda,K,\sigma,\tau) \| (1+r)^{\sigma+2} H(\cdot,0,0) \|_p \ , \end{aligned} \tag{5.9}$$

and if $\sigma < -2/p + \delta$,

$$\| (1+r)^{\sigma+2} \Delta u \|_p \leq C(p,\alpha,\lambda,\Lambda,K,\sigma,\delta) \| (1+r)^{\sigma+2} H(\cdot,0,0) \|_p \ , \tag{5.10}$$

$$|\gamma| \leq C(p,\alpha,\lambda,\Lambda,K,\sigma,\delta) \| (1+r)^{\sigma+2} H(\cdot,0,0) \|_p \ , \tag{5.11}$$

with, for $|x| \geq 1$,

$$|u(x)-\gamma\Gamma(x)| \leq C(p,\alpha,\lambda,\Lambda,K,\sigma,\delta) \|(1+r)^{\sigma+2}H(\cdot,0,0)\|_p |x|^{-\sigma - 2/p} . \tag{5.12}$$

Proof We point out once again that any function u in $W^{2,p}_{loc}(\mathbf{R}^n)$ with $p > n$ is continuous with continuous first order derivatives; since H is measurable in the first variable and continuous in the second two variables, $H(\cdot,u,Du)$ then is a measureable function on $\mathbf{R}^n$.

First we prove uniqueness. Let u and v be entire solutions of (5.1) in $W^{2,p}_{loc}(\mathbf{R}^n)$, and set $w := u - v$; then, by (5.1) and (5.7), w solves the equation

$$a \bullet D^2 w + B(\bullet,v,Du,Dv) \bullet Dw + C(\bullet,u,v,Du) \cdot w = 0 . \tag{5.13}$$

The conditions (N1) - (N5), (5.2), and (5.4) - (5.6) allow the application of Theorem 4, with $\mathbf{L}$ the operator

$$\mathbf{L}w := a \bullet D^2 w + B(\bullet,v,Du,Dv) \bullet Dw + C(\bullet,u,v,Du) \cdot w = 0 .$$

When $n = 3$, we apply (4.1) for any σ in the range $- n/p < \sigma < n - 2 - n/p$, noting that $w(\infty) = 0$, and conclude that $w \equiv 0$. When $n = 2$, if $u - \gamma_1 \Gamma$ and $v - \gamma_2 \Gamma$ vanish at infinity then so does $w - (\gamma_1 - \gamma_2) \Gamma$; then we apply (4.2) with $\tau < - 2/p < \sigma$ and again conclude $w \equiv 0$.

To establish the bounds, we let u be any entire solution of (5.1) in $W^{2,p}_{loc}(\mathbf{R}^n)$. Using (5.7), we may rewrite (5.1) as

$$\begin{aligned} &a \bullet D^2 u + [\ H(\bullet,u,Du) - H(\bullet,0,0)\ ] + H(\bullet,0,0) \\ &= a \bullet D^2 u + B(\bullet,0,Du,0) \bullet Du + C(\bullet,u,0,Du) \cdot u + H(\bullet,0,0) = 0 . \end{aligned}$$

Again we may apply Theorem 4, but now with $\mathbf{L}$ the operator

$$\mathbf{L}u := a \bullet D^2 u + B(\bullet,0,Du,0) \bullet Du + C(\bullet,u,0,Du) \cdot u = - H(\bullet,0,0), \tag{5.14}$$

where by (N5),

$$\| (1+r)^{\sigma+2} \mathbf{L}u \|_p = \| (1+r)^{\sigma+2} H(\bullet,0,0) \|_p < \infty . \tag{5.15}$$

Theorem 4 now yields the estimates (5.8) - (5.12).

Finally, we prove existence of a solution. We define, for $0 \le t \le 1$,

$$a_t := t\, a + (1 - t)\, I ,$$

and we consider an equation of the form

$$a_t \bullet D^2 u + t\, [\ H(\bullet,u,Du) - H(\bullet,0,0)\ ] = F , \tag{5.16}$$

where F is a measurable real valued function in $\mathbf{R}^n$ such that

$$\| (1+r)^{\sigma+2} F \|_p < \infty .$$

We show that we can solve (5.16), with the prescribed conditions on u at infinity, for any such F and any t in [0,1]; then the theorem is established by taking $t = 1$ and $F = - H(\bullet,0,0)$.

We may take λ smaller, if necessary, in (N1) so that $\lambda \le 1$. It is then easily checked that conditions (N1) - (N3) hold also for the matrix a_t , with the same constants λ, α, K, Λ, δ, σ and p. We may also take δ smaller, if necessry, so that $0 < \delta \leq 1$.

We first discuss the case $n \geq 3$. We likewise take σ smaller, if necessary, in (N5) so that $-n/p < \sigma < n - 2 - n/p$. We let $\mathbf{X}$ be the Banach space of all real valued functions u in $W^{2,p}_{loc}(\mathbf{R}^n)$ such that the norm

$$\| u \|_{\mathbf{X}} := \| (1+r)^{\sigma} u \|_p + \| (1+r)^{\sigma+1} Du \|_p + \| (1+r)^{\sigma+2} D^2u \|_p$$

is finite. We seek a solution of (5.16) in this space. Lemma 1B shows that for any function u in $\mathbf{X}$, $\| (1+r)^{\sigma + n/p} u \|_{\infty}$ is finite; hence $u(\infty) = 0$ since $\sigma + n/p > 0$.

When $t = 0$, (5.16) reduces to the equation $\Delta u = F$; then, by Theorem 3, there exists a solution u in $\mathbf{X}$. We assume now that, for some $t \geq 0$, (5.16) can be solved for any F with $\| (1+r)^{\sigma+2} F \|_p$ finite. We show that the same statement must hold for $t + \varepsilon$, where ε is independent of t; hence (5.16) can be solved for all t in [0,1].

Replacement of t by $t + \varepsilon$ in (5.16) gives the equation

$$\begin{aligned} &a_t \cdot D^2u + t \, [\, H(\cdot,u,Du) - H(\cdot,0,0) \,] \\ &= F - \varepsilon(a - I) \cdot D^2u - \varepsilon \, [\, H(\cdot,u,Du) - H(\cdot,0,0) \,] := \tilde{F} . \end{aligned} \tag{5.17}$$

When $u \in \mathbf{X}$, we may use (N3) to estimate the right side of this equation, obtaining

$$\begin{aligned} \| (1+r)^{\sigma+2} \tilde{F} \|_p &\leq \| (1+r)^{\sigma+2} F \|_p + \varepsilon \, \| (a - I) \|_{\infty} \, \| (1+r)^{\sigma+2} D^2u \|_p \\ &+ \varepsilon \, \Lambda \left(\| (1+r)^{\sigma+1} Du \|_p + \| (1+r)^{\sigma} u \|_p \right) , \end{aligned}$$

which is finite; thus we may define a mapping $T : \mathbf{X} \longrightarrow \mathbf{X}$ by setting $u = Tv$ whenever u is the solution in $\mathbf{X}$ of the equation

$$\begin{aligned} &a_t \cdot D^2u + t \, [\, H(\cdot,u,Du) - H(\cdot,0,0) \,] \\ &= F - \varepsilon(a - I) \cdot D^2v - \varepsilon \, [\, H(\cdot,v,Dv) - H(\cdot,0,0) \,] . \end{aligned} \tag{5.18}$$

We show that, for ε small enough, this mapping is a contraction; then it has a fixed point and (5.17) is solvable. Letting $u = Tv$, $w = Tz$, we have

$$\begin{aligned} &a_t \cdot D^2(u-w) + t \, [\, H(\cdot,u,Du) - H(\cdot,w,Dw) \,] = \\ &= - \varepsilon(a - I) \cdot (D^2v - D^2z) - \varepsilon \, [\, H(\cdot,v,Dv) - H(\cdot,z,Dz) \,] , \end{aligned}$$

or, by use of (5.7),

$$a_t \cdot D^2(u-w) + tB(\cdot,w,Du,Dw) \cdot D(u-w) + tC(\cdot,u,w,Du)\cdot(u-w) = G, \tag{5.19}$$

where

$$G := -\varepsilon(a - I) \cdot D^2(v - z)$$

$$- \varepsilon B(\cdot,z,Dv,Dz) \cdot D(v - z) - \varepsilon C(\cdot,v,z,Dv) \cdot (v - z) .$$

Inequalities (5.4) - (5.6) verify that the hypotheses of Theorem 4 are satisfied by the operator on the left of (5.19); hence Theorem 4 gives the bound

$$\| (1+r)^{\sigma} (u-w) \|_p + \| (1+r)^{\sigma+1} D(u-w) \|_p + \| (1+r)^{\sigma+2} D^2(u-w) \|_p$$

$$\le C(n,p,\alpha,\lambda,\Lambda,K,\sigma) \left[\| (1+r)^{\sigma+2} G \|_p \right] , \tag{5.20}$$

where we may use (N3), (5.4) and (5.5) to bound the norm of G,

$$\| (1+r)^{\sigma+2} G \|_p \le \varepsilon \, n \, \Lambda \left(\| (1+r)^{\sigma+2} D^2(v-z) \|_p \right.$$

$$\left. + \| (1+r)^{\sigma+1} D(v-z) \|_p + \| (1+r)^{\sigma} (v-z) \|_p \right)$$

$$= \varepsilon \, n \, \Lambda \| v - z \|_X .$$

Therefore, (5.20) yields

$$\| u - w \|_X = \| Tv - Tz \|_X \le \varepsilon \, n \, \Lambda \, C(n,p,\alpha,\lambda,\Lambda,K,\sigma) \| v - z \|_X ,$$

and T is a contraction for ε small enough, independently of t.

Now we consider the case n = 2. Again we may take σ smaller, if necessary, in (N5) so that $-2/p < \sigma < -2/p + \delta$. We let $\mathbf{Y}$ be the Banach space of all real valued functions u in $W^{2,p}_{loc}(\mathbf{R}^n)$ such that the norm

$$\| u \|_{\mathbf{Y}} := \| (1+r)^{\sigma+2} \Delta u \|_p$$

is finite, and such that $u - \gamma \Gamma$ vanishes at infinity for some real constant γ. We seek a solution of (5.16) in this space, but first we show that $\mathbf{Y}$ is a Banach space. For u in $\mathbf{Y}$ and related γ with $u - \gamma \Gamma$ vanishing at infinity, Theorem 3 gives the bounds

$$|\gamma| \le C(p,\sigma) \| (1+r)^{\sigma+2} \Delta u \|_p , \tag{5.21}$$

$$\| (1+r)^{\sigma-\delta} u \|_p + \| (1+r)^{\sigma-\delta+1} Du \|_p + \| (1+r)^{\sigma-\delta+2} D^2 u \|_p$$

$$\le C(p,\sigma,\delta) \| (1+r)^{\sigma+2} \Delta u \|_p , \tag{5.22}$$

and for $|x| \geq 1$,

$$|u(x) - \gamma\,\Gamma(x)| \leq C(p,\sigma)\ \|\ (1+r)^{\sigma+2}\ \Delta u\ \|_p\ |x|^{-\sigma - 2/p}\ . \tag{5.23}$$

If $\{u_m\}$ is a Cauchy sequence in the norm of $\mathbf{Y}$ with related constants $\{\gamma_m\}$, (5.21) and (5.22) show that $\{\gamma_m\}$ is a Cauchy sequence of real numbers and that $\{u_m\}$ is Cauchy also in the norm defined by the left side of (5.22). Hence there exists a function u in $W^{2,p}_{loc}(\mathbf{R}^n)$ such that $u_m \longrightarrow u$ in the norm defined by the left side of (5.22) and in the norm of $\mathbf{Y}$, and there exists a real constant γ such that $\gamma_m \longrightarrow \gamma$. We apply (5.23) to each u_m and γ_m, and let $m \longrightarrow \infty$ to conclude that it holds also for u and γ. Hence $u \in \mathbf{Y}$, and $\mathbf{Y}$ is complete.

We proceed as for $n \geq 3$, showing that (5.16) is solvable in the space $\mathbf{Y}$ for $0 \leq t \leq 1$. When $t = 0$, (5.16) reduces to $\Delta u = F$, and Theorem 3 guarantees a solution u in $\mathbf{Y}$. Assuming that, for some $t \geq 0$, (5.16) can be solved for any F with $\|\ (1+r)^{\sigma+2}\ F\ \|_p$ finite, we examine the case of $t + \varepsilon$.

Again we arrive at (5.17) and must verify that $\|\ (1+r)^{\sigma+2}\ \tilde{F}\ \|_p$ is finite; we use (N3) and (5.20) to derive

$$\begin{aligned}\|\ (1+r)^{\sigma+2}\ \tilde{F}\ \|_p &\leq \|(1+r)^{\sigma+2}\ F\|_p + \varepsilon\ \|(1+r)^{\delta}\ (a-I)\|_\infty\ \|(1+r)^{\sigma+2-\delta}\ D^2u\|_p \\ &\quad + \varepsilon\ \Lambda\ \|(1+r)^{\sigma+1-\delta}\ Du\|_p + \varepsilon\ \Lambda\ \|(1+r)^{\sigma-\delta}\ u\|_p \\ &\leq \|(1+r)^{\sigma+2}\ F\|_p + \varepsilon\ \Lambda\ C(p,\sigma,\delta)\ \|(1+r)^{\sigma+2}\ \Delta u\|_p\ ,\end{aligned}$$

which is finite. We define a mapping $T : \mathbf{Y} \longrightarrow \mathbf{Y}$ by setting $u = Tv$ where u is the solution in $\mathbf{Y}$ of (5.18). Letting $u = Tv$, $w = Tz$, we obtain again (5.19), and by Theorem 4,

$$\|\ (1+r)^{\sigma+2}\ \Delta(u-w)\ \|_p \leq C(p,\alpha,\lambda,\Lambda,K,\sigma,\delta)\ \Big[\ \|\ (1+r)^{\sigma+2}\ G\ \|_p\ \Big]\ . \tag{5.24}$$

We use (N3), (5.4), (5.5) and (5.22) applied to $v - z$ to derive

$$\begin{aligned}\|\ (1+r)^{\sigma+2}\ G\ \|_p &\leq \varepsilon\ n\ \Lambda\ \Big(\ \|\ (1+r)^{\sigma+2-\delta}\ D^2(v-z)\ \|_p \\ &\quad + \|\ (1+r)^{\sigma+1-\delta}\ D(v-z)\ \|_p + \|\ (1+r)^{\sigma-\delta}\ (v-z)\ \|_p\ \Big) \\ &\leq \varepsilon\ n\ \Lambda\ C(p,\sigma,\delta)\ \|\ v - z\ \|_{\mathbf{Y}}\ .\end{aligned}$$

Therefore, (5.24) gives

$$\|\ u - w\ \|_{\mathbf{Y}} = \|\ Tv - Tz\ \|_{\mathbf{Y}} \leq \varepsilon\ n\ \Lambda\ C(p,\sigma,\delta)\ C(p,\alpha,\lambda,\Lambda,K,\sigma,\delta)\ \|\ v - z\ \|_{\mathbf{Y}}\ ,$$

and T is a contraction for ε small enough, independently of t. ∎

We use Theorem 5 to prove a more general result :

Theorem 6 Assume conditions (N1) - (N5) on a and H.

(i) For any harmonic polynomial P of degree less than δ, there exists a unique entire solution u in $W^{2,p}_{loc}(\mathbf{R}^n)$ of (5.1) such that

(a) if $n \geq 3$, $u - P$ vanishes at infinity

(b) if $n = 2$, $u - P - \gamma\,\Gamma$ vanishes at infinity for some $\gamma \in \mathbf{R}$.

(ii) If u is an entire solution in $W^{2,p}_{loc}(\mathbf{R}^n)$ of (5.1) such that

$$|u(x)| = O(|x|)^{\tau} \quad \text{as } x \longrightarrow \infty$$

for some τ, $0 \leq \tau < \delta$, then there exists a unique harmonic polynomial P of degree no larger than τ such that (a) and (b) of (i) hold.

Proof (i) For any given harmonic polynomial P of degree m, $m < \delta$, we seek a solution of (5.1) of the form $u = v + P$. Substituting into (5.1), we find as our equation for v,

$$a \cdot D^2 v + a \cdot D^2 P + H(\cdot, v+P, Dv+DP) = 0 . \tag{5.25}$$

We define $\tilde{H}$ as the function

$$\tilde{H}(x,u,p) := a(x) \cdot D^2 P(x) + H(x, u+P(x), p+DP(x)) ,$$

and verify easily that $\tilde{H}$ satisfies conditions (N3) - (N4). We also find that

$$\begin{aligned} \tilde{H}(x,0,0) &= a(x) \cdot D^2 P(x) + H(x,P(x),DP(x)) \\ &= a(x) \cdot D^2 P(x) + H(x,P(x),DP(x)) - H(x,0,0) + H(x,0,0) . \end{aligned} \tag{5.26}$$

Since degree P = m, there exists a positive constant M such that, for $x \in \mathbf{R}^n$,

$$|P(x)| \leq M\,(1+|x|)^m , \quad |DP(x)| \leq M\,(1+|x|)^{m-1} , \quad |D^2 P(x)| \leq M\,(1+|x|)^{m-2} .$$

Since P is harmonic, we have $0 = \Delta P = I \cdot D^2 P$, and hence by (N3),

$$|a(x) \cdot D^2 P(x)| = |(a(x) - I) \cdot D^2 P(x)| \leq \Lambda\, M\, (1 + |x|)^{m-2-\delta} .$$

Using (N3) again, we also estimate

$$\begin{aligned} |H(x,P(x),DP(x)) - H(x,0,0)| &\leq \Lambda\,(1+|x|)^{-1-\delta}|DP(x)| + \Lambda\,(1+|x|)^{-2-\delta}|P(x)| \\ &\leq 2\,\Lambda\, M\,(1 + |x|)^{m-2-\delta} . \end{aligned}$$

Again we take σ smaller in (N5) so that $-n/p < \sigma < \delta - m - n/p$; then these estimates together with (5.26) and (N5) show that $\| (1+r)^{\sigma+2}\, \tilde{H}(\cdot,0,0) \|_p$ is finite. Therefore, we may apply Theorem 5 and conclude that there exists a unique solution v of (5.25) in $W^{2,p}_{loc}(\mathbf{R}^n)$ such that if $n \geq 3$, v vanishes at infinity, and if $n = 2$, $v - \gamma\,\Gamma$ vanishes at infinity for some real γ.

Setting u := v + P then gives our required solution u.

To establish uniqueness, we observe that the difference w of two solutions solves (5.13), as in the proof of Theorem 5, and $w(\infty) = 0$ when $n \geq 3$, $w - \gamma\,\Gamma$ vanishes at infinity for some real γ when $n = 2$. By the argument following (5.13), we conclude that $w \equiv 0$.

(ii) Uniqueness of P is clear. The difference of two such polynomials is a harmonic polynomial, vanishing at infinity when $n \geq 3$, and behaving like a multiple of log r at infinity when $n = 2$; in either case this difference must be zero.

To prove existence, we take u as described and choose σ smaller, if necessary, in (N5) so that $-n/p < \sigma < \delta - \tau - n/p$. Then $u(x) = O(|x|^{\tau})$ implies that $\| (1+r)^{\sigma-\delta} u \|_p$ is finite. As in the proof of Theorem 5, we have (5.14) and (5.15); hence, by (2.8) of Theorem 1, $\| (1+r)^{\sigma+2} \Delta u \|_p$ is finite. By Theorem 3, $\mathbf{S}(\Delta u) \in W^{2,p}_{loc}(\mathbf{R}^n)$, $\Delta(\mathbf{S}(\Delta u)) = \Delta u$; also, $\mathbf{S}(\Delta u)$ vanishes at infinity if $n \geq 3$, $\mathbf{S}(\Delta u) - \gamma\,\Gamma$ vanishes at infinity for some real γ if $n = 2$. The function $P := u - \mathbf{S}(\Delta u)$ is harmonic in $\mathbf{R}^n$. When $n \geq 3$, $u(x) = O(x^{\tau})$ as $x \longrightarrow \infty$ implies P is a polynomial of degree no larger than τ; moreover, $u - P = \mathbf{S}(\Delta u)$ vanishes at infinity. When $n = 2$, this reasoning applies when $\tau > 0$, and therefore P is a polynomial of degree no larger than τ. When $\tau = 0$, we have u bounded and P therefore growing no faster than a multiple of log r at infinity, implying that P is constant and hence a polynomial of degree no larger than 0. Note that $\gamma = 0$ also is implied in this case. ∎

A classical "Liouville Theorem" states that a bounded entire solution of an equation of a certain type is necessarily constant. For such a theorem to be possible, constant functions must solve the equation. For equation (5.1), we see that this requirement is equivalent to

$$H(x,c,0) = 0 \quad , \text{ for all } x \in \mathbf{R}^n \text{ and } c \in \mathbf{R} . \tag{5.27}$$

Corollary 6A Assume conditions (N1) - (N5) on a and H, and assume further that (5.27) holds. Then any bounded and entire solution u of (5.1) in $W^{2,p}_{loc}(\mathbf{R}^n)$ is constant.

Proof If u is a bounded solution, then, by Theorem 6(ii) with $\tau = 0$, there exists a constant c such that $u - c$ vanishes at infinity when $n \geq 3$, and $u - c - \gamma\,\Gamma$ vanishes at infinity for some real γ when $n = 2$. But the constant solution c has this property (with $\gamma = 0$ when $n = 2$); hence by the uniqueness statement of Theorem 6(i), $u \equiv c$. ∎

We point out the ramifications of Theorem 6 for the linear equation,

$$\mathbf{L}u = a \cdot D^2u + b \cdot Du + c \cdot u + f = 0 \quad . \tag{5.28}$$

This equation has the form (5.1), with

$$H(x,u,p) := b(x) \cdot p + c(x) \cdot u + f(x) \quad .$$

If we assume conditions (L1) - (L3), with $\delta > 0$, on a, b, and c, also that $c \leq 0$ when $n \geq 3$, $c \equiv 0$ when $n = 2$, and if we assume that $\| (1+r)^{\sigma+2} f \|_p$ is finite with $\sigma > - n/p$, then conditions (N1) - (N5) are easily verified for H. Theorems 5 and 6 then apply to solutions of (5.28). In particular, when $f \equiv 0$ we have the linear equation,

$$\mathbf{L}u = a \cdot D^2u + b \cdot Du + c \cdot u = 0 \quad , \tag{5.29}$$

in which case the solutions form a real vector space. Then Theorem 6 yields immediately the following :

Theorem 7 Suppose that conditions (L1) - (L3), with $\delta > 0$, hold on the coefficients of **L**, assume $c \leq 0$ if $n \geq 3$, $c \equiv 0$ if $n = 2$, and let τ lie in the range $0 \leq \tau < \delta$. Then the space of entire solutions of (5.29) in $W^{2,p}_{loc}(\mathbf{R}^n)$, with the property that

$$u(x) = O(|x|^{\tau}) \text{ as } x \longrightarrow \infty ,$$

form a finite dimensional vector space, of dimension the same as the space of harmonic polynomials of degree no larger than τ. There is a one-to-one correspondence between members of these spaces, as determined by statements (a) and (b) of Theorem 6. ∎

We also point out that, if we stipulate $c \equiv 0$, then constant functions solve the linear homogeneous equation (5.29), and Liouville's Theorem applies, asserting that bounded entire solutions are constant.

6. Bounds on the Newtonian Potential

In this section we derive weighted norm estimates on the Newtonian potential, and prove as a byproduct Theorem 2 of section 3.

Lemma 8A For positive integers n, real numbers ξ and η, and $x \in \mathbf{R}^n$, let $J = J(x,n,\xi,\eta)$, $K = K(x,n,\xi,\eta)$ be the integrals

$$J(x,n,\xi,\eta) := \int_{\mathbf{R}^n} |x-y|^{\xi} (1 + |y|)^{\eta} \, dy ,$$

$$K(x,n,\xi,\eta) := \int_{\mathbf{R}^n} | \log |x-y| |^{\xi} (1 + |y|)^{\eta} \, dy .$$

(a) If $\xi + n > 0$, $\eta + n > 0$, $\xi + \eta + n < 0$, then

$$J(x,n,\xi,\eta) \le C(n,\xi,\eta) (1 + |x|)^{\xi + \eta + n} .$$

(b) If $\xi \ge 0$, $\eta + n < 0$, then

$$K(x,n,\xi,\eta) \le C(n,\xi,\eta) \big(\log (e + |x|) \big)^{\xi} .$$

Proof (a) We write $J = J_1 + J_2 + J_3$, where

$$J_1 := \int_{|y-x| \le (1+|x|)/2} |x-y|^{\xi} (1 + |y|)^{\eta} \, dy ,$$

$$J_2 := \int_{(1+|x|)/2 \le |y-x| \le 2(1+|x|)} |x-y|^{\xi} (1 + |y|)^{\eta} \, dy ,$$

$$J_3 := \int_{|y-x| \ge 2(1+|x|)} |x-y|^{\xi} (1 + |y|)^{\eta} \, dy .$$

In J_1 , $|y-x| \le (1+|x|)/2$ implies that

$$(1 + |x|)/2 \le 1 + |y| \le 3(1 + |x|)/2 ,$$

giving

$$J_1 \le C(\eta)(1+|x|)^{\eta} \int_{|y-x| \le (1+|x|)} |x-y|^{\xi} \, dy = C(n,\xi,\eta)(1+|x|)^{\xi+\eta+n} ,$$

where we have used $\xi + n > 0$. In J_2, we can verify that

$$(1+|x|)/2 \le |y-x| \le 2(1+|x|) \Rightarrow |y| \le 2 + 3 |x| ,$$

and hence

$$J_2 \le C(\xi) (1+|x|)^{\xi} \int_{|y| \le 2+3|x|} (1+|y|)^{\eta} \, dy \le C(n,\xi,\eta) (1+|x|)^{\xi+\eta+n} ,$$

where we used the inequality $\eta + n > 0$. In J_3,

$$|y-x| \ge 2(1 + |x|) \Rightarrow |x| \le \frac{|x-y|}{2} \le |y| \le 1 + |y| ,$$

which, with use of $\eta < - n - \xi < 0$ and $\xi + \eta + n < 0$, gives

$$J_3 \le C(\eta) \int_{|y-x| \ge 1+|x|} |x-y|^{\xi} |x-y|^{\eta} \, dy = C(n,\xi,\eta) (1+|x|)^{\xi+\eta+n} .$$

Adding our estimates for the J_i's completes the proof of (a).

(b) Following the proof of (a), we set $K = K_1 + K_2 + K_3$, where the region of integration of each K_i is the same as that of J_i in (a).

For K_1, we obtain

$$K_1 \le C(\eta)\ (1 + |x|)^\eta \int_{|y-x| \le (1+|x|)} |\log|x-y||^\xi\, dy$$

$$= C(n,\eta)\ (1 + |x|)^\eta \int_0^{1+|x|} r^{n-1}\ |\log r|^\xi\, dr \quad .$$

Recalling that $\xi \ge 0$, and observing that r^{n-1} and $\log r$ are increasing and nonnegative on $[1,\infty)$, we estimate

$$\int_0^{1+|x|} r^{n-1}\ |\log r|^\xi\, dr = \int_0^1 r^{n-1}\ |\log r|^\xi\, dr + \int_1^{1+|x|} r^{n-1}\ |\log r|^\xi\, dr$$

$$= C(n,\xi) + \int_1^{1+|x|} r^{n-1}(\log r)^\xi\, dr \le C(n,\xi) + (1+|x|)^{n-1}[\log(1+|x|)]^\xi\ |x|$$

$$\le C(n,\xi) + (1+|x|)^n\ [\log(e+|x|)]^\xi \le C(n,\xi)\ (1+|x|)^n\ [\log(e+|x|)]^\xi \ .$$

Hence, using $n + \eta < 0$, we obtain

$$K_1 \le C(n,\xi,\eta)\ (1+|x|)^{n+\eta}\ [\log(e+|x|)]^\xi \le C(n,\xi,\eta)\ [\log(e+|x|)]^\xi \ .$$

To estimate K_2, we observe that in the interval $(1+|x|)/2 \le t \le 2(1+|x|)$ the function $|\log t|$, having a critical point only at $t = 1$ where it attains a minimum, attains its maximum at an endpoint of the interval. Thus, in K_2,

$$|\log|x-y|| \le \max\left\{ \left|\log\frac{1+|x|}{2}\right| , \left|\log 2(1+|x|)\right| \right.$$

$$= \max\left\{ |\log(1+|x|) - \log 2| , |\log(1+|x|) + \log 2| \right\}$$

$$= \log(1 + |x|) + \log 2 \le \log(e + |x|) + \log e \le 2\log(e + |x|).$$

Therefore, since we assume $\xi \ge 0$,

$$K_2 \le C(\xi)\ [\log(e + |x|)]^\xi \int_{|y| \le 2+3|x|} (1 + |y|)^\eta\, dy$$

$$\le C(n,\xi,\eta)\ [\log(e + |x|)]^\xi ,$$

where in the last step we used the inequality $\eta + n < 0$. For K_3, again using $\eta < -n$, $\xi \ge 0$, we obtain

$$K_3 \le C(\eta) \int_{|y-x| \ge 2(1+|x|)} [\log|x-y|]^\xi\ |x-y|^\eta\, dy$$

$$\leq C(n,\eta) \int_2^\infty r^{\eta+n-1} (\log r)^\xi \, dr = C(n,\xi,\eta) \leq C(n,\xi,\eta) \, [\log (e+|x|)]^\xi .$$

Adding the estimates for all K_i's completes the proof of (b). ∎

Lemma 8B Assume $n \geq 1$, $1 \leq p \leq s \leq \infty$, define q by $1/p + 1/q = 1$, and let λ, σ be real numbers such that

$$-\frac{1}{q} - \frac{1}{s} < \frac{\lambda}{n} < \frac{\sigma}{n} - \frac{1}{q} < 0 \quad . \tag{6.1}$$

For any real valued measurable function f on $\mathbf{R}^n$, let Vf be the function

$$Vf(x) := \int_{\mathbf{R}^n} |x-y|^\lambda \, |f(y)| \, dy \quad .$$

Then

$$\| (1 + r)^{\sigma - \lambda - n/q - n/s} \, Vf \|_s \leq C(n,p,s,\sigma,\lambda) \, \| (1 + r)^\sigma f \|_p \, . \tag{6.2}$$

Proof We may assume $\| (1+r)^\sigma f\|_p$ is neither zero nor infinite, as otherwise the inequality is clear. Note that the case $p = 1$, $s = \infty$, when also $q = \infty$, violates (6.1) and cannot occur. We consider case 1 below, and its alternative $p = s = \infty$, which is case 2.

CASE 1 : $1 \leq p \leq s < \infty$, or $1 < p < s = \infty$

Given p and s, let q, t be extended real numbers in the range $[1,\infty]$ determined by the relations

$$\frac{1}{p} + \frac{1}{q} = 1 \quad , \quad \frac{1}{s} + \frac{1}{t} = \frac{1}{p} \quad , \tag{6.3}$$

and further define real numbers α, β, γ, δ, τ, μ, ν, τ, ε according to

$$\left.\begin{aligned} &\alpha = \frac{q}{q+s} \quad , \quad \beta = \frac{s}{q+s} \quad , \quad \gamma = \frac{p}{s} \quad , \quad \delta = \frac{p}{t} \quad , \\ &\tau = \sigma p/t \quad , \quad \nu = -\frac{n + \sigma s}{q + s} = -n\alpha/q - \sigma\beta \quad , \\ &\mu = -\nu - \tau \quad , \quad \varepsilon = \sigma - \lambda - n/q - n/s \quad . \end{aligned}\right\} \tag{6.4}$$

We assume in Case 1 that p is finite; however s may be infinite, q may be infinite (when $p = 1$), and t may be infinite (when $s = p$). Moreover, q and s cannot be simultaneously infinite. Whenever $q = \infty$, in (6.4) we interpret $\alpha = 1$, $\beta = 0$, $\nu = 0$, and when s is infinite, we interpret $\alpha = 0$, $\beta = 1$, $\gamma = 0$, $\nu = -\sigma$. Note that when t is infinite, (6.4) gives $\delta = \tau = 0$. It is readily verified that, in all cases,

$$\alpha + \beta = 1 \quad , \quad \gamma + \delta = 1 \quad , \quad \mu + \nu + \tau = 0 \; , \quad \frac{1}{q} + \frac{1}{s} + \frac{1}{t} = 1 \quad . \tag{6.5}$$

Let Y, r be the functions determined by $Y(y) := y$, $r(x) := |x|$. By (6.5) and Hölder's inequality,

$$|Vf(x)| = \int_{\mathbf{R}^n} |x-y|^{\lambda(\alpha+\beta)} \; |f(y)|^{\gamma+\delta} \; (1 + |y|)^{\mu+\nu+\tau} \; dy$$

$$\leq \| \; |x-Y|^{\lambda\alpha} \; |f(Y)|^{\gamma} \; (1 + |Y|)^{\mu} \; \|_s$$

$$\cdot \; \| \; |x-Y|^{\lambda\beta} \; (1 + |Y|)^{\nu} \; \|_q \cdot \| \; (1 + |Y|)^{\tau} \; |f(Y)|^{\delta} \; \|_t \; .$$

If $t < \infty$, we use (6.4) to compute

$$\| \; (1 + Y)^{\tau} \; |f(Y)|^{\delta} \; \|_t = \left[\int_{\mathbf{R}^n} (1 + |y|)^{\tau t} \; |f(y)|^{\delta t} \; dy \right]^{1/t}$$

$$= \left[\int_{\mathbf{R}^n} (1 + |y|)^{\sigma p} \; |f(y)|^{p} \; dy \right]^{1/t} = \| \; (1 + r)^{\sigma} \; f \; \|_p^{\,p/t} \quad ,$$

while if $t = \infty$, then $\tau = \delta = 0$ and we achieve the same result,

$$\|(1 + Y)^{\tau} \; |f(Y)|^{\delta}\|_t = \operatorname{ess\,sup} \; (1+|y|)^0 \; |f(y)|^0 = 1 = \|(1+r)^{\sigma} \; f\|_p^{\,p/t} \; .$$

We let C denote a generic constant which can depend on some or all of the quantities n, p, s, σ, λ . If $q \neq \infty$, we apply Lemma 8A and obtain

$$\| \; |x-Y|^{\lambda\beta} \; (1 + |Y|)^{\nu} \; \|_q = \left[\int_{\mathbf{R}^n} |x-y|^{\lambda\beta q} \; (1 + |y|)^{\nu q} \; dy \right]^{1/q}$$

$$\leq C \; (1 + |x|)^{\lambda\beta \, + \, \nu \, + \, n/q} \quad ,$$

with the hypothesis of that lemma requiring that

$$\lambda \, \beta \, q + n > 0 \quad , \quad \nu \, q + n > 0 \quad , \quad \lambda \, \beta \, q + \nu \, q + n < 0 \; ,$$

inequalities justified by (6.4) and (6.1). If $q = \infty$, then $\beta = \nu = 0$, and we obtain again

$$\| \; |x-Y|^{\lambda\beta} \; (1 + |Y|)^{\nu} \; \|_q = \operatorname*{ess\,sup}_{y} \; |x-y|^0 \; (1 + |y|)^0 = 1$$

$$= 1 \cdot (1 + |x|)^0 = 1 \cdot (1 + |x|)^{\lambda\beta \, + \, \nu \, + \, n/q} \quad .$$

The inequalities of the preceding three paragraphs now yield

$$|Vf(x)| \leq C \; \| \; |x-Y|^{\lambda\alpha} \; |f(Y)|^{\gamma} \; (1 + |Y|)^{\mu} \; \|_s$$
$$\cdot \; \| \; (1 + r)^{\sigma} \; f \; \|_p^{\,p/t} \cdot (1 + |x|)^{\lambda\beta \, + \, \nu \, + \, n/q} \quad . \tag{6.6}$$

If $s \neq \infty$, we multiply this inequality by $(1 + |x|)^{\varepsilon}$, take both sides to the s power, integrate over $\mathbf{R}^n$ with respect to x , and change orders of integration on the right to obtain

$$\int_{\mathbf{R}^n} (1 + |x|)^{\varepsilon s} |Vf(x)|^s \, dx \leq \| (1 + r)^{\sigma} f \|_p^{ps/t} \cdot$$

$$\cdot \int_{\mathbf{R}^n} (1 + |y|)^{\mu s} |f(y)|^{\gamma s} \int_{\mathbf{R}^n} |x-y|^{\lambda\alpha s} (1 + |x|)^{s[\varepsilon+\lambda\beta+\nu+n/q]} \, dx \, dy .$$

Use of Lemma 8A and formulas (6.3) and (6.4) yield

$$\int_{\mathbf{R}^n} |x-y|^{\lambda\alpha s} (1 + |x|)^{s[\varepsilon+\lambda\beta+\nu+n/q]} \, dx$$

$$\leq C (1 + |y|)^{s[\lambda\alpha+\varepsilon+\lambda\beta+\nu+n/q]+n} = C (1 + |y|)^{s(\nu+\sigma)} ,$$

with Lemma 8A requiring the hypotheses

$\lambda\alpha s + n > 0$, $s(\varepsilon + \lambda\beta + \nu + n/q) > 0$, $s(\lambda\alpha + \varepsilon + \lambda\beta + \nu + n/q) + n < 0$, inequalities again justified by (6.4) and (6.1). Thus,

$$\int_{\mathbf{R}^n} (1 + |x|)^{\varepsilon s} |Vf(x)|^s \, dx \leq C \| (1 + r)^{\sigma} f \|_p^{ps/t} \cdot$$

$$\int_{\mathbf{R}^n} (1 + |y|)^{s(\mu+\nu+\sigma)} |f(y)|^{\gamma s} \, dy .$$

But (6.4) gives $\gamma s = p$, $s(\mu+\nu+\sigma) = \sigma p$; hence, taking both sides of our inequality to the 1/s power, we finally arrive at

$$\| (1 + r)^{\varepsilon} Vf \|_s \leq C \| (1 + r)^{\sigma} f \|_p^{p/t + p/s} = C \| (1 + r)^{\sigma} f \|_p ,$$

which is (6.2). If $s = \infty$, in (6.6) we have $\alpha = \gamma = \mu = 0$, and

$$\| |x-Y|^{\lambda\alpha} |f(Y)|^{\gamma} (1+|Y|)^{\mu} \|_s = \operatorname*{ess\,sup}_{y} |x-y|^0 |f(y)|^0 (1+|y|)^0 = 1 ;$$

then, since also $\beta = 1$, $t = p$ and $\nu = -\sigma$, (6.6) yields

$$(1 + |x|)^{\sigma - \lambda - n/q} |Vf(x)| \leq C \| (1 + r)^{\sigma} f \|_p ,$$

$$\| (1 + r)^{\sigma - \lambda - n/q} Vf \|_{\infty} \leq C \| (1 + r)^{\sigma} f \|_p ,$$

which again is (6.2).

<u>CASE</u> <u>2</u> : $p = s = \infty$

Using Lemma 8A we derive

$$|Vf(x)| \leq C \int_{\mathbf{R}^n} |x-y|^{\lambda} (1 + |y|)^{-\sigma} (1 + |y|)^{\sigma} |f(y)| \, dy$$

$$\leq C \, \| \, (1+r)^{\sigma} f \, \|_{\infty} \int_{\mathbf{R}^n} |x-y|^{\lambda} \, (1 + |y|)^{-\sigma} \, dy$$

$$\leq C \, \| \, (1+r)^{\sigma} f \, \|_{\infty} \, (1 + |x|)^{\lambda+n-\sigma} ,$$

provided that

$$\lambda + n > 0 \quad , \quad - \sigma + n > 0 \quad , \quad \lambda - \sigma + n < 0 .$$

These inequalities follow from (6.1); thus

$$\| \, (1+r)^{\sigma-\lambda-n} \, Vf \, \|_{\infty} \leq C \, \| \, (1+r)^{\sigma} f \, \|_{\infty} ,$$

which is (6.2). ∎

Lemma 8C Assume $n \geq 1$, $1 < p \leq \infty$, and let σ, ε be real numbers such that

$$\sigma > n - n/p \quad , \quad \varepsilon < - n/p .$$

For any real valued measurable function f in $\mathbf{R}^n$, let Wf be the function

$$Wf(x) := \int_{\mathbf{R}^n} | \log |x-y| \, | \, |f(y)| \, dy .$$

Then

$$\| \, (1+r)^{\varepsilon} \, Wf \, \|_p \leq C(n,p,\sigma,\varepsilon) \, \| \, (1+r)^{\sigma} f \, \|_p .$$

Proof As usual, we may assume the norm $\| (1+r)^{\sigma} f \|_p$ is finite. Let q be determined by $1/p + 1/q = 1$; then by Hölder's inequality,

$$Wf(x) = \int_{\mathbf{R}^n} | \log |x-y| \, | \, (1 + |y|)^{-\sigma} \, (1 + |y|)^{\sigma} \, |f(y)| \, dy$$

$$\leq \| \, (1+r)^{\sigma} f \, \|_p \left[\int_{\mathbf{R}^n} | \log |x-y| \, |^{q} \, (1 + |y|)^{-\sigma q} \, dy \right]^{1/q} .$$

We apply Lemma 8A to the last integral, requiring that $q \geq 0$, $- \sigma q + n < 0$, which hold by hypotheses; then

$$Wf(x) \leq C(n,p,\sigma) \, \| \, (1+r)^{\sigma} f \, \|_p \, \big(\log (e + |x|) \big) ,$$

$$\| \, (1+r)^{\varepsilon} \, Wf \, \|_p \leq C(n,p,\sigma) \, \| \, (1+r)^{\sigma} f \, \|_p \, \| \, (1+r)^{\varepsilon} \log (e+r) \, \|_p .$$

If $\varepsilon < - n/p$, then $\| (1+r)^{\varepsilon} \log (e+r) \|_p = C(n,p,\varepsilon) < \infty$, completing the proof of the theorem. ∎

Lemma 8D For positive integers n, real numbers ξ and η, and $x \in \mathbf{R}^n$, let $L = L(x,n,\xi,\eta)$ be the integral

$$L(x,n,\xi,\eta) := \int_{\mathbf{R}^n} | \log |x-y| - \log |x| \, |^{\xi} \, (1 + |y|)^{\eta} \, dy .$$

If $\eta + n < 0$ and $\xi + \eta + n > 0$, then for $|x| \geq 1$,

$$L(x,n,\xi,\eta) \leq C(n,\xi,\eta) \ |x|^{\eta + n} \ .$$

Proof We write $L = L_1 + L_2$, where

$$L_1 := \int_{|y|\leq|x|/2} \big| \log|x-y| - \log|x| \big|^{\xi} (1 + |y|)^{\eta} \, dy \ ,$$

$$L_2 := \int_{|y|\geq|x|/2} \big| \log|x-y| - \log|x| \big|^{\xi} (1 + |y|)^{\eta} \, dy \ ,$$

In L_1, we have $|x-y| \geq |x| - |y| \geq |x| - |x|/2 = |x|/2$, and we may use the mean value theorem to write

$$\log|x-y| - \log|x| = \frac{1}{t} \big(|x-y| - |x| \big)$$

for some t lying between $|x|$ and $|x-y|$. Thus $t \geq |x|/2$ and, noting that $\xi > -n - \eta > 0$, we obtain

$$\big|\log|x-y| - \log|x|\big| \leq \frac{2}{|x|} \big| |x-y| - |x| \big| \leq \frac{2}{|x|} |y| \ ,$$

$$L_1 \leq C(\xi) \ |x|^{-\xi} \int_{|y|\leq|x|/2} (1 + |y|)^{\xi + \eta} \, dy$$

$$\leq C(n,\xi) \ |x|^{-\xi} \int_0^{|x|} (1 + r)^{\xi + \eta + n - 1} \, dy \leq C(n,\xi,\eta) \ |x|^{\eta + n} \ ,$$

where in the last step we used $\xi + \eta + n > 0$, $|x| \geq 1$, $|x| \geq (1 + |x|)/2$.

We write L_2 as

$$L_2 = \int_{|y|\geq|x|/2} \left| \log \left| \frac{x}{|x|} - \frac{y}{|x|} \right| \right|^{\xi} (1+|y|)^{\eta} \, dy \quad [\text{ set } z = y/|x| \ ]$$

$$= \int_{|z|\geq 1/2} \left| \log \left| \frac{x}{|x|} - z \right| \right|^{\xi} \big(1 + |x| \, |z| \big)^{\eta} |x|^{n} \, dz \ .$$

But

$$|z| \geq 1/2 \ \Rightarrow \ |z| \geq (1+|z|)/3 \ \Rightarrow \ 1 + |x| \, |z| \geq |x| \, |z| \geq \frac{1}{3} |x|(1+|z|) \ ;$$

thus, since also $\eta < -n < 0$,

$$L_2 \leq C(\eta) \ |x|^{n+\eta} \int_{\mathbf{R}^n} \left| \log \left| \frac{x}{|x|} - z \right| \right|^{\xi} \big(1 + |z| \big)^{\eta} \, dz$$

$$= C(\eta) \ |x|^{n+\eta} K(x/|x|,n,\xi,\eta) \leq C(n,\xi,\eta) \ |x|^{n+\eta} \big(\log (e + 1) \big)^{\xi}$$

$$= C(n,\xi,\eta) \ |x|^{n+\eta} \ .$$

We combine the estimates for L_1 and L_2 to complete the proof. ∎

Lemma 8E Assume $n \geq 1$, $1 < p \leq \infty$, and let σ be a real number such that

$$n - n/p < \sigma < 1 + n - n/p \, . \tag{6.7}$$

For any real valued measurable function f in $\mathbf{R}^n$, let Zf be the function

$$Zf(x) := \int_{\mathbf{R}^n} \big| \, \log|x-y| - \log|x| \, \big| \, |f(y)| \, dy \, .$$

Then for $|x| \geq 1$,

$$Zf(x) \leq C(n,p,\sigma) \, \| (1+r)^{\sigma} f \|_p \, |x|^{n - \sigma - n/p} \, .$$

Proof We may assume $\| (1+r)^{\sigma} f \|_p$ is finite. Let q be determined by $1/p + 1/q = 1$; then by Hölder's inequality,

$$Zf(x) := \int_{\mathbf{R}^n} \big| \, \log|x-y| - \log|x| \, \big| \, (1+|y|)^{-\sigma} (1+|y|)^{\sigma} \, |f(y)| \, dy$$

$$\leq \| (1+r)^{\sigma} f \|_p \cdot \Big[\int_{\mathbf{R}^n} \big| \, \log|x-y| - \log|x| \, \big|^q (1 + |y|)^{-\sigma q} \, dy \Big]^{1/q}$$

$$= \| (1+r)^{\sigma} f \|_p \cdot \big(L(x,n,q,-\sigma q) \big)^{1/q} \, .$$

To use Lemma 8D, we check the conditions

$$-\sigma q + n < 0 \quad , \quad q - \sigma q + n > 0 \, ,$$

which follow from our assumptions (6.7); thus we obtain

$$\big(L(x,n,q,-\sigma q) \big)^{1/q} \leq \big(C(n,p,\sigma) \, |x|^{n - \sigma q} \big)^{1/q} = C(n,p,\sigma) \, |x|^{n - \sigma - n/p} \, ,$$

and consequently our desired estimate. ∎

Theorem 8 For $n \geq 2$, and f a measurable real valued function on $\mathbf{R}^n$, set

$$\mathbf{S}f(x) = \int_{\mathbf{R}^n} \Gamma(x-y) \, f(y) \, dy \, .$$

(a) If $n \geq 3$, $1 \leq p \leq s \leq \infty$, and

$$\frac{1}{p} - \frac{1}{s} < \frac{2}{n} < \frac{\sigma}{n} + \frac{1}{p} < 1 \, , \tag{6.8}$$

then

$$\| (1+r)^{\sigma + n/p - 2 - n/s} \, \mathbf{S}f \|_s \leq C(n,p,s,\sigma) \, \| (1+r)^{\sigma} f \|_p \, . \tag{6.9}$$

(b) If $n = 2$, $1 < p \leq \infty$, $2 - 2/p < \sigma$, $\varepsilon < -2/p$, then

$$\| (1+r)^{\varepsilon} \, \mathbf{S}f \|_p \leq C(n,p,\sigma,\varepsilon) \, \| (1+r)^{\sigma} f \|_p \, . \tag{6.10}$$

If moreover $\sigma < 3 - 2/p$, and if we define

$$f_o := \int_{\mathbf{R}^2} f(y)\, dy \quad ,$$

then

$$\left|\mathbf{S}f(x) - f_o\, \Gamma(x)\right| \le C(n,p,\sigma)\ \left\| (1+r)^{\sigma} f \right\|_p \left|x\right|^{2 - \sigma - 2/p} . \tag{6.11}$$

Proof (a) As usual, we may assume $\| (1+r)^{\sigma} f \|_p$ is finite. We estimate

$$\left|\mathbf{S}f(x)\right| \le C(n) \int_{\mathbf{R}^n} \left|x-y\right|^{2-n} \left|f(y)\right| dy = C(n)\ Vf(x) \ ,$$

where now $\lambda = 2-n$ in Lemma 8B. Conditions (6.1) in the hypotheses of Lemma 8B reduce to our assumptions; therefore we obtain the result of (a).

(b) We estimate

$$\left|\mathbf{S}f(x)\right| \le \frac{1}{2\pi}\ Wf(x) \quad ,$$

where W appears as in Lemma 8C. Our assumptions on σ, p and ε are the same as those of Lemma 8C; hence the result (6.10) follows. We have also

$$\mathbf{S}f(x) - f_o\ \Gamma(x) = 1/2\pi \cdot \int_{\mathbf{R}^2} \Big(\log\left|x-y\right| - \log\left|x\right| \Big)\ f(y)\ dy \ ,$$

giving

$$\left|\mathbf{S}f(x) - f_o\ \Gamma(x)\right| \le \frac{1}{2\pi} \cdot Zf(x) \quad .$$

If also $\sigma < 3 - 2/p$ then we may apply Lemma 8C and obtain 11. ∎

Finally, we may prove Theorem 2 of section 2.

Proof of Theorem 2 : We apply Theorem 8 with σ replaced by $\sigma + 2$. We apply (6.9) with $s = p$ to obtain (3.2), and with $s = \infty$ to obtain (3.3). We apply (6.10) with $\varepsilon = \tau$ to obtain (3.4), and apply (6.11) to obtain (3.5). ∎

References

1. R. A. Adams, Sobolev Spaces, (New York: Academic Press, 1975).

2. A. D. Aleksandrov, Uniqueness conditions and estimates for the solution of the Dirichlet problem, Vestnik Liningrad Univ. 18, no. 3 (1963), 5-29 [Russian]. English translation in Amer. Math. Soc. Transl. (2) 68 (1968), 89-119.

3. H. Begehr and G. N. Hile, Schauder estimates and existence theory for entire solutions of linear elliptic equations, Proc. Roy. Soc. Edinburgh, 110A (1988), 101-123.

4. E. Bohn and L. K. Jackson, The Liouville theorem for a quasilinear elliptic partial differential equation, Trans. Amer. Math. Soc. 104 (1962), 392-397.

5. J. M. Bony, Principe du maximum dans les espaces de Sobolev, C. R. Acad. Sci. Paris 265 (1967), 333-336.

6. A. Friedman, Bounded entire solutions of elliptic equations, Pacific J. Math. 44 (1973), 497-507.

7. D. Gilbarg and J. Serrin, On isolated singularities of solutions of second order elliptic differential equations, J. Analyse Math. 4 (1954-6), 309-340.

8. D. Gilbarg and N. S. Trudinger, Elliptic Partial Differential Equations of Second Order, Second Edition (Berlin: Springer, 1983).

9. S. Hildebrandt and K. Wildman, Sätze vom Liouvilleschen Typ für quasilineare elliptische Gleichungen und Systeme, Nachr. Akad. Wiss. Göttingen Math. Phys. Kl. II (1979), Nr. 4, 41-59.

10. G. N. Hile, Entire solutions of linear elliptic equations with Laplacian principal part, Pacific J. Math. 62 (1976), 127-140.

11. A. V. Ivanov, Local estimates for the first derivatives of solutions of quasilinear second order elliptic equations and their application to Liouville type theorems, Sem. Steklov Math. Inst. Leningrad 30 (1972), 40-50; Translated in J. Sov. Math. 4 (1975), 335-344.

12. C. P. Mawata, Schauder estimates and existence theory for entire solutions of linear parabolic equations, Differential and Integral Equations, vol. 2, No. 3 (1989), 251-274.

13. L. Peletier and J. Serrin, Gradient bounds and Liouville theorems for quasilinear elliptic equations, Ann. Scuola Norm. Sup. Pisa Cl. Sci., Ser. 4, vol. 5 (1978), 65-104.

14. R. Redheffer, On the inequality $\Delta u \geq f(u, |\text{grad } u|)$, J. Math. Anal. 1 (1960), 277-299.

15. J. Serrin, Entire solutions of nonlinear Poisson equations, Proc. London Math. Soc. 24 (1972), 348-366.

16. J. Serrin, Liouville theorems and gradient bounds for quasilinear elltiptic systems, Arch. Rational Mech. Anal. 66 (1977), 295-310.

17. N. Weck, Liouville theorems for linear elliptic systems, Proc. Roy. Soc. Edinburgh, 94A (1983), 309-322.

G. N. Hile
Mathematics Department
University of Hawaii
Honolulu, HI 96822 USA

C. P. Mawata
Mathematics Department
University of Tennessee at Chattanooga
Chattanooga, TN 37403 USA

H LI AND Z HOU

A coefficient inverse problem for an elliptic equation

Abstract In this paper, a coefficient inverse problem for an elliptic equation of second order is considered. By using Fourier transform, the inverse problem is reduced to a nonlinear integral equation in some class of functions. This equation is solved with the successive approximation method. It is proved that there exists a unique solution of the inverse problem, and a representation of this solution is given.

1 Introduction

Equations of mathematical physics describe, as a rule, physical processes, while their coefficients related to physical characteristics of the medium wherein the processes take place. In applied sense, many problems of determining properties of the medium are studied based on the results of observing these processes within the available range of measurements. These are the inverse problems of mathematical physics, i.e. of problems aimed at defining the coefficients of a differential equation through some functionals of its solutions. It has gained wide acceptance in gravitational, magnetic and electromagnetic surveying [1], [2], [3], et al.

In this paper, the solution of the following problem is sought:

In the domain $Q = \{(x,y,z) : (x,y,z) \in (-\infty,+\infty) \times D\}$, where D is a bounded domain of the (y,z)-plane, we find a function $g(y,z) \in C^1(\overline{D}), 0 < g(y,z)$, and a function $u(x,y,z) \in C^2(\overline{Q})$, satisfying the equation

$$g(y,z)\frac{\partial^2 u}{\partial x^2} + \frac{\partial^2 u}{\partial y^2} + \frac{\partial^2 u}{\partial z^2} = 0, \quad (x,y,z) \in Q \tag{1.1}$$

and the condition of determining the solution

$$u(x,y,z) = \varphi(x,y,z), \quad (x,y,z) \in (-\infty,+\infty) \times \Gamma_0, \tag{1.2}$$

$$u(x,y,z)|_{x=0} = \psi(y,z), \quad (y,z) \in D. \tag{1.3}$$

Here Γ_0 is the boundary of the domain $D, \varphi(x,y,z) \in C((-\infty,+\infty) \times \Gamma_0)$ and $\psi(y,z) \in C^2(\overline{D})$ are known functions, satisfying the compatibility condition

$$\varphi(0,y,z) = \psi(y,z), \quad (y,z) \in \Gamma_0. \tag{1.4}$$

In this paper, by using Fourier transform, we reduce problem (1.1) – (1.3) to some nonlinear integral equation. The latter is solvable, so we can solve problem (1.1) – (1.3).

2 Reduction of the inverse problem (1.1) – (1.2) to some nonlinear integral equation

Prior to delving into problem (1.1) – (1.3), we suppose function $g(y,z) > 0$ is known temporarily, and consider problem (1.1) – (1.2) by using Fourier transform with respect to the variable x. We have:

$$\Delta\widehat{u} - \lambda^2 g(y,z)\widehat{u}(\lambda,y,z) = 0\,, \quad (\lambda,y,z) \in Q\,, \tag{2.1}$$

$$\widehat{u}(\lambda,y,z)|_{\Gamma_0} = \widehat{\varphi}(\lambda,y,z)\,, \quad (\lambda,y,z) \in (-\infty,+\infty)\times\Gamma_0\,, \tag{2.2}$$

where $\Delta\widehat{u} = \frac{\partial^2\widehat{u}}{\partial y^2} + \frac{\partial^2\widehat{u}}{\partial z^2}$, $\widehat{u}(\lambda,y,z)$ is the Fourier transform of the function $u(x,y,z)$ with respect to the variable x,

$$\widehat{u}(\lambda,y,z) = \frac{1}{\sqrt{2\pi}}\int_{-\infty}^{+\infty} u(x,y,z)e^{-i\lambda x}dx\,, \tag{2.3}$$

$\widehat{\varphi}(\lambda,y,z)$ is the Fourier transform of the function $\varphi(x,y,z)$ with respect to x.

If problem (2.1) – (2.2) has a solution, the solution is unique. For proving this, we show that problem (2.1) – (2.2) has only the trivial solution when $\widehat{\varphi}$ is zero. In fact, let

$$\widehat{u}(\lambda,y,z) = \widehat{u}_1(\lambda,y,z) + i\widehat{u}_2(\lambda,y,z)\,,$$

where $\widehat{u}_1(\lambda,y,z)$ and $\widehat{u}_2(\lambda,y,z)$ are real functions, then they satisfy problem (2.1) – (2.2), i.e.

$$\Delta\widehat{u}_j - \lambda^2 g(y,z)\widehat{u}_j(\lambda,y,z) = 0\,, \quad (\lambda,y,z) \in Q\,, \tag{2.1'}$$

$$\widehat{u}_j|_{\Gamma_0} \equiv 0\,, \quad (\lambda,y,z) \in (-\infty,+\infty)\times\Gamma_0\,, \tag{2.2'}$$

where $j = 1,2,$. After multiplying equation (2.1') by $\widehat{u}_j(\lambda,y,z)$, we integrate about the domain D with respect to the variable (y,z), getting

$$\iint_D [\Delta\widehat{u}_j(\lambda,y,z)\widehat{u}_j(\lambda,y,z) - \lambda^2 g(y,z)\widehat{u}_j^2(\lambda,y,z)]\,dydz = 0\,. \tag{2.4}$$

By using Green formula, it follows from (8) that

$$\int_{\Gamma_0} \widehat{u}_j\frac{\partial\widehat{u}_j}{\partial n}\,ds - \iint_D [\widehat{u}_{jy}^2 + \widehat{u}_{jz}^2 + \lambda^2 g(y,z)\widehat{u}_j^2]\,dydz = 0 \tag{2.4'}$$

where n denotes the unit exterior normal vector to Γ_0, s denotes the arc length coordinate. Since on $\Gamma_0, u(y,z) \equiv 0$, from (2.2') – (2.4') we have:

$$\widehat{u}_j(\lambda,y,z) = C(\text{constant}) = 0\,, \quad (y,z) \in \overline{D}\,, j = 1,2\,.$$

So we gain

$$\widehat{u}(\lambda,y,z) = \widehat{u}_1(\lambda,y,z) + i\widehat{u}_2(\lambda,y,z) \equiv 0\,, \quad (y,z) \in \overline{D}\,.$$

By using the Green function $G(\xi,\eta;y,z)$ of the equation $\frac{\partial^2\widehat{u}}{\partial y^2}+\frac{\partial^2\widehat{u}}{\partial z^2}=0$ with respect to the domain D, we can reduce problem (2.1) – (2.2) to the equivalent integral equation, see [4],

$$\begin{aligned}\widehat{u}(\lambda,y,z) = & -\lambda^2\iint\limits_D g(\xi,\eta)G(\xi,\eta;y,z)\widehat{u}(\lambda,\xi,\eta)\,d\xi d\eta \\ & -\int\limits_{\Gamma_0}\widehat{\varphi}(\lambda,\xi,\eta)\frac{\partial G(\xi,\eta;y,z)}{\partial n}\,ds\,.\end{aligned} \tag{2.5}$$

Let

$$\mu(D)=\max_{\overline{D}}\iint\limits_D |G(\xi,\eta;y,z)|\,d\xi d\eta\,.$$

The conditions for problem (2.1) – (2.2) to be solvable are formulated in the following lemma.

Lemma 1. *Assume $\widehat{\varphi}(\lambda,y,z)\in C(\Gamma_0)$, and $\widehat{\varphi}\equiv 0$ for $|\lambda|>R$, and let $g(y,z)$ satisfy the conditions*

$$g(y,z)\in C^1(\overline{D}),\quad |g(y,z)|<\frac{1}{R^2\mu(D)}\,,$$

then the solution $\widehat{u}(\lambda,y,z)$ of problem (2.1) – (2.2) exists and is unique, $\widehat{u}(\lambda,y,z)\in C^2(\overline{D})$, and $\widehat{u}(\lambda,y,z)\equiv 0$ as $|\lambda|>R$, where R is a positive constant .

Proof. Since problem (2.1) – (2.2) is equivalent to equation (2.5), we only prove (2.5) to be solvable. As $|\lambda|\le R$, we solve (2.5) by using the method of successive approximations, iterating as follows.

Let

$$\begin{aligned}\widehat{u}_0(\lambda,y,z) &= -\int\limits_{\Gamma_0}\widehat{\varphi}(\lambda,\xi,\eta)\frac{\partial G(\xi,\eta;y,z)}{\partial n}\,ds\,,\\ \widehat{u}_n(\lambda,y,z) &= -\lambda^2\iint\limits_D g(\xi,\eta)\widehat{u}_{n-1}(\lambda,\xi,\eta)G(\xi,\eta;y,z)\,d\xi d\eta \\ &\quad -\int\limits_{\Gamma_0}\widehat{\varphi}(\lambda,\xi,\eta)\frac{\partial G(\xi,\eta;y,z)}{\partial n}\,ds,(n\ge 1)\,.\end{aligned}$$

Since $|g(y,z)|<\frac{1}{R^2\mu(D)}$, it is possible to show sequence $\{\widehat{u}_n(\lambda,y,z)\}$ to convege uniformly and absolutely to some function $\widehat{u}(\lambda,y,z)$. Hence, equation (2.5) has a solution $\widehat{u}(\lambda,y,z)$. From (2.1) we have

$$\Delta\widehat{u}(\lambda,y,z)=\lambda^2\widehat{u}(\lambda,y,z)g(y,z)\in C(\overline{D})\,,$$

so $\widehat{u}(\lambda,y,z)\in C^2(\overline{D})$. Problem (2.1) – (2.2) has only the trivial solution, when $|\lambda|>R$. We have shown the solution of problem (2.1) – (2.2) to be unique. Thus the function $\widehat{u}(\lambda,y,z)$ is the unique solution of problem (2.1) – (2.2). This concludes the proof.

From Lemma 1, we know that problem (2.1) – (2.2) has a unique solution $\widehat{u}(\lambda,y,z)$ and $\widehat{u}(\lambda,y,z)\equiv 0$ as $|\lambda|>R$. By using Fourier inverse transform, we have

$$u(x,y,z)=\frac{1}{\sqrt{2\pi}}\int_{-\infty}^{+\infty}\widehat{u}(\lambda,y,z)e^{-i\lambda x}d\lambda=\frac{1}{\sqrt{2\pi}}\int_{-R}^{R}\widehat{u}(\lambda,y,z)e^{-i\lambda x}d\lambda. \tag{2.6}$$

So for any $g(y,z)$ and $\varphi(x,y,z)$, which satisfy the conditions of Lemma 1, problem (1.1) – (1.2) has a unique solution $u(x,y,z)\in C^2(Q)$.

We now consider the inverse problem (1.1) – (1.3). First we reduce it to some nonlinear integral equation. From condition (1.3) and equality (2.6) we have

$$\Delta u|_{x=0}=\Delta\psi(y,z)=\frac{1}{\sqrt{2\pi}}\int_{-R}^{R}\Delta\widehat{u}(\lambda,y,z)\,d\lambda=\frac{1}{\sqrt{2\pi}}\int_{-R}^{R}\lambda^2 g(y,z)\widehat{u}(\lambda,y,z)\,d\lambda.$$

So we gain a representation of the function $g(y,z)$ by $\widehat{u}(\lambda,y,z)$

$$g(y,z)=\Delta\psi(y,z)\Big[\frac{1}{\sqrt{2\pi}}\int_{-R}^{R}\lambda^2\widehat{u}(\lambda,y,z)d\lambda\Big]^{-1}. \tag{2.7}$$

Thus problem (1.1) – (1.3) is reduced to the nonlinear integral equation for the function $\widehat{u}(\lambda,y,z)$

$$\begin{aligned}\widehat{u}(\lambda,y,z) &= -\lambda^2\iint_G G(\xi,\eta;y,z)\frac{\Delta\psi(\xi,\eta)\widehat{u}(\lambda,\xi,\eta)}{\frac{1}{\sqrt{2\pi}}\int_{-R}^{R}m^2\widehat{u}(m,\xi,\eta)dm}\,d\xi d\eta \\ &\quad -\int_{\Gamma_0}\widehat{\varphi}(\lambda,\xi,\eta)\frac{\partial G(\xi,\eta;y,z)}{\partial n}\,ds,\end{aligned} \tag{2.8}$$

under the condition

$$0<\Delta\psi(y,z)\Big[\frac{1}{\sqrt{2\pi}}\int_{-R}^{R}m^2\widehat{u}(m,y,z)dm\Big]^{-1}<\frac{1}{R^2\mu(D)}.$$

Let

$$\begin{aligned}w_0=\mu(D)\,\|\Delta\psi\|_{C(\overline{D})}R^2,\; A_1 &= \sup_{\substack{(y,z)\in\overline{D}\\ |\lambda|\le R}}\Big|\int_{\Gamma}\widehat{\varphi}(\lambda,\xi,\eta)\frac{\partial G(\xi,\eta;y,z)}{\partial n}\,ds\Big|,\\ A_2 &= \min_{(y,z)\in\overline{D}}\Big[-\frac{1}{\sqrt{2\pi}}\int_{-R}^{R}m^2\int_{\Gamma_0}\widehat{\varphi}(m,\xi,\eta)\frac{\partial G(\xi,\eta;y,z)}{\partial n}\,dsdm\Big].\end{aligned} \tag{2.9}$$

Lemma 2. *Assume*

(1) $\widehat{\varphi}(\lambda, y, z) \in C(\Gamma_0)$, *and* $\widehat{\varphi} \equiv 0$ *when* $|\lambda| > R$,

(2) $0 < \Delta\psi(y, z) \in C^1(D)$

(3) $A_2 > 0, A_2 \geq w_0 + R(8w_0 R A_1)^{1/2}$, $A_1 > 0$.

Then solving the inverse problem (1.1) – (1.3) is equivalent to solving equation (2.8) in the class H of functions, where

$$H = \Big\{\widehat{u}(\lambda, y, z) : \widehat{u}(\lambda, y, z) \in C^2(\overline{D}), \quad \text{and} \quad \widehat{u} \equiv 0 \quad \text{when} \quad |\lambda| > R,$$

$$\min_{\overline{D}} \Big[\frac{1}{\sqrt{2\pi}} \int_{-R}^{R} m^2 \widehat{u}(m, y, z) dm\Big] \geq B_0, B_0 = \frac{1}{2}\{w_0 + A_2 + [(A_2 - w_0)^2 - 8R^3 w_0 A_1]^{1/2}\}\Big\}.$$

Proof. As indicated above, we can reduce the inverse problem (1.1) – (1.3) to the equation (2.8) under condition (2.9). But $\Delta\psi(y, z) > 0$, so in order to satisfy condition (2.9), we must have

$$\min_{\overline{D}} \Big[\frac{1}{\sqrt{2\pi}} \int_{-R}^{R} m^2 \widehat{u}(m, y, z) dm\Big] \geq B_0 > 0. \tag{2.10}$$

Now we decide B_0. If $\widehat{u}(\lambda, y, z)$ is the solution of equation (2.8), then it satisfies

$$\begin{aligned} \widehat{u}(\lambda, y, z) &= -\lambda^2 \iint_D \frac{G(\xi, \eta; y, z)\Delta\psi(\xi, \eta)\widehat{u}(\lambda, \xi, \eta)}{\frac{1}{\sqrt{2\pi}} \int_{-R}^{R} m^2 \widehat{u}(m, \xi, \eta) dm} d\xi d\eta \\ &\quad - \int_{\Gamma_0} \widehat{\varphi}(\lambda, \xi, \eta) \frac{\partial G(\xi, \eta; y, z)}{\partial n} ds \\ &=: T\widehat{u}(\lambda, y, z). \end{aligned} \tag{2.11}$$

Thus we have

$$\sup_{\substack{|\lambda| \leq R \\ (y,z) \in \overline{D}}} |\widehat{u}(\lambda, y, z)| \leq \frac{B_0 A_1}{B_0 - w_0}. \tag{2.12}$$

The inequality

$$\min_{\overline{D}} \Big[\frac{1}{\sqrt{2\pi}} \int_{-R}^{R} m^2 T\widehat{u}(m, y, z) dm\Big] \geq B_0$$

leads to

$$\min_{\overline{D}} \Bigg[\frac{1}{\sqrt{2\pi}} \int_{-R}^{R} m^2 \Bigg\{ -m^2 \iint_D \frac{G(\xi, \eta; y, z)\Delta\psi(\xi, \eta)\widehat{u}(m, \xi, \eta)}{\frac{1}{\sqrt{2\pi}} \int_{-R}^{R} \widehat{u}(m, \xi, \eta) dm} d\xi d\eta$$

$$- \int_{\Gamma_0} \widehat{\varphi}(m, \xi, \eta) \frac{\partial G(\xi, \eta; y, z)}{\partial n} ds \Bigg\} dm\Bigg] \geq B_0. \tag{2.13}$$

If

$$A_2 - \frac{A_1 w_0}{B_0 - w_0} 2R^3 \geq B_0 \tag{2.14}$$

holds, then inequality (2.13) holds. Inequality (2.14) is written as

$$B_0^2 - B_0(w_0 + A_2) + A_2 w_0 + 2R^3 w_0 A_1 \leq 0\,. \tag{2.15}$$

For

$$(w_0 + A_2)^2 - 4(A_2 w_0 + 2R^3 A_1 w_0) = (A_2 - w_0)^2 - 8R^3 w_0 A_1$$

is nonnegative under condition (3) of Lemma 2, we can take

$$B_0 = \frac{1}{2}\{w_0 + A_2 + [(A_2 - w_0)^2 - 8R^3 w_0 A_1]^{1/2}\}\,,$$

which implies that

$$A_2 > B_0 > w_0\,.$$

It follows from Lemma 1 that solution $\widehat{u}(\lambda, y, z) \in C^2(\overline{D})$ and $\widehat{u}(\lambda, y, z) \equiv 0$ for $|\lambda| > R$. Thus we have proved our lemma.

Remark. *In Lemma 2,* $A_2 = \min\limits_{\overline{D}} \int\limits_{\Gamma_0} \varphi_{xx}(0, \xi, \eta) \dfrac{\partial G(\xi, \eta; y, z)}{\partial n}\, ds\,.$

3 Solving the nonlinear integral equation (2.8) in the class H

In section 2, we have reduced the inverse problem (1.1) – (1.3) to the nonlinear integral equation (2.8) in the class H. Hence, we have to solve (2.8) in the class H.

Lemma 3. *Assume the conditions of Lemma 2 are sastisfied. Then there exists a positive constant $\delta = \delta(R, A_1, A_2)$ such that for any given function $\psi(y, z)$ satisfying*

$$\Delta\psi(y, z) \in C^1(D)\,,\ 0 < \Delta\psi(y, z) < \delta\,,$$

equation (2.8) has a unique solution in the class H.

Proof. For the nonlinear integral equation (2.8), we construct a function sequence $\{\widehat{u}_n\}$ as follows. Let

$$\begin{aligned}
\widehat{u}_0(\lambda, y, z) &= -\int\limits_{\Gamma_0} \widehat{\varphi}(\lambda, \xi, \eta) \frac{\partial G(\xi, \eta; y, z)}{\partial n}\, ds \\
\widehat{u}_n(\lambda, y, z) &= T\widehat{u}_{n-1}(\lambda, y, z)\,, \quad n \geq 1\,.
\end{aligned}$$

Then we have with $C = C(\overline{D})$

$$\begin{aligned}
\|\widehat{u}_0\|_C &= A_1\,,\ \min_{\overline{D}} \Big[\frac{1}{\sqrt{2\pi}} \int_{-R}^{R} m^2 \widehat{u}_0(m, y, z) dm\Big] = A_2 > B_0\,, \\
\|\widehat{u}_1\|_C &\leq \frac{w_0}{A_2} A_1 + A_1\,,\ \min_{\overline{D}} \Big[\frac{1}{\sqrt{2\pi}} \int_{-R}^{R} m^2 \widehat{u}_1(m, y, z) dm\Big] \geq B_0\,, \\
\|\widehat{u}_n\|_C &\leq \Big(\frac{w_0^n}{B_0^n} + \frac{w_0^{n-1}}{B_0^{n-1}} + \ldots + 1\Big) A_1\,,\ \min_{\overline{D}} \Big[\frac{1}{\sqrt{2\pi}} \int_{-R}^{R} m^2 \widehat{u}_n(m, y, z) dm\Big] \geq B_0\,,
\end{aligned}$$

so for any integer $n \geq 0$, we have

$$\|\widehat{u}_n\| \leq \frac{B_0 A_1}{B_0 - w_0}, \ \min_{\overline{D}} \Big[\frac{1}{\sqrt{2\pi}} \int_{-R}^{R} m^2 \widehat{u}_n(m, y, z) dm\Big] \geq B_0 .$$

This means the functions $\widehat{u}_n(\lambda, y, z)$ belong to H.

Now we discuss the convergence of the sequence $\{\widehat{u}_n(\lambda, y, z).\}$ First, we estimate

$$\|\widehat{u}_n - \widehat{u}_{n-1}\|_C = \Big\| \lambda^2 \iint_D G(\xi, \eta; y, z) \Delta\psi(\xi, \eta)$$

$$\times \Bigg[\frac{\widehat{u}_{n-1}(\lambda, \xi, \eta)}{\frac{1}{\sqrt{2\pi}} \int_{-R}^{R} m^2 \widehat{u}_{n-1}(m, \xi, \eta) dm} - \frac{\widehat{u}_{n-2}(\lambda, \xi, \eta)}{\frac{1}{\sqrt{2\pi}} \int_{-R}^{R} m^2 \widehat{u}_{n-2}(m, \xi, \eta) dm} \Bigg] d\xi d\eta \Big\|_C$$

$$\leq w_0 \max_{\substack{(y,z)\in\overline{D} \\ |\lambda|\leq R}} \Bigg| \frac{\widehat{u}_{n-1}(\lambda, y, z)}{\frac{1}{\sqrt{2\pi}} \int_{-R}^{R} m^2 \widehat{u}_{n-1}(m, y, z) dm} - \frac{\widehat{u}_{n-2}(\lambda, y, z)}{\frac{1}{\sqrt{2\pi}} \int_{-R}^{R} m^2 \widehat{u}_{n-2}(m, y, z) dm} \Bigg|$$

$$\leq w_0 \max_{\substack{(y,z)\in\overline{D} \\ |\lambda|\leq R}} \Bigg[\Bigg| \frac{\widehat{u}_{n-1}(\lambda, y, z)}{\frac{1}{\sqrt{2\pi}} \int_{-R}^{R} m^2 \widehat{u}_{n-1}(m, y, z) dm} - \frac{\widehat{u}_{n-2}(\lambda, y, z)}{\frac{1}{\sqrt{2\pi}} \int_{-R}^{R} m^2 \widehat{u}_{n-1}(m, y, z) dm} \Bigg|$$

$$+ \Bigg| \frac{\widehat{u}_{n-2}(\lambda, y, z)}{\frac{1}{\sqrt{2\pi}} \int_{-R}^{R} m^2 \widehat{u}_{n-1}(m, y, z) dm} - \frac{\widehat{u}_{n-2}(\lambda, y, z)}{\frac{1}{\sqrt{2\pi}} \int_{-R}^{R} m^2 \widehat{u}_{n-2}(m, y, z) dm} \Bigg| \Bigg]$$

$$\leq w_0 \tfrac{1}{B_0} \|\widehat{u}_{n-1} - \widehat{u}_{n-2}\|_C$$

$$+ w_0 \tfrac{1}{B_0^2} \max_{\overline{D}} \Big| \frac{1}{\sqrt{2\pi}} \int_{-R}^{R} m^2 (\widehat{u}_{n-1}(m, y, z) - \widehat{u}_{n-2}(m, y, z)) dm \Big|$$

$$\leq \frac{w_0}{B_0} \|\widehat{u}_{n-1} - \widehat{u}_{n-2}\|_C + \frac{w_0}{B_0^2} \frac{B_0 A_1}{B_0 - w_0} \|\widehat{u}_{n-1} - \widehat{u}_{n-2}\|_C \frac{1}{\sqrt{2\pi}} \int_{-R}^{R} m^2 dm$$

$$= \Big[\frac{w_0}{B_0} + \frac{A_1 w_0}{B_0(B_0 - w_0)} \frac{2R^3}{3\sqrt{2\pi}} \Big] \|\widehat{u}_{n-1} - \widehat{u}_{n-2}\|_C$$

and

$$\|\widehat{u}_1 - \widehat{u}_0\| = T\widehat{u}_0 - \widehat{u}_0\|_C$$

$$= \Big\| - \lambda^2 \iint_D \frac{G(\xi, \eta; y, z) \Delta\psi(\xi, \eta) \widehat{u}_0(\lambda, \xi, \eta)}{\frac{1}{\sqrt{2\pi}} \int_{-R}^{R} m^2 \widehat{u}(m, \xi, \eta) dm} d\xi d\eta \Big\|_C \leq \frac{w_0}{B_0} .$$

Hence, the function sequence $\{\hat{u}_n(\lambda, y, z)\}$ converge uniformly and absolutely to some function $\hat{u}(\lambda, y, z)$ when

$$\frac{w_0}{B_0} + \frac{\sqrt{2}R^3 A_1 w_0}{3\sqrt{\pi}B_0(B_0 - w_0)} < 1\,, \tag{2.1}$$

and $\hat{u}(\lambda, y, z)$ is the solution of equation (2.8), $\hat{u}(\lambda, y, z) \in C(\overline{D})$. It is in the class H. In fact, it follows from equation (2.8) that $\hat{u}(\lambda, y, z) \in C^2(\overline{D})$, $\min\limits_{\overline{D}} \left[\frac{1}{\sqrt{2\pi}} \int\limits_{-R}^{R} m^2(\hat{u}(m, y, z)dm\right]$ $\geq B_0$ and $\hat{u}(\lambda, y, z) \equiv 0$ as $|\lambda| > R$.

It is easy to show the solution of equation (2.8) to be unique under condition (2.15). Secondly, we consider the condition which the function $\hat{u}(\lambda, y, z)$ must satisfy if the inequality (2.15) holds. Inserting

$$B_0 = \frac{1}{2}\{A_2 + w_0 + [(A_2 - w_0)^2 - 8R^3 w_0 A_1]^{1/2}\}$$

into inequality (2.15), gives

$$(A_2 - w_0)^2 + [(A_2 - w_0)^2 - 8R^3 w_0 A_1]^{1/2}(A_2 - w_0) - 8R^3 w_0 A_1 > 0 \tag{2.2}$$

or

$$[(A_2 - w_0)^2 - 8R^3 w_0 A_1]^{1/2}\{(A_2 - w_0)^2 - 8R^3 w_0 A_1]^{1/2} + A_2 - w_0\} > 0\,.$$

It means if

$$(A_2 - w_0)^2 > 8R^3 w_0 A\,,$$

then only

$$w_0 < A_2 + 4R^3 A_1 - 2R(4R^4 A_1^2 + 2RA_1 A_2)^{1/2}$$

or

$$w_0 > A_2 + 4R^3 A_1 + 2R(4R^4 A_1^2 + 2RA_1 A_2)^{1/2}$$

must hold. Hence, there exists a positive $\delta = \delta(R, A_1, A_2)$, namely

$$\delta(R, A_1, A_2) = \frac{1}{R^2 \mu(D)}[A_2 + 4R^3 A_1 - 2R(4R^4 A_1^2 + 2RA_1 A_2)^{1/2}] > 0\,,$$

such that inequality (2.15) holds for $0 < \Delta\psi < \delta$.

Remark. *Under the condition* $\hat{u}(m, y, z) \equiv 0$ *for* $|m| > R$, *in class* H *we have*

$$\min_{\overline{D}}[\frac{1}{\sqrt{2\pi}} \int_{-R}^{R} m^2 \hat{u}(m, y, z)dm] = \max_{\overline{D}}(-u_{xx}(0, y, z))\,.$$

3 Conclusion

In section 2, section 3, by using Fourier transform, we reduce the inverse problem (1.1) – (1.3) to the nonlinear integral equation (2.8) in class H, and prove Lemma 1 to Lemma 3. To sum up the whole process, we state the following result.

Theorem. *Suppose*

1. $\widehat{\varphi}(\lambda, y, z) \in C(\Gamma_0)$ *and* $\widehat{\varphi} \equiv 0$ *for* $|\lambda| > R$, *where* R *is a positive constant.*

2. $A_2 > 0, A_2 \geq w_0 + R(8w_0RA_1)^{1/2}, A_1 > 0$, *where*

$$A_1 = \sup_{\substack{(y,z)\in\overline{D} \\ |\lambda|\leq R}} \Big|\int_{\Gamma_0} \widehat{\varphi}(\lambda, \xi, \eta)\frac{\partial G(\xi, \eta; y, z)}{\partial n}\, ds\Big|, \; w_0 = \mu(D)\|\Delta\psi\|_C R^2 ,$$

$$A_2 = \min_{\overline{D}} \Big[-\frac{1}{\sqrt{2\pi}}\int_{-R}^{R} m^2 \int_{\Gamma_0} \widehat{\varphi}(m, \xi, \eta)\frac{\partial G(\xi, \eta; y, z)}{\partial n}\, ds dm\Big] .$$

3. $g(y, z) \in C^1(\overline{D})$, $g(y, z) > 0$.

Then there exists a positive constant $\delta(R, A_1A_2)$, *such that for any given function* $\psi(y, z), \Delta\psi(y, z) \in C^1(D)$ *and* $0 < \Delta\psi < \delta$, *the inverse problem (1.1) – (1.3) has a unique solution* $\{g(y, z), u(x, y, z)\}$ *given by*

$$g(y, z) = \Delta\psi(y, z)\Big[\frac{1}{\sqrt{2\pi}}\int_{-R}^{R} m^2 \widehat{u}(m, y, z)dm\Big]^{-1}, u(x, y, z) = \frac{1}{\sqrt{2\pi}}\int_{-R}^{R} \widehat{u}(m, y, z)e^{-imx}dm,$$

where $\widehat{u}(m, y, z)$ *is the unique solution of the nonlinear integral equation (2.8) in* H .

References

[1] Romanov, V.G., Inverse problem of mathematical physics. Translated by L.Ya. Yuzina, Utrecht, VNU Science Press BV, 1987.

[2] Engel, H.W. and Groetsch, C.W. (editors), Inverse and ill-posed problems, Academic Press, Inc. 1987.

[3] Anikonov, Iu.E. and Bubnov, B.A., Existence and uniqueness of the solution of inverse problem for a parabolic equation (in Russian), Dokl. Akad. Nauk SSSR 288, No. 4 (1988).

[4] Courant, R. and Hilbert, D., Methods of mathematical physics Vol. II, Inter–science, New York, 1962.

G C HSIAO[1] AND J SPREKELS[2]

On the identification of distributed parameters in hyperbolic equations*

In this note we consider the identification of spatially varying coefficients of viscosity and elasticity in hyperbolic equations. It is shown that the asymptotic regularization method introduced by Alt, Hoffmann and Sprekels can be applied to approximate the parameters. Convergence and stability of the method is proved.

1. Introduction

This paper is concerned with the identification of spatially varying coefficient functions in hyperbolic equations. Let us briefly indicate the problem in question. Assume that $\Omega \subset I\!R^N$ denotes an open and bounded set with (sufficiently smooth) boundary $\partial\Omega$. We consider the hyperbolic differential equation ($T > 0$)

$$\frac{\partial^2 \bar{u}}{\partial t^2} - \frac{\partial}{\partial t}\nabla \cdot (\bar{b}(x)\nabla \bar{u}) - \nabla \cdot (\bar{a}(x)\nabla \bar{u}) = \bar{f} \quad \text{in } \Omega \times (0,T) \tag{1.1}$$

together with the initial and boundary conditions:

$$\frac{\partial}{\partial t}\frac{\partial \bar{u}}{\partial n_{\bar{b}}} + \frac{\partial \bar{u}}{\partial n_{\bar{a}}} = \bar{F} \quad \text{on } \Gamma_1 \times (0,T), \tag{1.2}$$

$$\bar{u} = \bar{g} \quad \text{on } \Gamma_2 \times (0,T), \tag{1.3}$$

$$\bar{u}(0) = \bar{u}_0, \frac{\partial \bar{u}}{\partial t}(0) = \bar{u}_1 \quad \text{on } \bar{\Omega}. \tag{1.4}$$

Here, $\partial\Omega = \Gamma_1 \cup \Gamma_2$ with $\Gamma_1 \cap \Gamma_2 = \emptyset$ and meas $(\Gamma_2) > 0; \bar{a} = (\bar{a}_{ij}), \bar{b} = (\bar{b}_{ij})$ are spatially varying $(N \times N)$–matrices. The outer conormal derivative with respect to the matrix $a = (a_{ij})$ is defined by

$$\frac{\partial u}{\partial n_a} = n_i a_{ij}\frac{\partial u}{\partial x_j}, \tag{1.5}$$

where $n = (n_1, \ldots, n_N)$ denotes the outer normal at $\partial\Omega$. Note that we have used the summation convention which shall be maintained throughout the paper.

We consider the identification problem:

---

[1] Department of Mathematical Sciences, University of Delaware, Newark, DE 19716, USA.
[2] Fachberich 10 der Universität–GH Essen, Postfach 10 37 64, W–4300 Essen 1, FRG.
* Partially supported by the Deutsche Forschungsgemeinschaft (DFG).

$(\bar{P})$ *Given* $\bar{f}, \bar{F}, \bar{g}, \bar{u}$, *find positive definite matrix functions* $\bar{a}, \bar{b}$ *with entries in* $L^\infty(\Omega)$ *such that (1.1) - (1.4) hold.*

In physical terms, the entries $\bar{a}_{ij}$ and $\bar{b}_{ij}$, respectively, are related to coefficients of elasticity and viscosity, respectively. Without viscosity, i.e., if $\bar{b}$ is the null matrix, it is unnecessary to consider a dynamic problem in order to identify $\bar{a}$ as long as distributed measurements of $\bar{u}$ are available (see [1], [5], [6], for instance); the time-dependent dynamic problem is then of greater interest if the measurements are incomplete (say, if $\bar{u}$ is unknown in Ω, but over specified boundary data are available; see [3] and [9], for references). However, in the presence of viscosity we cannot restrict ourselves to a static situation.

To motivate the setting used in the subsequent sections, let us transform (1.1) - (1.4) into a variational form. To this end, assume the data are sufficiently smooth. Let

$$V = \{v \in H^1(\Omega) : v = 0 \quad \text{on } \Gamma_2\}, \tag{1.6}$$

take a test function $v \in H^1(0,T;V)$, and choose some $\Phi \in L^2(0,T;H^1(\Omega))$ such that $\Phi(t) = \bar{g}(t)$ on Γ_2, for a.e. $t \in (0,T)$, in the sense of traces. Replacing $\bar{u}$ by $\bar{u} - \Phi$ and keeping the notation $\bar{u}$, we obtain (where $'$ denotes time differentiation):

$$\begin{aligned}\int_\Omega (\bar{u}''(t) - \Phi''(t))v(t)dx &+ \int_\Omega \nabla v(t)\cdot \bar{b}\nabla(\bar{u}' - \Phi')(t)dx \\ &+ \int_\Omega \nabla v(t)\cdot \bar{a}\nabla(\bar{u} - \Phi)(t)dx = \int_\Omega \bar{f}(t)v(t)dx \\ &- \int_{\Gamma_1} \bar{F}(t)v(t)ds, \quad a.e. \text{ on } (0,T).\end{aligned} \tag{1.7}$$

Integration over $(0,T)$ yields via an integration by parts:

$$\begin{aligned}&\int_0^T \int_\Omega \nabla v \cdot \bar{b}\nabla(\bar{u}' - \Phi')dx\,dt + \int_0^T \int_\Omega \nabla v \cdot \bar{a}\nabla(\bar{u} - \Phi)dx\,dt \\ &= \int_0^T \int_\Omega \bar{f}v\,dx\,dt - \int_0^T \int_{\Gamma_1} \bar{F}v\,ds\,dt + \int_0^T \int_\Omega (\bar{u}' - \Phi')v'dx\,dt \\ &- \int_\Omega (\bar{u} - \Phi')(T)v(T)dx + \int_\Omega (\bar{u}' - \Phi')(0)v(0)dx.\end{aligned} \tag{1.8}$$

In the following sections we are concerned with the identification of $\bar{a}, \bar{b}$ in (1.8), where for the sake of simplicity we assume that $\bar{u}(0) = 0$ and $\bar{g} = 0$ (i.e., $\Phi = 0$). The general case can be treated rather similarly.

In Section 2, the method of asymptotic regularization by Alt, Hoffmann and Sprekels [1] (see also [2], [5], [6], for further applications) is adapted to our problem.

We prove that the resulting algorithm is convergent. In Section 3 we prove the stability of the algorithm.

2. Asymptotic Regularization

Let V be given as in (1.6). Then there holds the Poincaré inequality

$$\int_\Omega |u|^2 dx \leq \omega \int_\Omega |\nabla u|^2 dx, \forall u \in V \quad \text{for some} \quad \omega > 0. \tag{2.1}$$

We introduce the space

$$H_0^1(0,T;V) = \{u \in L^2(0,T;V) : u' \in L^2(0,T;V), u(0) = 0\}, \tag{2.2}$$

and denote its dual by $(H_0^1(0,T;V))'$. For the measurements we assume:

(A1) $\bar{F} \in L^2(0,T;L^2(\Gamma_1))$, $\bar{f} \in L^2(0,T;L^2(\Omega))$, $\bar{u} \in C([0,T];H^{1,\infty}(\Omega))$, $\bar{u}' \in C([0,T];H^{1,\infty}(\Omega)), \bar{u}(0) = 0$.

The assumption $\bar{u}(0) = 0$ is a normalization. Clearly, $\bar{u} \in H_0^1(0,T;V)$.

Now let $L^2_{\text{sym}}(\Omega)(L^\infty_{\text{sym}}(\Omega))$ define the space of pointwise a.e. symmetric matrix functions with entries in $L^2(\Omega)(L^\infty(\Omega))$. Recall that $L^2_{\text{sym}}(\Omega)$ becomes a Hilbert space when equipped with the inner product

$$< a, b >= \int_\Omega a_{ij} b_{ij} dx \quad \text{for } a = (a_{ij}), b = (b_{ij}) \in L^2_{\text{sym}}(\Omega). \tag{2.3}$$

In the sequel, the $L^2(\Omega)$ – norm in the spaces of scalar, vector, and matrix functions with entries in $L^2(\Omega)$ is always denoted by $||\cdot||$. In the $L^\infty(\Omega)$–spaces we use the notation $||\cdot||_{L^\infty(\Omega)}$. Furthermore, for convenience let us agree to omit the arguments of functions, if appropriate. For given $c_0 > 0, C_0 > 0$, we define

$$\begin{aligned} K(c_0, C_0) := \{a = (a_{ij}) \in L^\infty_{\text{sym}}(\Omega) : ||a_{ij}||_{L^\infty(\Omega)} \leq C_0, \forall i,j, \\ \xi \cdot a(x)\xi \geq c_0|\xi|^2, \forall \xi \in I\!R^N, \quad \text{a.e. in } \Omega\}. \end{aligned} \tag{2.4}$$

For the sets of admissible parameters we make the (physically motivated) assumption:

(A2) $K_i \subset K(c_0, C_0)$; K_i *is nonempty, convex and closed in* $L^2_{\text{sym}}(\Omega)$, $i = 1, 2$.

We consider the identification problem:

(P) *Find parameters* $(\bar{a}, \bar{b}) \in K_1 \times K_2$ *such that there holds:*

$$\begin{aligned} &\int_0^T \int_\Omega \nabla v \cdot \bar{b} \nabla \bar{u}\, dx\, dt + \int_0^T \int_\Omega \nabla v \cdot \bar{a} \nabla \bar{u}\, dx dt \\ &= \int_0^T \int_\Omega \bar{f} v\, dx\, dt - \int_0^T \int_{\Gamma_1} \bar{F} v\, ds dt + \int_0^T \int_\Omega \bar{u}' v' dx dt \\ &- \int_\Omega \bar{u}'(T) v(T) dx, \ \forall v \in H_0^1(0,T;V). \end{aligned} \tag{2.5}$$

We make the assumption (from the viewpoint of practical applications):

(A3) *(P) admits at least one solution* $(\bar{a}, \bar{b})$.

Of course $(\bar{a}, \bar{b})$ is unknown, a priori. But it is possible to use the mere knowledge of its existence to construct an algorithm to approximate a solution. To this end, we discuss an appropriate finite dimensional analogue of (P) first.

We choose n–dimensional subspaces U_n of $H_0^1(0,T;H^{1,\infty}(\Omega)\cap V)$ and finite dimensional subspaces A_n, B_n of $L^\infty_{\rm sym}(\Omega), n \in I\!N$ such that there hold:

(A4) $\bar{u} \in U_n \subset U_{n+1},\ \forall n \in I\!N$.

(A5) $\bigcup\limits_{n=1}^{\infty} U_n$ *is dense in* $H_0^1(0,T;V)$.

(A6) $\int_0^T \nabla u(t) \otimes \nabla(u(t)-v(t))dt \in A_n,\ \forall u, v \in U_n,\ n \in I\!N$.

(A7) $\int_0^T \nabla u'(t) \otimes \nabla(u(t)-v(t))dt \in B_n,\ \forall u, v \in U_n,\ n \in I\!N$.

(A8) $P_{A_n}(K_1) \subset K_1,\ P_{B_n}(K_2) \subset K_2,\ \forall n \in I\!N$, *with the* $<\cdot,\cdot>$ *- orthogonal projection operators* P_{A_n} *and* P_{B_n} *onto* A_n *and* B_n*, respectively.*

(A9) $K_1 \cap A_n \neq \emptyset,\ K_2 \cap B_n \neq \emptyset,\ \forall n \in I\!N$.

Here for vectors $\xi = (\xi_i), n = (n_i) \in I\!R^N$, the symmetric $(N \times N)$–matrix $\xi \otimes \eta$ is given by $(\xi \otimes \eta)_{ik} = \frac{1}{2}(\xi_i\eta_k + \xi_k\eta_i), i, k = 1, \ldots, N$. The assumptions (A6) and (A7) impose a compatibility between U_n and A_n, B_n; (A4) is trivial; (A5) can be satisfied since $H_0^1(0,T;H^{1,\infty}(\Omega)\cap V)$ is dense in $H_0^1(0,T;V)$; Finally, (A8), (A9) hold if $K_1 \subset A_n, K_2 \subset B_n$, for instance.

The finite dimensional analogue of (P) reads:

(P_n) *Find* $(\bar{a}, \bar{b}) \in K_1 \times K_2$ *such that (2.5) holds for any* $v \in U_n$.

By (A3), (P_n) has at least one solution. We now introduce a system of variational inequalities to approximate a solution of (P_n). More precisely, letting $h > 0$ be a chosen constant, we consider the problem:

$(\bar{P}_n)$ *Given* $(a_0, b_0, u_0) \in (K_1 \cap A_n) \times (K_2 \cap B_n) \times U_n$*, find* $(a_j, b_j, u_j) \in (K_1 \cap A_n) \times (K_2 \cap B_n) \times U_n, j \in I\!N$ *such that*

$$\begin{aligned}
&\int_0^T \int_\Omega \nabla v \cdot b_{j+1} \nabla u'_{j+1} dx\, dt + \int_0^T \int_\Omega \nabla v \cdot a_{j+1} \nabla u_{j+1} dx\, dt \\
&\quad + \int_0^T \int_\Omega \frac{\nabla u'_{j+1} - \nabla u'_j}{h} \cdot \nabla v' dx\, dt + \int_\Omega \frac{\nabla u_{j+1}(T) - \nabla u_j(T)}{h} \cdot \nabla v(T) dx \\
&\quad = \int_0^T \int_\Omega \bar{f} v\, dx\, dt - \int_0^T \int_{\Gamma_1} \bar{F} v\, ds\, dt - \int_\Omega \bar{u}'(T) v(T) dx \\
&\quad + \int_0^T \int_\Omega \bar{u}' v' dx\, dt,\ \forall v \in U_n,
\end{aligned} \tag{2.6}$$

$$\int_\Omega \left(\frac{a_{j+1} - a_j}{h} - \int_0^T \nabla(u_{j+1} - \bar{u}) \otimes \nabla u_{j+1} dt \right)_{ik} (\eta - a_{j+1})_{ik} dx \geq 0, \tag{2.7}$$
$$\forall \eta \in K_1 \cap A_n,$$

$$\int_\Omega \left(\frac{b_{j+1} - b_j}{h} - \int_0^T \nabla(u_{j+1} - \bar{u}) \otimes \nabla u'_{j+1} dt \right)_{ik} (\zeta - b_{j+1})_{ik} dx \geq 0, \tag{2.8}$$
$$\forall \zeta \in K_2 \cap B_n.$$

Postponing the question of solvability of $(\bar{P}_n)$, we assume that $\{(a_j, b_j, u_j)\}_{j\geq 0}$ solves $(\bar{P}_n)$. We derive a fundamental a priori estimate. We begin by letting $q_j = u_j - \bar{u}, r_j = a_j - \bar{a}, s_j = b_j - \bar{b}$, and note that $q_j \in U_n$ from (A4). Substitution of $v = q_{j+1}$ into (2.5) and (2.6) and subtraction of (2.5) from (2.6) yield, after an integration by parts,

$$\begin{aligned}
0 = & \int_0^T \int_\Omega \nabla q_{j+1} \cdot \bar{b} \nabla q'_{j+1} dx\, dt + \int_0^T \int_\Omega \nabla q_{j+1} \cdot s_{j+1} \nabla u'_{j+1} dx\, dt \\
& + \int_0^T \int_\Omega \nabla q_{j+1} \cdot \bar{a} \nabla q_{j+1} dx\, dt + \int_0^T \int_\Omega \nabla q_{j+1} \cdot r_{j+1} \nabla u_{j+1} dx\, dt \\
& + \int_0^T \int_\Omega \nabla q'_{j+1} \cdot \frac{\nabla q'_{j+1} - \nabla q'_j}{h} dx\, dt + \int_\Omega \nabla q_{j+1}(T) \cdot \frac{\nabla q_{j+1}(T) - \nabla q_j(T)}{h} dx \\
\geq & \frac{c_0}{2} ||\nabla q_{j+1}(T)||^2 + c_0 \int_0^T ||\nabla q_{j+1}(T)||^2 dt \\
& + \int_\Omega \nabla q_{j+1}(T) \cdot \frac{\nabla q_{j+1}(T) - \nabla q_j(T)}{h} dx \\
& + \int_0^T \int_\Omega \nabla q_{j+1} \cdot \left\{ s_{j+1} \nabla u'_{j+1} + r_{j+1} \nabla u_{j+1} + \frac{\nabla q'_{j+1} - \nabla q'_j}{h} \right\} dx\, dt.
\end{aligned} \tag{2.9}$$

Here we have used that $\bar{a}, \bar{b} \in K(c_0, C_0)$. Next substitute $\eta = P_{A_n}\bar{a}$ into (2.7). Note that $\eta - a_{j+1} = (P_{A_n}\bar{a} - \bar{a}) - r_{j+1}$, and since $\bar{a} - P_{A_n}\bar{a} \in A_n^\perp$, we obtain from (A6):

$$0 \geq \int_\Omega (\frac{1}{h}(r_{j+1} - r_j) - \int_0^T (\nabla q_{j+1} \otimes \nabla u_{j+1}) dt)_{ik} (r_{j+1})_{ik} dx\, dt. \tag{2.10}$$

Similarly we obtain from (2.8) with $\zeta = P_{B_n}\bar{b}$:

$$0 \geq \int_\Omega \left(\frac{1}{h}(s_{j+1} - s_j) - \int_0^T \nabla q_{j+1} \otimes \nabla u'_{j+1} dt \right)_{ik} (s_{j+1})_{ik} dx\, dt. \tag{2.11}$$

Adding (2.9) – (2.11), we obtain

$$
\begin{aligned}
0 \geq\ & \frac{c_0}{2}||\nabla q_{j+1}(T)||^2 + c_0 \int_0^T ||\nabla q_{j+1}(t)||^2 dt \\
&+ \frac{1}{h}\int_\Omega \nabla q_{j+1}(T)\cdot(\nabla q_{j+1}(T) - \nabla q_j(T))dx \\
&+ \frac{1}{h}\Bigg[\int_0^T \int_\Omega \{(\nabla q'_{j+1} - \nabla q'_j)\cdot \nabla q'_{j+1} + (r_{j+1} - r_j)_{ik}(r_{j+1})_{ik} \\
&+ (s_{j+1} - s_j)_{ik}(s_{j+1})_{ik}\}dx\, dt\Bigg].
\end{aligned}
\tag{2.12}
$$

From standard arguments there follows that

$$
\Phi_j = \int_0^T ||\nabla q'_j(t)||^2 dt + ||\nabla q_j(T)||^2 + ||r_j||^2 + ||s_j||^2 \tag{2.13}
$$

is decreasing for $j \geq 0$, and, via summation, that there holds the a priori estimate

$$
\begin{aligned}
\sup_{j\in I\!N} \Phi_j + c_0 h \sum_{j=1}^{\infty} ||\nabla q_j(T)||^2 + 2c_0 h \sum_{j=1}^{\infty} \int_0^T ||\nabla q_j(t)||^2 dt \\
\leq \Phi_0 < +\infty.
\end{aligned}
\tag{2.14}
$$

Note that the bound Φ_0 depends neither on h nor on the dimension n of U_n.

Next we show

Theorem 2.1. *Let (A1) – (A9) hold, and let $n \in I\!N$ be arbitrary. Then $(\bar{P}_n)$ has at least one solution $\{(a_j, b_j, u_j)\}_{j\geq 0}$. Moreover, there exists some $h_0 > 0$ such that for $0 < h \leq h_0$ the solution is unique.*

Proof. Let $B = A_n \times B_n \times U_n$ and $C = (K_1 \cap A_n) \times (K_2 \cap B_n) \times U_n$. Then $C \neq \emptyset$ is closed and convex, and dim $B < +\infty$. Let $[\cdot,\cdot]$ denote the dual pairing between elements of B' and B, where B' is the dual of B. We conclude by induction and assume that some suitable $(a_k, b_k, u_k) \in C, 0 \leq k \leq j$, have already been constructed. To find $(a_{j+1}, b_{j+1}, u_{j+1})$ with (2.6) – (2.8), we have to show the existence of $(a, b, u) \in C$ with

$$
[T_j(a, b, u), (\eta, \zeta, v) - (a, b, u)] \geq 0,\ \forall(\eta, \zeta, v) \in C, \tag{2.15}
$$

where the operator $T_j : C \to B'$ is given by

$$
\begin{aligned}
[T_j(a,b,u),(\eta,\zeta,v)] &= \int_0^T \int_\Omega \nabla v \cdot b \nabla u' \, dx dt \\
&+ \int_0^T \int_\Omega \nabla v \cdot a \nabla u \, dx \, dt + \int_\Omega \nabla v(T) \cdot \frac{\nabla u(T) - \nabla u_j(T)}{h} dx \\
&+ \int_0^T \int_\Omega \nabla v' \cdot \frac{\nabla u' - \nabla u'_j}{h} dx \, dt - \int_0^T \int_\Omega \bar{f} v \, dx \, dt + \int_0^T \int_{\Omega_1} \bar{F} v \, ds \, dt \\
&+ \int_0^T \int_\Omega \bar{u}' v' \, dx \, dt - \int_\Omega \bar{u}'(T) v(T) dx \\
&+ \int_\Omega \left(\frac{a - a_j}{h} \right)_{ik} \eta_{ik} dx - \int_0^T \int_\Omega \nabla(u - \bar{u}) \cdot \eta \nabla u \, dx \, dt \\
&+ \int_\Omega \left(\frac{b - b_j}{h} \right)_{ik} \zeta_{ik} dx - \int_0^T \int_\Omega \nabla(u - \bar{u}) \cdot \zeta \nabla u' \, dx \, dt.
\end{aligned}
\tag{2.16}
$$

We show that T_j is coercive with respect to $(\bar{u}, P_{A_n}\bar{a}, P_{B_n}\bar{b}) \in C$. Since T_j is obviously continuous, the solvability of (2.15) then follows from standard results on variational inequalities (e.g. [8], p.14, Corollary 4.3).

Let $(a,b,u) \in C$. Recall that $P_{A_n}\bar{a} = \bar{a} + (P_{A_n}\bar{a} - \bar{a})$, where $P_{A_n}\bar{a} - \bar{a} \in A_n^\perp$, and similarly, $P_{B_n}\bar{b} = \bar{b} + (P_{B_n}\bar{b} - \bar{b})$, where $P_{B_n}\bar{b} - \bar{b} \in B_n^\perp$. Hence, owing to (A5) and (A6), we obtain from (2.16), with $q := u - \bar{u}, r := a - \bar{a}, s := b - \bar{b}$,

$$
\begin{aligned}
&[T_j(a,b,u) - T_j(\bar{u}, P_{A_n}\bar{a}, P_{B_n}\bar{b}), (a,b,u) - (\bar{u}, P_{A_n}\bar{a}, P_{B_n}\bar{b})] \\
&= [T_j(a,b,u) - T_j(\bar{u},\bar{a},\bar{b}), (a,b,u) - (\bar{u},\bar{a},\bar{b})] \\
&= \frac{1}{h}(||\nabla q(T)||^2 + \int_0^T ||\nabla q'(t)||^2 dt + ||r||^2 + ||s||^2) \\
&+ \int_0^T \int_\Omega \nabla q \cdot \bar{b} \nabla q' dx \, dt + \int_0^T \int_\Omega \nabla q \cdot \bar{a} \nabla q \, dx \, dt.
\end{aligned}
\tag{2.17}
$$

Obviously, we may define the norm of B by

$$
||(a,b,u)||_B^2 = ||\nabla u(T)||^2 + \int_0^T ||\nabla u'(t)||^2 dt + ||b||^2 + ||a||^2. \tag{2.18}
$$

But

$$
||r||^2 + ||s||^2 \geq ||a - P_{A_n}\bar{a}||^2 + ||b - P_{B_n}\bar{b}||^2, \tag{2.19}
$$

and thus (2.17) implies the coercivity, and the existence is proved.

For the uniqueness, it suffices to show that (2.15) admits for any $j \geq 0$ at most one solution. Assume $j \geq 0$ is fixed, and $(a_\nu, b_\nu, u_\nu), \nu = 1,2$ are any solutions of

(2.15). Let $a = a_1 - a_2, b = b_1 - b_2, u = u_1 - u_2$. Obviously, from (2.6),

$$\begin{aligned} 0 = &\int_0^T \int_\Omega \nabla u \cdot (b_1 \nabla u_1' - b_2 \nabla u_2') dx\, dt \\ &+ \int_0^T \int_\Omega \nabla u \cdot (a_1 \nabla u_1 - a_2 \nabla u_2) dx\, dt \\ &+ \frac{1}{h} \int_0^T ||\nabla u'(t)||^2 dt + \frac{1}{h} ||\nabla u(T)||^2. \end{aligned} \tag{2.20}$$

Equations (2.7) and (2.8) give

$$\begin{aligned} 0 \geq &\frac{1}{h} ||a||^2 - \int_0^T \int_\Omega \nabla(u_1 - \bar{u}) \cdot a \nabla u_1\, dx\, dt \\ &+ \int_0^T \int_\Omega \nabla(u_2 - \bar{u}) \cdot a \nabla u_2\, dx\, dt, \end{aligned} \tag{2.21}$$

and

$$\begin{aligned} 0 \geq &\frac{1}{h} ||b||^2 - \int_0^T \int_\Omega \nabla(u_1 - \bar{u}) \cdot b \nabla u_1'\, dx\, dt \\ &+ \int_0^T \int_\Omega \nabla(u_2 - \bar{u}) \cdot b \nabla u_2'\, dx\, dt, \end{aligned} \tag{2.22}$$

respectively. Addition of (2.20) – (2.22) and rearrangement of terms lead to

$$\begin{aligned} 0 \geq &\frac{1}{h} \left[\int_0^T ||\nabla u'(t)||^2 dt + ||\nabla u(T)||^2 + ||b||^2 \right] \\ &+ \int_0^T \int_\Omega \{ \nabla u \cdot b_1 \nabla u' + \nabla u \cdot a_1 \nabla u + \nabla u \cdot b \nabla u_2' \\ &+ \nabla u \cdot a \nabla u_2 - \nabla u \cdot a \nabla (u_1 + u_2) + \nabla u \cdot a \nabla \bar{u} \\ &+ \nabla u' \cdot b \nabla \bar{u} - \frac{1}{2} \frac{d}{dt} (\nabla u \cdot b \nabla (u_1 + u_2)) \}\, dx\, dt. \end{aligned} \tag{2.23}$$

Clearly,

$$\int_0^T \int_\Omega \{ \nabla u \cdot b_1 \nabla u' + \nabla u \cdot a_1 \nabla u \}\, dx\, dt \geq 0. \tag{2.24}$$

Moreover, for any $\eta, \xi \in I\!R^N$ and any $(N \times N)$–matrix $a = (a_{ik})$ there holds the estimate

$$|\xi \cdot a\eta| \leq \frac{1}{2} |\xi|^2 + \frac{N^2}{2} \max_{1 \leq i,k \leq N} |a_{ik}|^2 |\eta|^2, \tag{2.25}$$

and, due to the finite dimension of A_n and B_n, there exists some $C_n > 0$ such that

$$||a||^2_{L^\infty(\Omega)} \leq C_n ||a||^2, \ \forall a \in A_n \cup B_n. \tag{2.26}$$

Finally, since (a_ν, b_ν, u_ν) solve $(\bar{P}_n)$ at the index j, they satisfy the general a priori estimate (2.14). In particular, there exists some constant $\sigma > 0$ such that, for $\nu = 1, 2,$

$$\int_0^T ||\nabla u'_\nu(t)||^2 dt + ||\nabla u_\nu(T)||^2 + ||a_\nu||^2 + ||b_\nu||^2 \leq \sigma. \tag{2.27}$$

Moreover, note that $u(0) = 0$ implies that

$$\int_0^T ||\nabla u(t)||^2 dt \leq T^2 \int_0^T ||\nabla u'(t)||^2 dt. \tag{2.28}$$

Thus,

$$\int_0^T ||\nabla u_\nu(t)||^2 dt \leq T^2 \sigma, \; \nu = 1, 2. \tag{2.29}$$

We have the estimate:

$$\begin{aligned}\int_0^T \int_\Omega \nabla u \cdot b \nabla u'_2 \, dx \, dt \geq & -\frac{1}{2} \int_0^T ||\nabla u(t)||^2 dt \\ & - \frac{N^2}{2} ||b||^2_{L^\infty(\Omega)} \int_0^T ||\nabla u'_2(t)||^2 dt \\ \geq & -\frac{T^2}{2} \int_0^T ||\nabla u'(t)||^2 dt - \frac{N^2}{2} C_n \sigma ||b||^2.\end{aligned} \tag{2.30}$$

The remaining terms can be estimated similarly; since this is straightforward but lengthy, we omit the details. Finally one arrives at an inequality of the form

$$\begin{aligned}0 \geq & \left(\frac{1}{h} - \delta_1\right) \int_0^T ||\nabla u'(t)||^2 dt + \left(\frac{1}{h} - \delta_2\right) ||\nabla u(T)||^2 \\ & + \left(\frac{1}{h} - \delta_3\right) ||a||^2 + \left(\frac{1}{h} - \delta_4\right) ||b||^2,\end{aligned} \tag{2.31}$$

where the finite constants $\delta_i > 0, i = 1, 2, 3, 4$, do not depend on the index j. Taking $h > 0$ appropriately small, we obtain the result. ∎

We have also the following convergence result.

Theorem 2.2. *Let (A1) – (A9) hold, and let* $\{(a_j, b_j, u_j)\}_{j \geq 0}$ *be any solution of* $(\bar{P}_n)$. *Then we have*

$$||\nabla(u_j - \bar{u})(T)||^2 + \int_0^T ||\nabla(u'_j - \bar{u}')(t)||^2 dt \to 0 \text{ as } j \to \infty, \tag{2.32}$$

$$(a_j, b_j) \to (a_\infty, b_\infty) \text{ as } j \to \infty, \text{ where } (a_\infty, b_\infty) \text{ solves } (P_n). \tag{2.33}$$

Proof. The a priori estimate (2.14) implies (2.32) and the boundedness of $\{a_j\}$ and $\{b_j\}$ in $L^2_{\text{sym}}(\Omega)$. By the finite dimension of $A_n \times B_n$, there is some subsequence $\{(a_{j_k}, b_{j_k})\}$ with $(a_{j_k}, b_{j_k}) \to (a_\infty, b_\infty) \in A_n \times B_n$. Due to the closedness of K_1, K_2 in $L^2_{\text{sym}}(\Omega)$, we have $(a_\infty, b_\infty) \in K_1 \times K_2$. Replacing the index $j+1$ in (2.6) by j_k and passing to the limit as $k \to \infty$, we obtain, in view of (2.32), that (a_∞, b_∞) indeed satisfies (2.5) for any $v \in U_n$; i.e., (a_∞, b_∞) solves (P_n). It remains now to show that the whole sequence converges to (a_∞, b_∞), i.e., that (a_∞, b_∞) is the only limit point.

First, note that the conclusions leading to the a priori estimates (2.13), (2.14) will remain unchanged, if we weaken the assumption of$(\bar{a}, \bar{b})$ solving (P) to that of $(\bar{a}, \bar{b})$ merely solving (P_n). Then, for any such solution $(\bar{a}, \bar{b})$ of (P_n), we can conclude that the sequence $\{\Phi_j\}$ is decreasing and converges to some $\delta > 0$. By (2.32), we see that

$$||a_j - \bar{a}||^2 + ||b_j - \bar{b}||^2 \to \delta \quad \text{as} \quad j \to \infty, \tag{2.34}$$

and hence

$$||a_\infty - \bar{a}||^2 + ||b_\infty - \bar{b}||^2 = \delta \tag{2.35}$$

for any limit point (a_∞, b_∞) of $\{(a_j, b_j)\}$. Thus, if $(a^i_\infty, b^i_\infty), i = 1, 2$, are any limit points, it follows that

$$||a^1_\infty - \bar{a}||^2 + ||b^1_\infty - \bar{b}||^2 = ||a^2_\infty - \bar{a}||^2 + ||b^2_\infty - \bar{b}||^2 \tag{2.36}$$

for any solution $(\bar{a}, \bar{b})$ of (P_n). But we have just proved that any limit point, and thus (a^1_∞, b^1_∞), solves (P_n). Hence we obtain

$$0 = ||a^2_\infty - a^1_\infty||^2 + ||b^2_\infty - b^1_\infty||^2, \tag{2.37}$$

which proves the assertion. ■

Next we discuss the situation for $n \to \infty$. By Theorem 2.2, for any $n \in I\!N$ there exists an $(a^n_\infty, b^n_\infty) \in (K_1 \cap A_n) \times (K_2 \cap B_n)$ such that, for each $v \in U_n$,

$$\begin{aligned}\int_0^T \int_\Omega \nabla v \cdot b^n_\infty \nabla \bar{u}' \, dx \, dt + \int_0^T \int_\Omega \nabla v \cdot a^n_\infty \nabla \bar{u} \, dx \, dt \\ = \int_0^T \int_\Omega \bar{f} v \, dx \, dt - \int_0^T \int_{\Gamma_1} \bar{F} v \, ds \, dt + \int_0^T \int_\Omega \bar{u}' v' \, dx \, dt \\ - \int_\Omega \bar{u}'(T) v(T) \, dx.\end{aligned} \tag{2.38}$$

Since the constant Φ_0 in the a priori estimate (2.14) does not depend on n, $\{a^n_\infty\}$, $\{b^n_\infty\}$ are bounded in $L^2_{\text{sym}}(\Omega)$. Hence for a subsequence, again denoted by $\{a^n_\infty, b^n_\infty\}$, we have

$$a^n_\infty \to a^*, b^n_\infty \to b^* \quad \text{weakly in } L^2_{\text{sym}}(\Omega). \tag{2.39}$$

By (A2), K_i is weakly closed in $L^2_{\text{sym}}(\Omega)$, whence

$$(a^*, b^*) \in K_1 \times K_2. \tag{2.40}$$

Now let $v \in H^1_0(0,T;V)$ be arbitrary. By (A5), there are $v_n \in U_n$ such that $v_n \to v$ in $H^1_0(0,T;V)$. From this and the fact that by (A1) we have $\nabla \bar{u} \in L^\infty(0,T;L^\infty(\Omega))$ and $\nabla \bar{u}' \in L^\infty(0,T;L^\infty(\Omega))$, respectively, one readily deduces by passing to the limit as $n \to \infty$ that

$$\begin{aligned}\int_0^T \int_\Omega \nabla v \cdot b^* \nabla \bar{u}'\, dx\, dt &+ \int_0^T \int_\Omega \nabla v \cdot a^* \nabla \bar{u}\, dx\, dt \\ &= \int_0^T \int_\Omega \bar{f} v\, dx\, dt - \int_0^T \int_{\Gamma_1} \bar{F} v\, ds\, dt + \int_0^T \int_\Omega \bar{u}' v'\, dx\, dt \\ &- \int_\Omega \bar{u}'(T) v(T)\, dx.\end{aligned} \tag{2.41}$$

Thus we have proved

Theorem 2.3. *Let (A1) – (A9) hold, and let (a^n_∞, b^n_∞) denote a limit point of the solution of $(\bar{P}_n), n \in I\!N$. Then each subsequence of $\{(a^n_\infty, b^n_\infty)\}$ has a weak cluster point in $L^2_{\text{sym}}(\Omega)$ which is a solution of (P).*

Remarks:

1. Results corresponding to Theorems 2.1 – 2.3 for elliptic and parabolic problems have been proved in [1], [2], [6]. In neither of these papers viscosity terms were considered. However, the methods of proof presented here are close to those used in [1], [2], [6].
2. The proofs of Theorems 2.1 – 2.3 can be modified if instead of h in (2.6) – (2.8) different step sizes $h_i > 0, i = 1,2,3$, are used. In practical computations this use of different scales has proved to be very useful in order to accelerate the convergence in the parameters (a, b) at the expense of that in the state u.

3. Stability Results

Let $n \in I\!N$ be fixed, and let (A1) – (A9) hold. Again we consider the finite dimensional identification problem (P_n). We investigate the dependence of the limit points $(a_\infty, b_\infty) = (a_\infty, b_\infty)(a_0, b_0, u_0, \bar{u}, \bar{f}, \bar{F})$ upon the initial data (a_0, b_0, u_0) and the measurements $(\bar{u}, \bar{f}, \bar{F})$. We do this without information about the asymptotic behavior of (a_j, b_j), i.e., without estimates of $||a_j - a_\infty|| + ||b_j - b_\infty||$ by the data. We exploit techniques similar to those employed in [2] in a somewhat different context.

Let $\bar{M} \subset (C^1([0,T]; V) \cap U_n) \times L^2(0,T; L^2(\Omega)) \times L^2(0,T; L^2(\Gamma_1))$ denote some bounded set of measurements $(\bar{u}, \bar{f}, \bar{F})$. To any $(\bar{u}, \bar{f}, \bar{F}) \in \bar{M}$, the corresponding set of solutions of (P_n) is denoted by $L_n(\bar{u}, \bar{f}, \bar{F})$. Furthermore, let $M_0 \subset (K_1 \cap A_n) \times (K_2 \cap B_n) \times U_n$ denote some bounded set. We assume:

(A10) $L_n(\bar{u}, \bar{f}, \bar{F}) \neq \emptyset, \quad \forall (\bar{u}, \bar{f}, \bar{F}) \in \bar{M}$.

(A11) *For any sequence* $(\bar{u}_k, \bar{f}_k, \bar{F}_k) \in \bar{M}$ *such that* $\bar{u}_k \to \bar{u}, \bar{f}_k \to \bar{f}$ *weakly in* $L^2(0,T; L^2(\Omega)), \bar{F}_k \to \bar{F}$ *weakly in* $L^2(0,T; L^2(\Gamma_1))$, *there holds* $(\bar{u}, \bar{f}, \bar{F}) \in \bar{M}$, *and*

$$\operatorname{dist}((\hat{a}, \hat{b}), L_n(\bar{u}_k, \bar{f}_k, \bar{F}_k)) \to 0 \text{ as } k \to \infty, \ \forall (\hat{a}, \hat{b}) \in L_n(\bar{u}, \bar{f}, \bar{F}). \tag{3.1}$$

The last assumption means that (P_n) itself is stable. We want to show that this implies the stability of the algorithm. The result is mainly based upon

Lemma 3.1. *To To every* $\epsilon > 0$ *there exist* $\delta > 0$ *and* $N_0 \in I\!N$ *such that every solution* $\{(a_j, b_j, u_j)\}_{j \geq 0}$ *of* $(\bar{P}_n)$ *corresponding to some* $((a_0, b_0, u_0), (\bar{u}, \bar{f}, \bar{F})) \in M_0 \times \bar{M}$ *satisfies: Whenever we have*

$$\begin{aligned} \int_0^T \{||\nabla(u_j' - \bar{u}')(t)||^2 + ||\nabla(u_{j+1}' - \bar{u}')(t)||^2\}\, dt \\ + \{||\nabla(u_j - \bar{u})(T)||^2 + ||\nabla(u_{j+1} - \bar{u})(T)||^2\} \leq \delta \end{aligned} \tag{3.2}$$

for some $j \geq N_0$, *then there exists some* $(\hat{a}, \hat{b}) \in L_n(\bar{u}, \bar{f}, \bar{F})$ *with*

$$||a_{j+1} - \hat{a}||^2 + ||b_{j+1} - \hat{b}||^2 \leq \epsilon. \tag{3.3}$$

Proof. Assume the contrary. Then there exists some $\epsilon > 0$ so that to any $k \in I\!N$ there exist some solution $\{(a_{k,j}, b_{k,j}, u_{k,j})\}_{j \geq 0}$ of $(\bar{P}_n)$, corresponding to some $((a_{k,0}, b_{k,0}, u_{k,0}), (\bar{u}_k, \bar{f}_k, \bar{F}_k)) \in M_0 \times \bar{M}$, and an integer $j(k) \geq k$ such that

$$\begin{aligned} \int_0^T \{||\nabla(u_{k,j(k)}' - \bar{u}_k')(t)||^2 + ||\nabla(u_{k,j(k)+1}' - \bar{u}_k')(t)||^2\} dt \\ + \{||\nabla(u_{k,j(k)} - \bar{u}_k)(T)||^2 + ||\nabla(u_{k,j(k)+1} - \bar{u}_k)(T)||^2\} \leq \frac{1}{k}, \end{aligned} \tag{3.4}$$

but

$$||a_{k,j(k)+1} - \hat{a}||^2 + ||b_{k,j(k)+1} - \hat{b}||^2 \geq \epsilon \tag{3.5}$$

for any $(\hat{a},\hat{b}) \in L_n(\bar{u}_k, \bar{f}_k, \bar{F}_k)$. since $\bar{M}$ is bounded, we may without loss of generality assume that $\bar{u}_k \to \bar{u}$ in U_n, $\bar{f}_n \to \bar{f}$ weakly in $L^2(0,T;L^2(\Omega))$, $\bar{F}_n \to \bar{F}$ weakly in $L^2(0,T;L^2(\Gamma_1))$. By (A11), $(\bar{u},\bar{f},\bar{F}) \in \bar{M}$. By assumption (A10), $L_n(\bar{u},\bar{f},\bar{F}) \neq \emptyset$.

Now note that $\{(a_{k,j(k)+1}, b_{k,j(k)+1})\} \subset (K_1 \cap A_n) \times (K_2 \cap B_n)$ is, due to (A2), bounded. Hence, without loss of generality we may assume that

$$||a_{k,j(k)+1} - \bar{a}||^2 + ||b_{k,j(k)+1} - \bar{b}||^2 \to 0 \quad \text{as } k \to \infty \tag{3.6}$$

for suitable $(\bar{a},\bar{b}) \in (K_1 \cap A_n) \times (K_2 \cap B_n)$. We will show that

$$(\bar{a},\bar{b}) \in L_n(\bar{u},\bar{f},\bar{F}). \tag{3.7}$$

Once this is shown, the proof is complete; indeed, if (3.7) holds, there exist, by (A11), suitable $(\tilde{a}_k,\tilde{b}_k) \in L_n(\bar{u}_k,\bar{f}_k,\bar{F}_k)$ such that $(\tilde{a}_k,\tilde{b}_k) \to (\bar{a},\bar{b})$ as $k \to \infty$. But this implies, for sufficiently large k, that

$$\begin{aligned} ||a_{k,j(k)+1} - \tilde{a}_k||^2 + ||b_{k,j(k+1)} - \tilde{b}_k||^2 &\le 2||a_{k,j(k)+1} - \bar{a}||^2 \\ &+ 2||\tilde{a}_k - \bar{a}||^2 + 2||b_{k,j(k)+1} - \bar{b}||^2 + 2||\tilde{b}_k - \bar{b}||^2 < \epsilon, \end{aligned}$$

which contradicts (3.5).

To establish (3.7), we first have to show that (2.5) holds for any $v \in U_n$. By assumption, we obtain from (2.6) for any $k \in I\!N$ and $v \in U_n$ the identity

$$\begin{aligned} &\int_0^T \int_\Omega \nabla v \cdot b_{k,j(k)+1} \nabla u'_{k,j(k)+1}\, dx\, dt \\ &\quad + \int_0^T \int_\Omega \nabla v \cdot a_{k,j(k)+1} \nabla u_{k,j(k)+1}\, dx\, dt \\ &\quad + \int_0^T \int_\Omega \frac{\nabla u'_{k,j(k)+1} - \nabla u'_{k,j(k)}}{h} \cdot \nabla v'\, dx\, dt \\ &\quad + \int_\Omega \frac{\nabla u_{k,j(k)+1} - \nabla u_{k,j(k)}}{h}(T) \cdot \nabla v(T)\, dx \\ &\quad = \int_0^T \int_\Omega \bar{f}_k v\, dx\, dt - \int_0^T \int_{\Gamma_1} \bar{F}_k v\, ds\, dt - \int_\Omega \bar{u}'_k(T) v(T)\, dx \\ &\quad + \int_0^T \int_\Omega \bar{u}'_k v'\, dx\, dt. \end{aligned} \tag{3.8}$$

Now, from (3.6) and the finite dimension of A_n, B_n, it follows that

$$||a_{k,j(k)+1} - \bar{a}||^2_{L^\infty(\Omega)} + ||b_{k,j(k)+1} - \bar{b}||^2_{L^\infty(\Omega)} \to 0 \text{ as } k \to \infty. \tag{3.9}$$

Moreover, by (3.4) and since $\bar{u}_k \to \bar{u}$,

$$\int_0^T ||\nabla u'_{k,j(k)+1} - \nabla \bar{u}'(t)||^2 dt \leq 2 \int_0^T ||\nabla \bar{u}'_k(t) - \nabla \bar{u}'(t)||^2 dt$$
$$\cdot + \int_0^T ||\nabla u'_{k,j(k)+1}(t) - \nabla \bar{u}'_k(t)||^2 dt \to 0 \text{ as } k \to \infty. \tag{3.10}$$

Similarly,

$$\int_0^T ||\nabla (u_{k,j(k)} - \bar{u}')(t)||^2 dt \to 0 \text{ as } k \to \infty, \tag{3.11}$$

$$||\nabla (u)_{k,j(k)+1} - \bar{u})(T)||^2 + ||\nabla (u_{k,j(k)} - \bar{u})(T)||^2 \to 0 \text{ as } k \to \infty. \tag{3.12}$$

Finally, recall that $\bar{f}_k \to \bar{f}$ weakly in $L^2(0,T;L^2(\Omega))$, and $\bar{F}_k \to \bar{F}$ weakly in $L^2(0,T;L^2(\Gamma_1))$, respectively. Thus, passing to the limit as $k \to \infty$ in (3.8), we obtain (3.7), and the Lemma is proved. ■

We need a further preliminary:

Lemma 3.2. *There There exists some $\hat{h} \in (0, h_0)$ such that for $0 < h \leq \hat{h}$ the solution of $(\bar{P}_n)$(unique, by Theorem 2.1) depends continuously on the data $((a_0, b_0, u_0), (\bar{u}, \bar{f}, \bar{F})) \in M_0 \times \bar{M}$ up to each finite index j.*

Proof. Let $\{(a_{k,j}, b_{k,j}, u_{k,j})\}_{j \geq 0}$ solve $(\bar{P}_n)$ with respect to the data $((a_{k,0}, b_{k,0}, u_{k,0}))$, $(\bar{u}_k, \bar{f}_k, \bar{F}_k)) \in M_0 \times \bar{M}, k = 1, 2$. We put:

$$u_j = u_{1,j} - u_{2,j},\ a_j = a_{1,j} - a_{2,j},\ b_j = b_{1,j} - b_{2,j},\ j \geq 0$$

and

$$\bar{u} = \bar{u}_1 - \bar{u}_2,\ \bar{f} = \bar{f}_1 - \bar{f}_2,\ \bar{F} = \bar{F}_1 - \bar{F}_2.$$

Then for arbitrary $j \in I\!N$, one easily deduces, from (2.6)–(2.8), the inequality

$$\begin{aligned}
&\int_0^T \int_\Omega \nabla u_{j+1} \cdot (b_{1,j+1} \nabla u'_{1,j+1} - b_{2,j+1} \nabla u'_{2,j+1}) dx\, dt \\
&\quad + \int_0^T \int_\Omega \nabla u_{j+1} \cdot (a_{1,j+1} \nabla u_{1,j+1} - a_{2,j+1} \nabla u_{2,j+1}) dx\, dt \\
&\quad + \int_0^T \int_\Omega \nabla u'_{j+1} \cdot \frac{\nabla u'_{j+1} - \nabla u'_j}{h} dx\, dt \\
&\quad + \int_\Omega \nabla u_{j+1}(T) \cdot \frac{\nabla u_{j+1}(T) - \nabla u_j(T)}{h} dx \\
&\quad + \int_\Omega \left(\frac{b_{j+1} - b_j}{h}\right)_{ik} (b_{j+1})_{ik} dx - \int_0^T \int_\Omega \nabla(u_{1,j+1} - \bar{u}_1) \cdot b_{j+1} \nabla u'_{1,j+1} dx\, dt \\
&\quad + \int_0^T \int_\Omega \nabla(u_{2,j+1} - \bar{u}_2) \cdot b_{j+1} \nabla u'_{2,j+1} dx\, dt \\
&\quad + \int_\Omega \left(\frac{a_{j+1} - a_j}{h}\right)_{ik} (a_{j+1})_{ik} dx - \int_0^T \int_\Omega \nabla(u_{1,j+1} - \bar{u}_1) \cdot a_{j+1} \nabla u_{1,j+1} dx\, dt \\
&\quad + \int_0^T \int_\Omega \nabla(u_{2,j+1} - \bar{u}_2) \cdot a_{j+1} \nabla u_{2,j+1} dx\, dt \\
&\quad \le \int_0^T \int_\Omega f u_{j+1} dx\, dt - \int_0^T \int_{\Gamma_1} \bar{F} u_{j+1} ds\, dt - \int_\Omega \bar{u}'(T) u_{j+1}(T) dx \\
&\quad + \int_0^T \int_\Omega \dot{\bar{u}}' u'_{j+1} dx\, dt.
\end{aligned} \tag{3.13}$$

In view of the global a priori estimate (2.14) and due to the finite dimension of the spaces A_n, B_n, U_n, we conclude that there exists a $C_1 > 0$ (which does not depend on j) such that

$$\begin{aligned}
&\int_0^T \int_\Omega \nabla u_{j+1} \cdot (b_{1,j+1} \nabla u'_{1,j+1} - b_{2,j+1} \nabla u'_{2,j+1} dx\, dt \\
&\quad = \int_0^T \int_\Omega \nabla u_{j+1} \cdot b_{1,j+1} \nabla u'_{j+1} dx\, dt + \int_0^T \int_\Omega \nabla u_{j+1} \cdot b_{j+1} \nabla u'_{2,j+1} dx\, dt \\
&\quad \ge \frac{c_0}{2} ||\nabla u_{j+1}(T)||^2 - C_1 ||b_{j+1}||^2.
\end{aligned} \tag{3.14}$$

Similarly,

$$\begin{aligned}
&\int_0^T \int_\Omega \nabla u_{j+1} \cdot (a_{1,j+1} \nabla u_{1,j+1} - a_{2,j+1} \nabla u_{2,j+1} dx\, dt \\
&\quad \ge c_0 \int_0^T ||\nabla u_{j+1}(t)||^2 dt - C_2 ||a_{j+1}||^2,
\end{aligned} \tag{3.15}$$

and, moreover,

$$-\int_0^T \int_\Omega \nabla(u_{1,j+1} - \bar{u}_1) \cdot b_{j+1} \nabla u'_{1,j+1} dx\, dt \geq -C_3 ||b_{j+1}||^2, \tag{3.16}$$

$$\int_0^T \int_\Omega \nabla(u_{2,j+1} - \bar{u}_2) \cdot b_{j+1} \nabla u'_{2,j+1} dx\, dt \geq -C_4 ||b_{j+1}||^2, \tag{3.17}$$

$$-\int_0^T \int_\Omega \nabla(u_{1,j+1} - \bar{u}_1) \cdot a_{j+1} \nabla u_{1,j+1} dx\, dt \geq -C_5 ||a_{j+1}||^2, \tag{3.18}$$

$$\int_0^T \int_\Omega \nabla(u_{2,j+1} - \bar{u}_2) \cdot a_{j+1} \nabla u_{2,j+1} dx\, dt \geq -C_6 ||a_{j+1}||^2. \tag{3.19}$$

Thus we obtain from (3.13):

$$\begin{aligned} &\frac{1}{2h} \int_0^T ||\nabla u'_{j+1}(t)||^2 dt - \frac{1}{2h} \int_0^T ||\nabla u'_j(t)||^2 dt \\ &\quad + \frac{1}{2h} ||\nabla u_{j+1}(T)||^2 - \frac{1}{2h} ||\nabla u_j(T)||^2 \\ &\quad + \left(\frac{1}{2h} - C_1 - C_3 - C_4\right) ||b_{j+1}||^2 - \frac{1}{2h} ||b_j||^2 \\ &\quad + \left(\frac{1}{2h} - C_2 - C_5 - C_6\right) ||a_{j+1}||^2 - \frac{1}{2h} ||a_j||^2 \\ &\leq \frac{1}{2} \int_0^T ||\bar{f}(t)||^2 dt + \frac{1}{2} \int_0^T \int_{\Gamma_1} |\bar{F}|^2 ds\, dt \\ &\quad + \frac{1}{2} ||\bar{u}'(T)||^2 + \frac{1}{2} \int_0^T ||\bar{u}'(t)||^2 dt \\ &\quad + C_7 \int_0^T ||\nabla u'_{j+1}(t)||^2 dt + C_8 ||\nabla u_{j+1}(T)||^2, \end{aligned} \tag{3.20}$$

where we have applied Poincaré's and Young's inequality to the right hand side of (3.13).

Now define

$$\psi_j = \int_0^T ||\nabla u'_j(t)||^2 dt + ||\nabla u_j(T)||^2 + ||b_j||^2 + ||a_j||^2 \tag{3.21}$$

and

$$\begin{aligned} \psi = &\int_0^T ||\bar{f}(t)||^2 dt + \int_0^T \int_{\Gamma_1} |\bar{F}|^2 ds\, dt \\ &+ ||\bar{u}'(t)||^2 + \int_0^T ||\bar{u}'(t)||^2 dt. \end{aligned} \tag{3.22}$$

Then it follows that, with some $C_9 > 0$ which does not depend on j,

$$(1 - C_9 h)\psi_{j+1} \leq \psi_j + h\psi. \tag{3.23}$$

Choosing $h < \frac{1}{C_9}$, the assertion now follows from the discrete version of Gronwall's Lemma.

We are now ready to prove the main result of this section.

Theorem 3.3. *Let (A1) – (A11) hold, and let $h \in (0, h_0]$ be sufficiently small. Then the limit point (a_∞, b_∞) in Theorem 2.2 depends continuously upon the data $((a_0, b_0, u_0), (\bar{u}, \bar{f}, \bar{F})) \in M_0 \times \bar{M}$.*

Proof. Assume that $h < \frac{1}{C_9}$ (compare the proof of Lemma 3.2), and let $\{(a_{k,0}, b_{k,0}, u_{k,0})\} \subset M_0$ and $\{(\bar{u}_k, \bar{f}_k, \bar{F}_k)\} \subset \bar{M}$ denote sequences such that

$$\begin{aligned} &(a_{k,0}, b_{k,0}, u_{k,0}) \to (a_0, b_0, u_0) \in M_0, \\ &(\bar{u}_k, \bar{f}_k, \bar{F}_k) \to (\bar{u}, \bar{f}, \bar{F}) \in \bar{M} \text{ as } k \to \infty. \end{aligned} \tag{3.24}$$

The corresponding (unique) solutions of $(\bar{P}_n)$ are denoted by $\{(a_{k,j}, b_{k,j}, u_{k,j})\}_{j\geq 0}$ and $\{(a_j, b_j, u_j)\}_{j\geq 0}$, respectively; the corresponding (unique) limit points are denoted by $(a_{k,\infty}, b_{k,\infty})$, $k \in I\!N$ and (a_∞, b_∞), respectively. We have to show that

$$||a_{k,\infty} - a_\infty||^2 + ||b_{k,\infty} - b_\infty||^2 \to 0 \text{ as } k \to \infty.$$

Assume the contrary. Then there exist an $\hat{\epsilon} > 0$ and some subsequence, without loss of generality $\{(a_{k,\infty}, b_{k,\infty})\}$ itself, such that

$$||a_{k,\infty} - a_\infty||^2 + ||b_{k,\infty} - b_\infty||^2 \geq \hat{\epsilon} \quad \text{for } k \in I\!N. \tag{3.25}$$

Let $\epsilon > 0$ be arbitrary. Then there exist $\delta > 0$ and $N_0 \in I\!N$ as asserted in Lemma 3.1.

Now $u_j \to \bar{u}$ and $(a_j, b_j) \to (a_\infty, b_\infty)$, and hence there is some $N_1(\epsilon) \in I\!N, N_1(\epsilon) \geq N_0$ such that

$$\begin{aligned} &\int_0^T \{||\nabla(u_j' - \bar{u})(t)||^2 + ||\nabla(u_{j+1}' - \bar{u}')(t)||^2\}\, dt \\ &+ \{||\nabla(u_j - \bar{u})(T)||^2 + ||\nabla(u_{j+1} - \bar{u})(T)||^2\} + ||a_{j+1} - a_\infty||^2 \\ &+ ||b_{j+1} - b_\infty||^2 \leq \frac{1}{2} \min(\delta, \epsilon), \forall j \geq N_1(\epsilon). \end{aligned} \tag{3.26}$$

From Lemma 3.2, there follows that for sufficiently large $k \in I\!N$, say $k \geq k_0$, we have

$$\begin{aligned}\int_0^T &\{||\nabla(u'_{k,j} - \bar{u}'_k)(t)||^2 + ||\nabla(u'_{k,j+1} - \bar{u}'_k)(t)||^2\}dt \\ &+ \{||\nabla(u_{k,j} - \bar{u})(T)||^2 + ||\nabla(u_{k,j+1} - \bar{u})(T)||^2\} \\ &+ ||a_{k,j+1} - a_\infty||^2 + ||b_{k,j+1} - b_\infty||^2 \\ &\leq \min(\delta, \epsilon) \quad \text{for } j = N_1(\epsilon).\end{aligned} \tag{3.27}$$

From Lemma 3.1, there follows for $k \geq k_0$ the existence of some $(\hat{a}_k, \hat{b}_k) \in L_n(\bar{u}_k, \bar{f}_k, \bar{F}_k)$ such that

$$||a_{k,j+1} - \hat{a}_k||^2 + ||b_{k,j+1} - \hat{b}_k||^2 \leq \epsilon \qquad \text{for } j = N_1(\epsilon). \tag{3.28}$$

Thus, for $j = N_1(\epsilon)$ and $k \geq k_0$,

$$\begin{aligned}\Phi_{k,j+1} = \int_0^T &||\nabla(u'_{k,j+1} - \bar{u}'_k)(t)||^2 dt \\ &+ ||\nabla(u_{k,j+1} - \bar{u})(T)||^2 + ||a_{k,j+1} - \hat{a}_k||^2 \\ &+ ||b_{k,j+1} - \hat{b}_k||^2 \leq 2\epsilon.\end{aligned} \tag{3.29}$$

Due to (2.13), for any fixed $k \geq k_0$ the sequence $\{\Phi_{k,j}\}_{j \in I\!N}$ is decreasing. Hence (3.29) holds for any $k \geq k_0$ and for any $j \geq N_1(\epsilon)$. Letting $j \to \infty$, we have, for any $k \geq k_0$,.

$$||a_{k,\infty} - \hat{a}_k||^2 + ||b_{k,\infty} - \hat{b}_k||^2 \leq 2\epsilon. \tag{3.30}$$

Since $a_{k,\infty} \in A_n$ and $b_{k,\infty} \in B_n$, respectively, there follows

$$||a_{k,\infty} - P_{A_n}\hat{a}_k||^2 + ||b_{k,\infty} - P_{B_n}\hat{b}_k||^2 \leq 2\epsilon \tag{3.31}$$

with the orthogonal projections onto A_n and B_n.

Clearly, $\{(P_{A_n}\hat{a}_k, P_{B_n}\hat{b}_k)\}$ is bounded in $A_n \times B_n$. Hence, for some subsequence,

$$(P_{A_n}\hat{a}_{k_\ell}, P_{B_n}\hat{b}_{k_\ell}) \to (\hat{a}, \hat{b}) \in (K_1 \cap A_n) \times (K_2 \cap B_n). \tag{3.32}$$

From (3.31), for sufficiently large ℓ,

$$||a_{k_\ell,\infty} - \hat{a}||^2 + ||b_{k_\ell,\infty} - \hat{b}||^2 \geq 3\epsilon. \tag{3.33}$$

Moreover, due to (3.29), for sufficiently large ℓ,

$$||a_{k_\ell,j+1} - P_{A_n}\hat{a}_{k_\ell}||^2 + ||b_{k_\ell,j+1} - P_{B_n}\hat{b}_{k_\ell}||^2 \leq 2\epsilon,\ j \geq N_1(\epsilon). \tag{3.34}$$

Letting $\ell \to \infty$ for fixed $j \geq N_1(\epsilon)$ leads to

$$||a_{j+1} - \hat{a}||^2 + ||b_{j+1} - \hat{b}||^2 \leq 2\epsilon, \tag{3.35}$$

whence by letting $j \to \infty$,

$$||a_\infty - \hat{a}||^2 + ||b_\infty - \hat{b}||^2 \leq 2\epsilon. \tag{3.36}$$

Hence, by (3.33), for sufficiently large ℓ,

$$||a_{k_\ell,\infty} - a_\infty||^2 + ||b_{k_\ell,\infty} - b_\infty||^2 \leq 10\epsilon. \tag{3.37}$$

Taking $\epsilon < \frac{\hat{\epsilon}}{10}$, this contradicts (3.25) which concludes the proof. ■

4. Concluding Remarks

3. The rate of convergence of (a_j, b_j) to (a_∞, b_∞) is as yet unknown. In particular, it is still an open problem how many steps of the iteration in $(\bar{P}_n)$ have to be performed in order to guarantee the desired accuracy.
4. It is as yet unknown under which circumstances the conditions (A10), (A11) (which are crucial for the stability theory) are satisfied. For elliptic and parabolic identification problems sufficient conditions have been stated in Alt, Hoffmann & Sprekels [1] and in Hsiao & Sprekels [7], respectively.
5. The techniques presented in the preceding sections carry over to other hyperbolic identification problems of the same structure. For instance, they may be applied to the problem of reconstructing the coefficients of viscosity and elasticity from measurements of the displacement field for visco–elastic solids (see [4], for the relevant equations). The obtained results are analogous.

Acknowledgment

The second author gratefully acknowledges the stimulating atmosphere and the financial support during his visit at the Department of Mathematical Sciences of the University of Delaware.

References

[1] H.W. Alt, K.–H. Hoffmann & J. Sprekels: A numerical procedure to solve certain identification problems. In: *Intern. Ser. Numer. Math. Vol. 68*, 11–43, Birkhäuser Verlag, Basel 1984.

[2] H.W. Alt, H.–H. Hoffmann & J. Sprekels: Convergence and stability of the asymptotic regularization method for restricted parameter identification problems. Institut für Mathematik der Universität Augsburg, Preprint No. 60, Augsburg 1985.

[3] K.P. Bube & R. Burridge: The one–dimensional inverse problem of reflection seismology. *SIAM Rev.* 25 (1983), 497–559.

[4] G. Duvaut & J.L. Lions: *Inequalities in Mechanics and Physics*, Springer–Verlag, Berlin–Heidelberg–New York, 1976.
[5] K.–H. Hoffmann & J. Sprekels: On the identification of elliptic problems by asymptotic regularization, *Numer. Funct. Anal. Optim*, 7 (1984/85), 157–178.
[6] K.–H. Hoffmann & J. Sprekels: On the identification of parameters in general variational inequalities by asymptotic regularization, *SIAM J. Math. Anal*, 17 (1986), 1198–1217.
[7] G.C. Hsiao & J. Sprekels: A stability result for distributed parameter identification in bilinear systems, *Math. Meth. Appl. Sci.*, 10 (1988), 447–456.
[8] D. Kinderlehrer & G. Stampacchia: *An Introduction to Variational Inequalities and Their Applications*, Academic Press, New York–London–Toronto 1980.
[9] G.Q. Xie: A new iterative method for solving the coefficient inverse problem of the wave equation, *Comm. Pure Appl. Math.*, 39 (1986), 307–322.

E MEISTER, F PENZEL, F-O SPECK AND F S TEIXEIRA[(*)]

Two-media scattering problems in a half-space

We consider mixed boundary-transmission problems for the Helmholtz equation in a half-space, taking different wave numbers in each quadrant. Dirichlet, Neumann or mixed type boundary conditions are imposed on the half-planes $\pm x_1 > 0$, $x_2 = 0$, $x_3 \in \mathbb{R}$ and transmission conditions are prescribed in the half-plane $x_1 = 0$, $x_2 > 0$, $x_3 \in \mathbb{R}$. Those problems are seen to be well posed in the setting of finite energy norm spaces H_1, and explicit solutions are given for the Dirichlet and Neumann problems.

1 Introduction

This paper deals with the study of (canonical) diffraction problems of electromagnetic (or acoustic) waves incident from, say, a free space quadrant onto a quadrant filled with a different dielectric medium, in the presence of a ground plane. Such plane will be supposed to be either an electric (soft) or magnetic (hard) perfectly conducting surface, or still a composition of half-planes of each type.

Assuming a time-harmonic incident field, linearly polarized and parallel to the common edge of the configuration, the problems, by the use of Maxwell's equations, are formulated as mixed boundary-transmission problems for two-dimensional homogeneous Helmholtz equations, where the dependence on x_3 is dropped. Different wave numbers are given in each quadrant of $\mathbb{R}^2_+$, transmission conditions are prescribed on the interface between the two different media ($x_1 = 0$, $x_2 > 0$), and boundary conditions of Dirichlet, Neumann or mixed type are assumed on the conducting screen ($x_2 = 0$).

These problems, formulated by E. Meister in 1987 (see [7], pp. 232), remained unsolved until now. Quite recently, R. Hurd presented a closed-form solution for one particular problem of this class, the diffraction problem of a plane electromagnetic wave incident from a free space quadrant onto an anisotropic ferrite quadrant, in the presence of a perfectly conducting plane [5]. In that paper a function-theoretic method was used, and led to a scalar Wiener-Hopf equation with a rather complicated symbol. By a totally different approach, T. von Petersdorff studied a more general class of boundary-transmission problems in a Sobolev space setting, which includes the problems

(*)Sponsored by the Deutsche Forschungsgemeinschaft under grant number KO 634/32-1

stated above. This author used another method, based on boundary integral equation techniques, and proved already existence and uniqueness results to the problems under consideration.

We shall present a different operator-theoretic method for solving the boundary -value problems correspondent to the diffraction problems mentioned before, also looking for solutions in the energy-norm space H_1, but with the aim of obtaining explicit analytical representations for the scattered fields. This objective is achieved for the Dirichlet and Neumann problems (see sections 3 and 4), while for the mixed problem (section 5) we only succeed to reduce it to a vectorial Riemann-Hilbert problem in $[L_2^+(\mathbb{R})]^2$ (see (5.32)), whose explicit solution depends on the determination of a canonical generalized factorization [1], [12] of its 2×2 matrix-valued piecewise continuous symbol, for which no method seems to be available presently. Nevertheless, for a plane wave excitation function, it turned out to be possible to get the asymptotic behaviour near the origin of the solution to such a system of singular integral equations, despite of not having an explicit factorization. This is worked out in the last section, as an application of the method earlier developed by F. Penzel [13]. From the results obtained therein, a straightforward procedure, based on Abel's type theorems, could be used to determine the asymptotic behaviour of the scattered fields, which is known to be of special interest for the applications (cf. [5]).

2 Formulation of the problems and preliminary results

We shall introduce now some notation and present basic results which will be useful in the sequel.

For a well-posed formulation of the mixed boundary-transmission problems, we will use the Sobolev-Slobodeckii spaces, which naturally incorporate the finite energy condition.

Let $\mathbb{R}^2_+ = \{(x_1, x_2) \in \mathbb{R}^2 : x_2 > 0\}$ denote the upper-half space of $\mathbb{R}^2$ and consider the space $H_1(\mathbb{R}^2)$ endowed with the usual norm [3]. Furthermore, let Q_j $(j = 1, 2)$ stand for the first and second quadrants of $\mathbb{R}^2$, and consider the spaces $H_1(Q_j)$ formed by the restrictions to Q_j of all functions in $H_1(\mathbb{R}^2)$. It is well known that the trace operators

$$T_0 : H_1(Q_j) \to H_{1/2}(\partial Q_j) \quad , \quad T_0 u_j = u_{j|_{\partial Q_j}} \qquad (j = 1, 2) \tag{2.1}$$

are continuous and right-invertible operators [4].

Let $\mathbb{R}^{\pm} = \{x \in \mathbb{R} : \pm x > 0\}$ and split the boundaries ∂Q_j into their straight parts (see Figure 1)

$$\partial Q_1 = \Gamma_1 \cup \Gamma_2 \quad , \quad \partial Q_2 = \Gamma_2 \cup \Gamma_3 \quad , \tag{2.2}$$

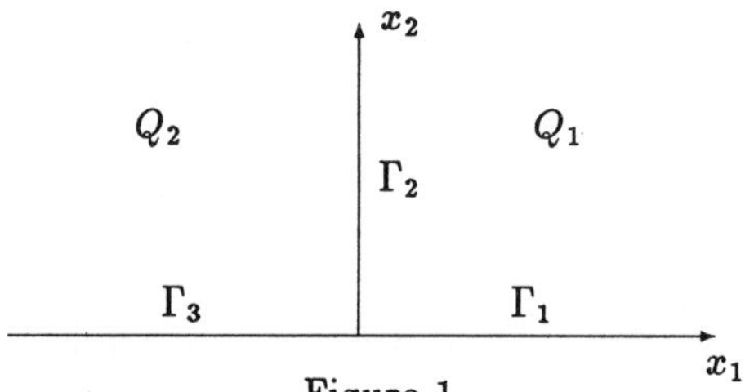

Figure 1.

taking Γ_1, Γ_2 and Γ_3 as copies of $\mathbb{R}^+$ and $\mathbb{R}^-$, respectively. We shall use the notation T_{0,Γ_i} for the (partial) trace operators defined by

$$T_{0,\Gamma_i} u = r_{\Gamma_i} T_0 u \quad , \quad i = 1, 2, 3$$

where r_{Γ_i} denotes the restriction operator to Γ_i.

Given $u_1 \in H_1(Q_1)$, let

$$f_1 = T_{0,\Gamma_1} u_1 \quad , \quad f_2^+ = T_{0,\Gamma_2} u_1 \tag{2.3}$$

We have the following characterization for the trace space $H_{1/2}(\partial Q_1)$ (see [2], [4], [14])

$$H_{1/2}(\partial Q_1) = \{(f_1, f_2^+) \in [H_{1/2}(\mathbb{R}^+)]^2 : f_1 - f_2^+ \in \widetilde{H}_{1/2}(\mathbb{R}^+)\} \tag{2.4}$$

where $\widetilde{H}_{1/2}(\mathbb{R}^+)$ denotes the subspace of $H_{1/2}(\mathbb{R}^+)$ (cf. [3]) formed by all the functions whose zero extension to $\mathbb{R}$ are in $H_{1/2}(\mathbb{R})$.

Similarly, taking $u_2 \in H_1(Q_2)$ and having

$$f_2^- = T_{0,\Gamma_2} u_2 \quad , \quad f_3 = T_{0,\Gamma_3} u_2 \tag{2.5}$$

the space $H_{1/2}(\partial Q_2)$ can be understood as

$$H_{1/2}(\partial Q_2) = \{(f_2^-, f_{3(-)}) \in [H_{1/2}(\mathbb{R}^+)]^2 : f_2^- - f_{3(-)} \in \widetilde{H}_{1/2}(\mathbb{R}^+)\} \tag{2.6}$$

(here $f_{3(-)}$ denotes the reflection of the function f_3 defined by $f_3(-x_1)$, $x_1 \in \mathbb{R}^+$ a.e.).

In the spaces $H_1(Q_j)$ it is not possible to define the traces of the normal derivatives on the boundary, directly by the trace theorem. However, if $\kappa \in \mathbb{C} \backslash \mathbb{R}$ and if we consider the subspaces

$$H_1(Q_j; \Delta; \kappa) = \{u_j \in H_1(Q_j) : (\Delta + \kappa^2) u_j \in L_2(Q_j)\}$$

for $\Delta = \frac{\partial^2}{\partial x_1^2} + \frac{\partial^2}{\partial x_2^2}$, endowed with the norm

$$\|u_j\|_{H_1(Q_j;\Delta;\kappa)} = \|u_j\|_{H_1(Q_j)} + \|(\Delta + \kappa^2)u_j\|_{L_2(Q_j)} \tag{2.7}$$

then we can define the continuous trace operators (see [14])

$$T_1 : H_1(Q_j;\Delta;\kappa) \to H_{-1/2}(\partial Q_j) \quad , \quad T_1 u_j = \frac{\partial u_j}{\partial n}\Big|_{\partial Q_j} \tag{2.8}$$

where n denotes the outer normal vector. As before, if

$$g_1 = T_{1,\Gamma_1} u_1 = \frac{\partial}{\partial(-x_2)} u_1\big|_{\Gamma_1} \quad \textit{and} \quad g_2^+ = T_{1,\Gamma_2} u_1 = \frac{\partial}{\partial(-x_1)} u_1\big|_{\Gamma_2} \tag{2.9}$$

we have the following characterization for $H_{-1/2}(\partial Q_1)$ (see also [2]):

$$H_{-1/2}(\partial Q_1) = \{(g_1, g_2^+) \in [H_{-1/2}(\mathbb{R}^+)]^2 : g_1 + g_2^+ \in \widetilde{H}_{-1/2}(\mathbb{R}^+)\} \tag{2.10}$$

where $\widetilde{H}_{-1/2}(\mathbb{R}^+)$ denotes the subspace of $H_{-1/2}(\mathbb{R}^+)$ (cf. [3]) formed by all distributions which are extendible by zero onto $\mathbb{R}$ in $H_{-1/2}(\mathbb{R})$. Analogously,

$$H_{-1/2}(\partial Q_2) = \{(g_2^-, g_{3(-)}) \in [H_{-1/2}(\mathbb{R}^+)]^2 : g_2^- + g_{3(-)} \in \widetilde{H}_{-1/2}(\mathbb{R}^+)\} \tag{2.11}$$

where $g_2^- = T_{1,\Gamma_2} u_2 = \frac{\partial u_2}{\partial x_1}\big|_{\Gamma_2}$, $g_3 = T_{1,\Gamma_3} u_2 = \frac{\partial}{\partial(-x_2)} u_2\big|_{\Gamma_3}$ and $g_{3(-)}$ represents the reflection of the distribution g_3.

Now we can formulate rigorously our mixed boundary-transmission problems. Let κ_1, κ_2 denote two complex wave numbers, such that

$$\operatorname{Re} \kappa_j > 0 \quad , \quad \operatorname{Im} \kappa_j > 0 \qquad (j = 1, 2) \tag{2.12}$$

and denote by ρ_j the transmission coefficients, which are supposed to fulfill the physically natural conditions

$$\operatorname{Re} \rho_j > 0 \quad , \quad \operatorname{Im} \rho_j \geq 0 \qquad (j = 1, 2) \tag{2.13}$$

Moreover, let $u_j(x_1, x_2)$ $(j = 1, 2)$ denote the scalar potentials associated to the diffracted fields in each quadrant Q_j. Then, we look for weak solutions $u_j \in H_1(Q_j)$ to the Helmholtz equations

$$(\Delta + \kappa_j^2)u_j = 0 \quad \text{in} \quad Q_j \qquad (j = 1, 2) \tag{2.14}$$

which satisfy the transmission conditions (at the interface Γ_2 of the two media)

$$\begin{aligned} T_{0,\Gamma_2} u_1 - T_{0,\Gamma_2} u_2 &= f_2 \\ \frac{1}{\rho_1} T_{1,\Gamma_2} u_1 + \frac{1}{\rho_2} T_{1,\Gamma_2} u_2 &= g_2 \end{aligned} \quad \text{in} \quad \Gamma_2 , \tag{2.15}$$

where $f_2 \in H_{1/2}(\mathbb{R}^+)$, $g \in H_{-1/2}(\mathbb{R}^+)$ are given distributions (related to the traces of the incident field and its normal derivative on Γ_2, respectively, (cf. [7])).

Furthermore, dependent on the type of the screen at $x_2 = 0$, we impose one of the following boundary conditions [7]:

(I) Dirichlet conditions

$$T_{0,\Gamma_1} u_1 = f_1 \quad , \quad T_{0,\Gamma_3} u_2 = f_3 \tag{2.16}$$

where $f_1 \in H_{1/2}(\mathbb{R}^+)$, $f_3 \in H_{1/2}(\mathbb{R}^-)$ are given functions.

(II) Neumann conditions

$$T_{1,\Gamma_1} u_1 = g_1 \quad , \quad T_{1,\Gamma_3} u_2 = g_3 \tag{2.17}$$

where $g_1 \in H_{-1/2}(\mathbb{R}^+)$, $g_3 \in H_{-1/2}(\mathbb{R}^-)$ are given distributions.

(III) Mixed Dirichlet/Neumann conditions

$$T_{0,\Gamma_1} u_1 = f_1 \quad , \quad T_{1,\Gamma_3} u_2 = g_3 \tag{2.18}$$

with $f_1 \in H_{1/2}(\mathbb{R}^+)$ and $g_3 \in H_{-1/2}(\mathbb{R}^-)$ given.

In the last problem, f_1 and g_3 are arbitrary data functions (distributions) in $H_{1/2}(\mathbb{R}^+)$ and $H_{-1/2}(\mathbb{R}^-)$, respectively. However, for the first two boundary-transmission problems there are some compatibility conditions which must be fulfilled, involving the transmission data f_2, g_2 and the boundary data in (2.16) and (2.17). Indeed, as a consequence of the characterization of the spaces $H_{\pm 1/2}(\partial Q_j)$ $(j = 1, 2)$ in (2.4) - (2.11), we have:

Proposition 2.1: *If there exists a solution ot the Dirichlet boundary-transmission problem* (2.14) - (2.16) *then the following compatibility condition holds*

(i) $f_1 - f_2 - f_{3(-)} \in \widetilde{H}_{-1/2}(\mathbb{R}^+)$

Similarly, if there exists a solution to the Neumann boundary-transmission problem (2.14), (2.15), (2.17), *then*

(ii) $\dfrac{1}{\rho_1} g_1 + g_2 + \dfrac{1}{\rho_2} g_{3(-)} \in \widetilde{H}_{-1/2}(\mathbb{R}^+)$

Proof: (i) Let $(u_1, u_2) \in H_1(Q_1) \times H_1(Q_2)$ be a solution to (2.14) – (2.16). Then, by use of (2.13), we have

$$f_1 - f_2 - f_{3(-)} = (f_1 - T_{0,\Gamma_2} u_1) + (T_{0,\Gamma_2} u_2 - f_{3(-)})$$

where $f_1 = T_{0,\Gamma_1} u_1$ and $f_3 = T_{0,\Gamma_3} u_2$, with $(f_1, T_{0,\Gamma_2} u_1) \in H_{1/2}(\partial Q_1)$ and $(T_{0,\Gamma_2} u_2, f_3) \in H_{1/2}(\partial Q_2)$. The result follows directly from (2.4), (2.6).
(ii) The proof is analogous to the previous one, using (2.10) and (2.11). ■

We will consider the Dirichlet, Neumann and mixed boundary-transmission problems separately, in sections 3, 4 and 5, respectively. Therein we shall use the following basic results, whose proofs can be found in [10] or [15]. For a fixed complex constant $\kappa \in \mathbb{C} \backslash \mathbb{R}$, let t denote the squareroot function

$$t(\xi) = (\xi^2 - \kappa^2)^{1/2} \quad , \quad \xi \in \mathbb{R} \tag{2.19}$$

with branch cut $\Gamma = \Gamma_+ \cup \Gamma_-$, $\Gamma_\pm = \{z \in \mathbb{C} : z = \pm\kappa \pm i\tau , \tau > 0\}$ $(\mathrm{Re}\, t(\xi) > 0 , \xi \in \mathbb{R})$.
If $f \in H_{1/2}(\mathbb{R})$ is given, the function $u \in H_1(\mathbb{R}^2_+)$ represented by

$$u(x_1, x_2) = \mathcal{F}^{-1}_{\xi \to x_1} \{\hat{f}(\xi) e^{-t(\xi) x_2}\} \quad , \quad (x_1, x_2) \in \mathbb{R}^2_+ \tag{2.20}$$

where $\hat{f}$ denote the Fourier transform

$$\hat{f}(\xi) = \mathcal{F}_{x_1 \to \xi} f(x_1) = \int_{\mathbb{R}} e^{ix_1\xi} f(x_1) dx_1 \tag{2.21}$$

is the (unique) $H_1(\mathbb{R}^2_+)$ solution of the Dirichlet problem for the Helmholtz equation $(\Delta + \kappa^2)u = 0$ in $\mathbb{R}^2_+$, with trace f on $x_1 \in \mathbb{R}$, $x_2 = 0$ (see [10], [15]). Formula (2.20) defines a continuous potential operator

$$K_{D,\mathbb{R}^2_+}(\kappa) : H_{1/2}(\mathbb{R}) \to H_1(\mathbb{R}^2_+), \ \left(K_{D,\mathbb{R}^2_+}(\kappa) f\right)(x_1, x_2) = \mathcal{F}^{-1}_{\xi \to x_1} \left\{\hat{f}(\xi)\, e^{-t(\xi) x_2}\right\} \tag{2.22}$$

which is left invertible by the trace operator
$\gamma_0 : u \in H_1(\mathbb{R}^2_+) \mapsto \gamma_0 u = u(\cdot, x_2)\big|_{x_2=0} \in H_{1/2}(\mathbb{R})$ (cf. [3]).

Analogously, if $\mathbb{R}^2_r = \{(x_1, x_2) \in \mathbb{R}^2 : x_1 > 0\}$ and $\mathbb{R}^2_\ell = \mathbb{R}^2 \backslash \overline{\mathbb{R}^2_r}$ denote the right and left half-planes, respectively, we introduce the potential operators

$$K_{D,\mathbb{R}^2_r}(\kappa) : H_{1/2}(\mathbb{R}) \to H_1(\mathbb{R}^2_r), \ \left(K_{D,\mathbb{R}^2_r}(\kappa) f\right)(x_1, x_2) = \mathcal{F}^{-1}_{\xi \to x_2} \left\{\hat{f}(\xi)\, e^{-t(\xi) x_1}\right\} \tag{2.23}$$

$$K_{D,\mathbb{R}^2_\ell}(\kappa) : H_{1/2}(\mathbb{R}) \to H_1(\mathbb{R}^2_\ell), \ \left(K_{D,\mathbb{R}^2_\ell}(\kappa) f\right)(x_1, x_2) = \mathcal{F}^{-1}_{\xi \to x_2} \left\{\hat{f}(\xi)\, e^{t(\xi) x_1}\right\} \tag{2.24}$$

which give the H_1–solutions of the Helmholtz equation $(\Delta+\kappa^2)u=0$ in $\mathbb{R}^2_r$ and $\mathbb{R}^2_\ell$, respectively, with Dirichlet data f on $x_1=0$, $x_2\in\mathbb{R}$.

Moreover, we shall also use the Neumann potential operators (see [10], [15])

$$K_{N,\mathbb{R}^2_+}(\kappa):H_{-1/2}(\mathbb{R})\to H_1(\mathbb{R}^2_+),\ \left(K_{N,\mathbb{R}^2_+}(\kappa)g\right)(x_1,x_2)=\mathcal{F}^{-1}_{\xi\to x_1}\left\{\frac{\hat{g}(\xi)}{t(\xi)}e^{-t(\xi)x_2}\right\} \quad (2.25)$$

which maps continuously the Neumann data $g\in H_{-1/2}(\mathbb{R})$ into the $H_1(\mathbb{R}^2_+)$ solution of the Helmholtz equation in $\mathbb{R}^2_+$. Further

$$K_{N,\mathbb{R}^2_r}(\kappa):H_{-1/2}(\mathbb{R})\to H_1(\mathbb{R}^2_r),\ \left(K_{N,\mathbb{R}^2_r}(\kappa)g\right)(x_1,x_2)=\mathcal{F}^{-1}_{\xi\to x_2}\left\{\frac{\hat{g}(\xi)}{t(\xi)}e^{-t(\xi)x_1}\right\} \quad (2.26)$$

$$K_{N,\mathbb{R}^2_\ell}(\kappa):H_{-1/2}(\mathbb{R})\to H_1(\mathbb{R}^2_\ell),\ \left(K_{N,\mathbb{R}^2_\ell}(\kappa)g\right)(x_1,x_2)=\mathcal{F}^{-1}_{\xi\to x_2}\left\{\frac{\hat{g}(\xi)}{t(\xi)}e^{t(\xi)x_1}\right\} \quad (2.27)$$

give the H_1–solution of the Helmholtz equation in $\mathbb{R}^2_r$ and $\mathbb{R}^2_\ell$, respectively, in dependence of the Neumann data g on $x_1=0$, $x_2\in\mathbb{R}$.

3 The Dirichlet boundary-transmission problems

With the notations introduced in the previous section, let us consider the Dirichlet boundary-transmission problem (2.12) – (2.16).

Problem $\mathcal{P}_D$: Find a weak solution $(u_1,u_2)\in H_1(Q_1)\times H_1(Q_2)$ to the Helmholtz equations

$$(\Delta+k_j^2)u_j=0 \quad \text{in} \quad Q_j \qquad (j=1,2) \quad (3.1)$$

which satisfies the boundary conditions

$$T_{0,\Gamma_1}u_1=f_1 \quad , \quad T_{0,\Gamma_3}u_2=f_3 \quad (3.2)$$

and the transmission conditions

$$f_2^+-f_2^- = f_2 \quad (3.3)$$

$$\frac{1}{\rho_1}g_2^+ +\frac{1}{\rho_2}g_2^- = g_2 \quad (3.4)$$

where $f_2^+=T_{0,\Gamma_2}u_1$, $f_2^-=T_{0,\Gamma_2}u_2$, $g_2^+=\Gamma_{1,\Gamma_2}u_1$ and $g_2^-=T_{1,\Gamma_2}u_2$.

The known data functions

$$f_1,f_2\in H_{1/2}(\mathbb{R}^+) \quad , \quad f_3\in H_{1/2}(\mathbb{R}^-)$$

are assumed to fulfil the compatibility condition (see Proposition 2.1 (i))

$$f_1 - f_2 - f_{3(-)} \in \widetilde{H}_{1/2}(\mathbb{R}^+) \quad , \tag{3.5}$$

and $g_2 \in H_{-1/2}(\mathbb{R}^+)$ is a given distribution.

We will prove the existence and uniqueness of solution to the above problem, getting simultaneously an explicit analytical formula for its solution. For this purpose, we start with reducing Problem $\mathcal{P}_D$ to a simpler equivalent one, introducing the functions (see (2.22))

$$\widetilde{u}_1 = u_1 - r_{Q_1} K_{D,\mathbb{R}^2_+}(\kappa_1)\,\ell^e f_1 \tag{3.6}$$

$$\widetilde{u}_2 = u_2 - r_{Q_2} K_{D,\mathbb{R}^2_+}(\kappa_2)\ell^e f_3 \tag{3.7}$$

where r_{Q_j} $(j = 1, 2)$ are the operators of restriction to Q_j and $\ell^e : H_{1/2}(\mathbb{R}^{\pm}) \to H_{1/2}(\mathbb{R})$ denote the operators of even extension, given almost everywhere by

$$\ell^e f(x) = \begin{cases} f(x) & , \quad x \in \mathbb{R}^{\pm} \\ f(-x) & , \quad x \in \mathbb{R}^{\mp} \end{cases} \tag{3.8}$$

which are well defined and continuous [9].

If (u_1, u_2) is a solution of Problem $\mathcal{P}_D$, then it is easily seen that

$$T_{0,\Gamma_1}\widetilde{u}_1 = 0 \quad \text{and} \quad T_{0,\Gamma_3}\widetilde{u}_2 = 0 \tag{3.9}$$

Furthermore, let

$$\widetilde{f}_2^+ = T_{0,\Gamma_2}\widetilde{u}_1 = f_2^+ - C_0(\kappa_1)\ell^e f_1 \tag{3.10}$$

and

$$\widetilde{f}_2^- = T_{0,\Gamma_2}\widetilde{u}_2 = f_2^- - C_0(\kappa_2)\ell^e f_1 \tag{3.11}$$

where $f_2^+ = T_{0,\Gamma_2} u_1$, $f_2^- = T_{0,\Gamma_2} u_2$ and $C_0(\kappa_j)$ $(j = 1, 2)$ denotes the so-called operator "around the corner" (cf. [8])

$$C_0(\kappa_j) = T_{0,\Gamma_2} r_{Q_j} K_{D,\mathbb{R}^2_+}(\kappa_j) : H_{1/2}(\mathbb{R}) \to H_{1/2}(\mathbb{R}^+) \tag{3.12}$$

given by (see (2.22))

$$C_0(\kappa_j)\, f(x) = \frac{1}{2\pi}\int_{\mathbb{R}} \hat{f}(\xi)\, e^{-t_j(\xi)x}\, d\xi \quad , \quad x > 0 \tag{3.13}$$

with $t_j(\xi) = (\xi^2 - \kappa_j^2)^{1/2}$ (see (2.19)). We note that $\widetilde{f}_2^{\pm} \in \widetilde{H}_{1(2}(\mathbb{R}^+)$, due to (3.9), (2.4) and (2.6), as $(0, \widetilde{f}_2^+) \in H_{1/2}(\partial Q_1)$ and $(\widetilde{f}_2^-, 0) \in H_{1/2}(\partial Q_2)$.

Similarly, we define (see (3.6), (3.7))

$$\widetilde{g}_2^+ = T_{1,\Gamma_2}\, \widetilde{u}_1 = g_2^+ - T_{1,\Gamma_2}\, r_{Q_1} K_{D,\mathbb{R}^2_+}(\kappa_1)\ell^e f_1 = g_2^+ \tag{3.14}$$

and

$$\widetilde{g}_2^- = T_{1,\Gamma_2}\, \widetilde{u}_2 = g_2^- - T_{1,\Gamma_2}\, r_{Q_2} K_{D,\mathbb{R}^2_+}(\kappa_2)\ell^e f_1 = g_2^- \tag{3.15}$$

The last equality in both expressions is a consequence of the even property of $\ell^e f_1$. Indeed, taking f_1 in a dense subspace of smooth functions, it follows from (2.22) that

$$(T_{1,\Gamma_2} r_{Q_j} K_{D,\mathbb{R}^2_+}(\kappa_j)\ell^e f_1)(x_2) = \frac{(-1)^{j+1}}{2\pi} \int_{\mathbb{R}} i\xi\, \widehat{\ell^e f_1}(\xi)\, e^{-t_j(\xi)x_2}\, d\xi = 0$$

for $j = 1, 2$.

Therefore, we have obtained the following equivalent boundary-transmission problem:

Problem $\mathcal{P}_D^0$: Find $(\widetilde{u}_1, \widetilde{u}_2) \in H_1(Q_1) \times H_1(Q_2)$ such that

$$(\Delta + \kappa_j^2)\, \widetilde{u}_j = 0 \quad \text{in} \quad Q_j \qquad (j = 1, 2) \tag{3.16}$$

which satisfies the boundary conditions (see (3.9))

$$T_{0,\Gamma_1}\, \widetilde{u}_1 = 0 \quad \text{and} \quad T_{0,\Gamma_3}\, \widetilde{u}_2 = 0 \tag{3.17}$$

and the transmission conditions (see (3.10) – (3.15))

$$\widetilde{f}_2^+ - \widetilde{f}_2^- \;=\; \widetilde{f}_2 \tag{3.18}$$

$$\frac{1}{\rho_1}\widetilde{g}_2^+ + \frac{1}{\rho_2}\widetilde{g}_2^- \;=\; g_2 \tag{3.19}$$

where $\widetilde{f}_2 = f_2 - \Big(C_0(\kappa_1)\ell^e f_1 - C_0(\kappa_2)\ell^e f_3\Big) \in \widetilde{H}_{1/2}(\mathbb{R}^+)$ and $g_2 \in H_{-1/2}(\mathbb{R}^+)$ are given functions (distributions).

It is a crucial fact to have homogeneous boundary conditions in Problem $\mathcal{P}_D^0$. Taking, for instance, $\widetilde{u}_1 \in H_1(Q_1)$, because we impose that $\widetilde{u}_1$ has a zero trace on Γ_1, its odd extension to the fourth quadrant is a function in $H_1(\mathbb{R}^2_r)$, which also satisfies the Helmholtz equation in $\mathbb{R}^2_r$ (cf. [18, §3]). Moreover, its trace on $x_1 = 0$, $x_2 \in \mathbb{R}$ coincides with the odd extension of $\widetilde{f}_2^+$ onto $\mathbb{R}$ (see (3.10)), which we will denote by $\ell^o \widetilde{f}_2^+$. This function is an element of $H_{1/2}(\mathbb{R})$, since $\widetilde{f}_2 \in \widetilde{H}_{1/2}(\mathbb{R}^+)$. In fact, by writing $\ell^o = \ell^e - 2\ell_0$, where $\ell_0 : \widetilde{H}_{1/2}(\mathbb{R}^+) \to H_{1/2}(\mathbb{R})$ is the operator of extension by zero, it comes out from (3.8) that $\ell^o \widetilde{f}_2^+ \in H_{1/2}(\mathbb{R})$.

Remark: In general, the odd extension of an element of $H_{1/2}(\mathbb{R}^+)$ is not in $H_{1/2}(\mathbb{R})$. However, the odd extension operator

$$\ell^o g(x) = \begin{cases} g(x) & , \quad x \in \mathbb{R}^\pm \\ -g(-x) & , \quad x \in \mathbb{R}^\mp \end{cases} \tag{3.20}$$

is well defined in $H_{-1/2}(\mathbb{R}^\pm)$, and continuous (see [9] for details).

Then, from the above considerations, it is clear that $\widetilde{u}_1$ has the following representation formula

$$\widetilde{u}_1 = r_{Q_1} K_{D,\mathbb{R}^2_\tau}(\kappa_1)\ell^o \widetilde{f}_2^+ \tag{3.21}$$

By analogous considerations for $\widetilde{u}_2$, we also have

$$\widetilde{u}_2 = r_{Q_2} K_{D,\mathbb{R}^2_\ell}(\kappa_2)\ell^o \widetilde{f}_2^- \tag{3.22}$$

with the potential operators defined in (2.23), (2.24).

The explicit representation for the solution of Problem $\mathcal{P}_D^0$ (and then, by (3.6), (3.7)), to our original problem $\mathcal{P}_D$) can now be obtained by noting that (3.21), (3.22) imply (see (3.14), (3.15))

$$\ell^o \widetilde{g}_2^+ = A_1^{-1} \ell^o \widetilde{f}_2^+ \quad \text{and} \quad \ell^o \widetilde{g}_2^- = A_2^{-1} \ell^o \widetilde{f}_2^- \tag{3.23}$$

where

$$A_j = \mathcal{F}^{-1} \frac{1}{t_j} \mathcal{F} : H_{-1/2}(\mathbb{R}) \to H_{1/2}(\mathbb{R}) \qquad (j = 1, 2) \tag{3.24}$$

are invertible convolution (or pseudodifferential) operators with inverses given by $A_j^{-1} = \mathcal{F}^{-1} t_j \mathcal{F}$.

From the transmission conditions (3.18) – (3.19) and (3.23) it follows that:

$$\begin{pmatrix} \ell^o \widetilde{f}_2 \\ \\ \ell^o g_2 \end{pmatrix} = B_D \begin{pmatrix} \ell^o \widetilde{f}_2{}^+ \\ \\ \ell^o \widetilde{f}_2{}^- \end{pmatrix} = \begin{bmatrix} 1 & -1 \\ \\ \dfrac{1}{\rho_1} A_1^{-1} & \dfrac{1}{\rho_2} A_2^{-1} \end{bmatrix} \begin{pmatrix} \ell^o \widetilde{f}_2{}^+ \\ \\ \ell^o \widetilde{f}_2{}^- \end{pmatrix} \tag{3.25}$$

where the boundary operator $B_D : [H_{1/2}(\mathbb{R})]^2 \to H_{1/2}(\mathbb{R}) \times H_{-1/2}(\mathbb{R})$ is an invertible pseudodifferential operator on the line, with inverse given by

$$B_D^{-1} = \mathcal{F}^{-1} \begin{bmatrix} \dfrac{t_2/\rho_2}{t_1/\rho_1 + t_2/\rho_2} & \dfrac{1}{t_1/\rho_1 + t_2/\rho_2} \\ \dfrac{-t_1/\rho_1}{t_1/\rho_1 + t_2/\rho_2} & \dfrac{1}{t_1/\rho_1 + t_2/\rho_2} \end{bmatrix} \mathcal{F} \tag{3.26}$$

(note that $t_1/\rho_1 + t_2/\rho_2 \neq 0$, $\xi \in \mathbb{R}$ due to (2.12), (2.13) and (2.19)).

Therefore, we obtained the following

Theorem 3.1: *Problem $\mathcal{P}_D$ is uniquely solvable. Its solution is given by*

$$\begin{pmatrix} u_1 \\ u_2 \end{pmatrix} = \begin{bmatrix} r_{Q_1} K_{D,\mathbb{R}^2_+}(\kappa_1)\ell^e & 0 \\ 0 & r_{Q_2} K_{D,\mathbb{R}^2_+}(\kappa_2)\ell^e \end{bmatrix} \begin{pmatrix} f_1 \\ f_3 \end{pmatrix} +$$

$$+ \begin{bmatrix} r_{Q_1} K_{D,\mathbb{R}^2_\tau}(\kappa_1) & 0 \\ 0 & r_{Q_2} K_{D,\mathbb{R}^2_\ell}(\kappa_2) \end{bmatrix} B_D^{-1}\ell^o \begin{pmatrix} \widetilde{f}_2 \\ g_2 \end{pmatrix} \tag{3.27}$$

with $\widetilde{f}_2 = f_2 - (C_0(\kappa_1)\ell^e f_1 - C_0(\kappa_2)\ell^e f_3) \in \widetilde{H}_{1/2}(\mathbb{R}^+)$.

Proof: By direct computation, using (3.6), (3.7) and the representation formulas (3.21), (3.22) together with (3.25) one assembles that final formula. ∎

To conclude this section we like to point out that (3.27) defines the resolvent operator to Problem $\mathcal{P}_D$, which maps continuously the data space into $H_1(Q_1) \times H_1(Q_2)$, yielding the well-posedness of the problem in this natural space setting.

Thus we see afterwards that the spaces $H_1(Q_j; \Delta; \kappa)$ can be avoided, since the representation of weak solutions by potential operators yields sufficient regularity for the existence of normal derivatives in the above sense.

Moreover, having an explicit representation for the solution, it is possible to obtain easily further information on it, for instance the asymptotics for the near and far-fields, which is of great interest in applications [5]. Such behaviour could be carried out by use of standard methods of asymptotic analysis.

4 The Neumann boundary-transmission problem

In this section we consider the Neumann boundary-transmission problem (2.12) – (2.15), (2.17), and prove the well-posedness of this problem, when the compatibility condition in Proposition 2.1 is fulfilled.

Problem $\mathcal{P}_N$: Find a weak solution $(u_1, u_2) \in H_1(Q_1) \times H_1(Q_2)$ to the Helmholtz equations

$$(\Delta + k_j^2)u_j = 0 \quad \text{in} \quad Q_j \qquad (j = 1, 2) \tag{4.1}$$

which satisfies the boundary conditions

$$T_{1,\Gamma_1} u_1 = g_1 \quad , \quad T_{1,\Gamma_3} u_2 = g_3 \tag{4.2}$$

and the transmission conditions

$$f_2^+ - f_2^- = f_2 \tag{4.3}$$

$$\frac{1}{\rho_1} g_2^+ + \frac{1}{\rho_2} g_2^- = g_2 \tag{4.4}$$

where $f_2^+ = T_{0,\Gamma_2} u_1$, $f_2^- = T_{0,\Gamma_2} u_2$, $g_2^+ = T_{1,\Gamma_2} u_1$ and $g_2^- = T_{1,\Gamma_2} u_2$.

The known distributions

$$g_1, g_2 \in H_{-1/2}(\mathbb{R}^+) \quad , \quad g_3 \in H_{-1/2}(\mathbb{R}^-) \tag{4.5}$$

are assumed to satisfy the compatibility condition

$$\frac{1}{\rho^1} g_1 + g_2 + \frac{1}{\rho^2} g_3(-) \in \widetilde{H}_{-1/2}(\mathbb{R}^+) \tag{4.6}$$

and $f_2 \in H_{1/2}(\mathbb{R}^+)$ is a given function.

To obtain an explicit analytical representation for the solution, we start, as before, by reducing Problem $\mathcal{P}_N$ to an equivalent one, with homogeneous boundary conditions in Γ_1 and Γ_3.

Let

$$\widetilde{u}_1 = u_1 - r_{Q_1} K_{N,\mathbb{R}^2_+}(\kappa_1)\ell^o g_1 \tag{4.7}$$

$$\widetilde{u}_2 = u_2 - r_{Q_2} K_{N,\mathbb{R}^2_+}(\kappa_2)\ell^o g_3 \tag{4.8}$$

define the substituted functions (see (2.25), (3.20)). If (u_1, u_2) is a solution to Problem $\mathcal{P}_D$, we clearly have $(\widetilde{u}_1, \widetilde{u}_2) \in H_1(Q_1) \times H_1(Q_2)$ and

$$T_{1,\Gamma_1} \widetilde{u}_1 = 0 \quad , \quad T_{1,\Gamma_3} \widetilde{u}_2 = 0 \tag{4.9}$$

Now, denote by $\widetilde{f_2^\pm}$ the traces of $\widetilde{u}_1, \widetilde{u}_2$, on Γ_2, respectively. We obtain from (4.7), (4.8)

$$\widetilde{f_2^+} = T_{0,\Gamma_2} \widetilde{u}_1 = f_2^+ - T_{0,\Gamma_2} r_{Q_1} K_{N,\mathbb{R}^2_+}(\kappa_1) \ell^o g_1 = f_2^+ \tag{4.10}$$

and

$$\widetilde{f_2^-} = T_{0,\Gamma_2} \widetilde{u}_2 = f_2^- - T_{0,\Gamma_2} r_{Q_2} K_{N,\mathbb{R}^2_+}(\kappa_2) \ell^o g_3 = f_2^- \tag{4.11}$$

because of the odd character of $\ell^o g_1, \ell^o g_3$ (see (2.25)).

Similarly, we consider the distributions

$$\widetilde{g_2^+} = T_{1,\Gamma_2} \widetilde{u}_1 = g_2^+ - C_1(\kappa_1) \ell^o g_1 \tag{4.12}$$

$$\widetilde{g_2^-} = T_{1,\Gamma_2} \widetilde{u}_2 = g_2^- + C_1(\kappa_2) \ell^o g_3 \tag{4.13}$$

where $C_1(\kappa_j)$ $(j = 1, 2)$ denotes the operator "around the corner" (cf. [8])

$$C_1(\kappa_j) = T_{1,\Gamma_2} r_{Q_j} K_{N,\mathbb{R}^2_+}(\kappa_j) : H_{-1/2}(\mathbb{R}) \to H_{-1/2}(\mathbb{R}^+) \tag{4.14}$$

given by (see (2.25), (3.13))

$$C_1(\kappa_j)\, g(x) = \frac{i}{2\pi} \int_{\mathbb{R}} \frac{\xi}{t_j(\xi)} \hat{g}(\xi)\, e^{-t_j(\xi) x}\, d\xi \quad , \quad x > 0 \tag{4.15}$$

Note that $\widetilde{g_2^\pm} \in \widetilde{H}_{-1/2}(\mathbb{R}^+)$ due to (4.9), (2.10) and (2.11), since $(0, \widetilde{g_2^+}) \in H_{-1/2}(\partial Q_1)$ and $(\widetilde{g_2^-}, 0) \in H_{-1/2}(\partial Q_2)$.

The boundary-transmission Problem $\mathcal{P}_N$ is then equivalent to the following one:

Problem $\mathcal{P}_N^0$: Find $(\widetilde{u}_1, \widetilde{u}_2) \in H_1(Q_1) \times H_1(Q_2)$ such that

$$(\Delta + \kappa_j^2) \widetilde{u}_j = 0 \quad \text{in} \quad Q_j \qquad (j = 1, 2) \tag{4.16}$$

which satisfies the boundary conditions (see (4.9))

$$T_{1,\Gamma_1} \widetilde{u}_1 = 0 \quad \text{and} \quad T_{1,\Gamma_3} \widetilde{u}_2 = 0 \tag{4.17}$$

and the transmission conditions (see (4.10) - (4.15))

$$\widetilde{f}_2^+ - \widetilde{f}_2^- = f_2 \tag{4.18}$$

$$\frac{1}{\rho_1}\widetilde{g}_2^+ + \frac{1}{\rho_2}\widetilde{g}_2^- = \widetilde{g}_2 \tag{4.19}$$

where $f_2 \in H_{1/2}(\mathbb{R}^+)$ and $\widetilde{g}_2 = g_2^- - \left(\frac{1}{\rho_1} C_1(\kappa_1)\ell^o g_1 - \frac{1}{\rho_2} C_1(\kappa_2)\ell^o g_3\right) \in \widetilde{H}_{-1/2}(\mathbb{R}^+)$ are given functions (distributions).

Analogous considerations to those made immediately after the statement of Problem $\mathcal{P}_D^0$ yield the following representation formula for the solution $(\widetilde{u}_1, \widetilde{u}_2)$ to Problem $\mathcal{P}_N^0$ (see also (2.26), (2.27)):

$$\widetilde{u}_1 = r_{Q_1} K_{N,\mathbb{R}_r^2}(\kappa_1)\ell^e \widetilde{g}_2^+ \tag{4.20}$$

$$\widetilde{u}_2 = r_{Q_2} K_{N,\mathbb{R}_\ell^2}(\kappa_2)\ell^e \widetilde{g}_2^- \tag{4.21}$$

We note that $\ell^e \widetilde{g}_2^\pm \in H_{-1/2}(\mathbb{R})$ because $\widetilde{g}_2^\pm \in \widetilde{H}_{-1/2}(\mathbb{R}^+)$. Moreover the homogeneous boundary conditions (4.17) are automatically satisfied by $\widetilde{u}_1, \widetilde{u}_2$ in (4.20), (4.21) due to the even property of $\ell^e \widetilde{g}_2^\pm$. Taking the trace on $x_1 = 0$ of $\widetilde{u}_1, \widetilde{u}_2$ given by (4.20), (4.21), it is straightforward to verify that

$$\ell^e \widetilde{f}_2^+ = A_1 \ell^e \widetilde{g}_2^+ \quad \text{and} \quad \ell^e \widetilde{f}_2^- = A_2 \ell^e \widetilde{g}_2^- \tag{4.22}$$

hold (see (3.24)).

Therefore, by using the transmission conditions (4.18), (4.19) and the above relations, we get

$$\begin{pmatrix} \ell^e f_2 \\ \ell^e \widetilde{g}_2 \end{pmatrix} = B_N \begin{pmatrix} \ell^e \widetilde{g}_2^+ \\ \ell^e \widetilde{g}_2^- \end{pmatrix} = \begin{bmatrix} A_1 & -A_2 \\ \dfrac{1}{\rho_1} & \dfrac{1}{\rho_2} \end{bmatrix} \begin{pmatrix} \ell^e \widetilde{g}_2^+ \\ \ell^e \widetilde{g}_2^- \end{pmatrix} \tag{4.23}$$

where the boundary operator $B_N : [H_{-1/2}(\mathbb{R})]^2 \to H_{1/2}(\mathbb{R}) \times H_{-1/2}(\mathbb{R})$ is an invertible pseudodifferential operator on the line, with inverse given by

$$B_N^{-1} = \mathcal{F}^{-1} \begin{bmatrix} \dfrac{t_1 t_2/\rho_2}{t_1/\rho_1 + t_2/\rho_2} & \dfrac{t_1}{t_1/\rho_1 + t_2/\rho_2} \\ \dfrac{-t_1 t_2/\rho_1}{t_1/\rho_1 + t_2/\rho_2} & \dfrac{t_2}{t_1/\rho_1 + t_2/\rho_2} \end{bmatrix} \mathcal{F} \tag{4.24}$$

Thus, the (unique) solution of Problem $\mathcal{P}_N^0$ is given by (4.20), (4.21), with $(\ell^e\widetilde{g_2^+}, \ell^e\widetilde{g_2^-})^T = B_N^{-1}(\ell^e f_2, \ell^e\widetilde{g}_2)$. Consequently, from (4.7) and (4.8) we have the following result:

Theorem 4.1: *Problem $\mathcal{P}_N$ is uniquely solvable. Its solution is given by*

$$\begin{pmatrix} u_1 \\ u_2 \end{pmatrix} = \begin{bmatrix} r_{Q_1} K_{N,\mathbb{R}^2_+}(\kappa_1)\ell^o & 0 \\ 0 & r_{Q_2} K_{N,\mathbb{R}^2_+}(\kappa_2)\ell^o \end{bmatrix} \begin{pmatrix} g_1 \\ g_3 \end{pmatrix} +$$

$$+ \begin{bmatrix} r_{Q_1} K_{N,\mathbb{R}^2_r}(\kappa_1) & 0 \\ 0 & r_{Q_2} K_{N,\mathbb{R}^2_\ell}(\kappa_2) \end{bmatrix} B_N^{-1}\ell^e \begin{pmatrix} f_2 \\ \widetilde{g}_2 \end{pmatrix} \qquad (4.25)$$

with $\widetilde{g}_2 = g_2^- - \left(\frac{1}{\rho_1} C_1(\kappa_1)\ell^o g_1 - \frac{1}{\rho_2} C_1(\kappa_2)\ell^o g_3\right) \in \widetilde{H}_{-1/2}(\mathbb{R}^+)$.

Again we emphasize that (4.25) defines a continuous mapping from the data space into $H_1(Q_1) \times H_1(Q_2)$, which shows the well-posedness of Problem $\mathcal{P}_N$.

5 The mixed boundary-transmission problem

In the last two sections we solved explicitly the Dirichlet and Neumann boundary-transmission problems $\mathcal{P}_D$ and $\mathcal{P}_N$, by reducing them to certain boundary pseudodifferential (or convolution) operators on the real line (see (3.25), (4.22)) whose inverse operators were obtained just be the computation of the inverses of the correspondent 2×2 symbols.

In this section we study the mixed boundary-transmission problem (2.12) – (2.15), (2.18). For such a problem we prove the possibility of reducing it to a particular Riemann-Hilbert problem in $[L_2^+(\mathbb{R})]^2$ with a piecewise continuous 2×2 matrix-valued presymbol. Its explicit solution can be obtained through the determination of a generalized factorization [1] of the presymbol, which is seen to be L_2-non-singular, with zero total index (see Proposition 5.2). However, no method seems to be available presently to determine the explicit factors, with exception to the particular case where we consider equal wave numbers ($\kappa_1 = \kappa_2$) (see Remark 5.3).

Problem $\mathcal{P}_M$: Find a weak solution $(u_1, u_2) \in H_1(Q_1) \times H_1(Q_2)$ to the Helmholtz equations

$$(\Delta + \kappa_j^2) u_j = 0 \quad \text{in} \quad Q_j \tag{5.1}$$

which satisfies the boundary conditions (of mixed type)

$$T_{0,\Gamma_1} u_1 = f_1 \quad , \quad T_{1,\Gamma_3} u_2 = g_3 \tag{5.2}$$

for given $f_1 \in H_{1/2}(\mathbb{R}^+)$, $g_3 \in H_{-1/2}(\mathbb{R}^-)$, and the transmission conditions

$$f_2^+ - f_2^- = f_2 \tag{5.3}$$

$$\frac{1}{\rho_1} g_2^+ + \frac{1}{\rho_2} g_2^- = g_2 \tag{5.4}$$

where $f_2^+ = T_{0,\Gamma_2} u_1$, $f_2^- = T_{0,\Gamma_2} u_2$, $g_2^+ = T_{1,\Gamma_2} u_1$, $g_2^- = T_{1,\Gamma_2} u_2$ and f_2, g_2 are given functions or distributions in $H_{\pm 1/2}(\mathbb{R}^+)$, respectively.

As in the preceeding sections, we begin by obtaining an equivalent problem, with homogeneous boundary conditions. For this, let (see (2.22), (2.25))

$$\widetilde{u}_1 = u_1 - r_{Q_1} K_{D,\mathbb{R}^2_+}(\kappa_1) \ell^e f_1 \tag{5.5}$$

$$\widetilde{u}_2 = u_2 - r_{Q_2} K_{N,\mathbb{R}^2_+}(\kappa_2) \ell^o g_3 \tag{5.6}$$

define the substituted functions. If $(u_1, u_2) \in H_1(Q_1) \times H_1(Q_2)$ is a solution to Problem $\mathcal{P}_M$, then from (5.2), (5.5) and (5.6) it is clear that

$$T_{0,\Gamma_1} \widetilde{u}_1 = 0 \quad \text{and} \quad T_{1,\Gamma_3} \widetilde{u}_2 = 0 \tag{5.7}$$

Moreover, a procedure analogous to that followed in sections 3 and 4, shows that the transmission conditions (5.3), (5.4) for (u_1, u_2) are transformed into the following ones for $(\widetilde{u}_1, \widetilde{u}_2)$:

$$\widetilde{f}_2^+ - \widetilde{f}_2^- = \widetilde{f}_2 \tag{5.8}$$

$$\frac{1}{\rho_1} \widetilde{g}_2^+ + \frac{1}{\rho_2} \widetilde{g}_2^- = \widetilde{g}_2 \tag{5.9}$$

with (see (3.12))

$$\widetilde{f}_2^+ = T_{0,\Gamma_2} \widetilde{u}_1 = f_2^+ - C_0(\kappa_1) \ell^e f_1 \in \widetilde{H}_{1/2}(\mathbb{R}^+) \tag{5.10}$$

due to $(0, \widetilde{f}_2^+) \in H_{1/2}(\partial Q_1)$, see (2.4), and

$$\widetilde{f}_2^- = T_{0,\Gamma_2} \widetilde{u}_2 = f_2^- - T_{0,\Gamma_2} r_{Q_2} K_{N,\mathbb{R}^2_+}(\kappa_2)\ell^o g_3 = f_2^- \tag{5.11}$$

because of the odd character of $\ell^o g_3$, see (2.25). Furthermore, we have, by similar arguments:

$$\widetilde{g}_2^+ = T_{1,\Gamma_2} \widetilde{u}_1 = g_2^+ - T_{1,\Gamma_2} r_{Q_1} K_{D,\mathbb{R}^2_+}(\kappa_1)\ell^e f_1 = g_2^+ \tag{5.12}$$

and (see (4.14))

$$\widetilde{g}_2^- = T_{1,\Gamma_2} \widetilde{u}_2 = g_2^- + C_1(\kappa_2)\ell^o g_3 \in \widetilde{H}_{-1/2}(\mathbb{R}^+) \tag{5.13}$$

Hence, by means of (5.5), (5.6), Problem $\mathcal{P}_M$ is equivalent to

Problem $\mathcal{P}_M^0$: Find $(\widetilde{u}_1, \widetilde{u}_2) \in H_1(Q_1) \times H_1(Q_2)$ such that

$$(\Delta + \kappa_j^2)\widetilde{u}_j = 0 \quad \text{in} \quad Q_j \qquad (j = 1, 2) \tag{5.14}$$

which satisfies the homogeneous boundary conditions (5.7) and the transmission conditions (5.8), (5.9) with $\widetilde{f}_2 = f_2 - C_0(\kappa_1)\ell^e f_1 \in H_{1/2}(\mathbb{R}^+)$ and $\widetilde{g}_2 = g_2 + \frac{1}{\rho_2} C_1(\kappa_2)\ell^o g_3 \in H_{-1/2}(\mathbb{R}^+)$.

Because we have homogeneous boundary conditions in Γ_1, Γ_3, an easy reasoning shows that the following representation formulas hold (cf. [18, §3])

$$\widetilde{u}_1 = r_{Q_1} K_{D,\mathbb{R}^2_r}(\kappa_1)\ell^o \widetilde{f}_2^+ \tag{5.15}$$

$$\widetilde{u}_2 = r_{Q_2} K_{N,\mathbb{R}^2_\ell}(\kappa_2)\ell^e \widetilde{g}_2^- \tag{5.16}$$

where we used the fact that $\widetilde{f}_2^+ \in \widetilde{H}_{1/2}(\mathbb{R}^+)$ and $\widetilde{g}_2^- \in \widetilde{H}_{-1/2}(\mathbb{R}^+)$ (see (5.10), (5.13)).

Using the above formulas and (2.23), (2.27), we also have, by direct computation,

$$\ell^o \widetilde{g}_2^+ = A_1^{-1} \ell^o \widetilde{f}_2^+ \quad \text{and} \quad \ell^e \widetilde{f}_2^- = A_2 \ell^e \widetilde{g}_2^- \tag{5.17}$$

(see (3.24)). These two relations, together with the transmission conditions (5.8), (5.9), will allow to express the unknowns $\ell^o \widetilde{f}_2^+$ and $\ell^e \widetilde{g}_2^-$ in (5.15), (5.16) in terms of the known data $\widetilde{f}_2$ and $\widetilde{g}_2$. Taking the even extension in (5.8) and the odd extension in (5.9), we get

$$\begin{cases} \ell^e \widetilde{f}_2^+ - \ell^e \widetilde{f}_2^- = \ell^e \widetilde{f}_2 \\ \dfrac{1}{\rho_1} \ell^o \widetilde{g}_2^+ + \dfrac{1}{\rho_2} \ell^o \widetilde{g}_2^- = \ell^o \widetilde{g}_2 \end{cases}$$

or yet, by (5.17):

$$\begin{cases} \ell^e \widetilde{f}_2^+ - A_2\, \ell^e \widetilde{g}_2^- = \ell^e \widetilde{f}_2 \\ \dfrac{1}{\rho_1} A_1^{-1} \ell^o \widetilde{f}_2^+ + \dfrac{1}{\rho_2} \ell^o \widetilde{g}_2^- = \ell^o \widetilde{g}_2 \end{cases} \tag{5.18}$$

This system of pseudodifferential equations can be reformulated as a special Riemann-Hilbert problem. Indeed, if we denote by $\ell_0 : \widetilde{H}_{\pm 1/2}(\mathbb{R}^+) \to H^+_{\pm 1/2}(\mathbb{R})$ the operator of extension by zero, it follows from (5.10), (5.13) that

$$\varphi^+ = \ell_0 \widetilde{f}_2^+ \in H^+_{1/2}(\mathbb{R}) \quad \text{and} \quad \psi^+ = \ell_0 \widetilde{g}_2^- \in H^+_{-1/2}(\mathbb{R}) \tag{5.19}$$

where $H^+_{\pm 1/2}(\mathbb{R})$ denote the subspace of $H_{\pm 1/2}(\mathbb{R})$ formed by all the functions (distributions) with support contained in $\overline{\mathbb{R}}^+$. Then, introducing the reflection operator J in $H_s(\mathbb{R})$ $(s \in \mathbb{R})$, defined by (for smooth functions in a dense subspace)

$$J\varphi(x) = \varphi(-x) \quad , \quad x \in \mathbb{R} \quad , \tag{5.20}$$

it turns out immediately that the following decompositions hold

$$\begin{cases} \ell^e \widetilde{f}_2^+ &= \varphi^+ + J\varphi^+ \\ \ell^o \widetilde{f}_2^+ &= \varphi^+ - J\varphi^+ \end{cases} \quad \text{and} \quad \begin{cases} \ell^e \widetilde{g}_2^- &= \psi^+ + J\psi^+ \\ \ell^o \widetilde{g}_2^- &= \psi^+ - J\psi^+ \end{cases} \tag{5.21}$$

Thus, inserting (5.21) into system (5.18) we obtain

$$\begin{cases} \varphi^+ - A_2 \psi^+ &= \ell^e \widetilde{f}_2 - J\varphi^+ + A_2 J\psi^+ \\ \dfrac{1}{\rho_1} A_1^{-1} \varphi^+ + \dfrac{1}{\rho_2} \psi^+ &= \ell^o \widetilde{g}_2 + \dfrac{1}{\rho_1} A_1^{-1} J\varphi^+ + \dfrac{1}{\rho_2} J\psi^+ \end{cases} \tag{5.22}$$

After a direct computation, this system of equations can be written in the matrix form

$$\mathcal{F}^{-1} G \mathcal{F} \begin{pmatrix} \varphi^+ \\ \psi^+ \end{pmatrix} = \begin{pmatrix} f \\ g \end{pmatrix} + \begin{pmatrix} J\varphi^+ \\ J\psi^+ \end{pmatrix} \tag{5.23}$$

where the matrix-valued function G is given by (see (3.24))

$$G = \frac{t_2}{t_1/\rho_1 + t_2/\rho_2} \begin{bmatrix} -\dfrac{t_2/\rho_2 - t_1/\rho_1}{t_2} & \dfrac{2}{\rho_2} \dfrac{1}{t_2} \\ \dfrac{2}{\rho_1} t_1 & \dfrac{t_2/\rho_2 - t_1/\rho_1}{t_2} \end{bmatrix} \tag{5.24}$$

(with $t_1/\rho_1 + t_2/\rho_2 \neq 0$, $\xi \in \mathbb{R}$, see (2.12), (2.13), (2.19)) and $(f,g)^T \in H_{1/2}(\mathbb{R}) \times H_{-1/2}(\mathbb{R})$ is the vector defined by

$$\begin{pmatrix} f \\ g \end{pmatrix} = \begin{pmatrix} -\mathcal{F}^{-1}\dfrac{t_2/\rho_2}{t_1/\rho_1+t_2/\rho_2}\mathcal{F}\ell^e\widetilde{f}_2 + \mathcal{F}^{-1}\dfrac{1}{t_1/\rho_1+t_2/\rho_2}\mathcal{F}\ell^o\widetilde{g}_2 \\ \mathcal{F}^{-1}\dfrac{t_1t_2/\rho_1}{t_1/\rho_1+t_2/\rho_2}\mathcal{F}\ell^e\widetilde{f}_2 + \mathcal{F}^{-1}\dfrac{t_2}{t_1/\rho_1+t_2/\rho_2}\mathcal{F}\ell^o\widetilde{g}_2 \end{pmatrix} \tag{5.25}$$

Clearly (5.23) represents a special Riemann-Hilbert problem (where the "minus" vector has to fulfil the additional condition of being the reflection of the unknown "plus" vector $(\varphi^+, \psi^+)^T$), which is equivalent to Problem $\mathcal{P}_M^0$, and, through (5.5), (5.6), to Problem $\mathcal{P}_M$. We summarize in the next theorem the results obtained so far.

Theorem 5.1: *Problem $\mathcal{P}_M$ is uniquely solvable if and only if the Riemann-Hilbert problem* (5.22)-(5.24) *is uniquely solvable. Moreover, we have:*

(i) *If $(\varphi^+, \psi^+)^T \in H^+_{1/2}(\mathbb{R}) \times H^+_{-1/2}(\mathbb{R})$ is a solution of* (5.22), *then*

$$\begin{pmatrix} u_1 \\ u_2 \end{pmatrix} = \begin{bmatrix} r_{Q_1} K_{D,\mathbf{R}^2_+}(\kappa_1)\ell^e & 0 \\ 0 & r_{Q_2} K_{N,\mathbf{R}^2_+}(\kappa_2)\ell^o \end{bmatrix} \begin{pmatrix} f_1 \\ g_3 \end{pmatrix} +$$

$$+ \begin{bmatrix} r_{Q_1} K_{D,\mathbf{R}^2_\tau}(\kappa_1) & 0 \\ 0 & r_{Q_2} K_{N,\mathbf{R}^2_\ell}(\kappa_2) \end{bmatrix} \begin{pmatrix} \varphi^+ - J\varphi^+ \\ \psi^+ + J\psi^+ \end{pmatrix} \tag{5.26}$$

is a solution to Problem $\mathcal{P}_M$ (see (5.5), (5.6), (5.14), (5.15) *and* (5.20)*).*

(ii) *If $(u_1, u_2) \in H_1(Q_1) \times H_1(Q_2)$ is a solution to Problem $\mathcal{P}_M$, then*

$$\begin{pmatrix} \varphi^+ \\ \psi^+ \end{pmatrix} = \ell_0 \begin{pmatrix} T_{0,\Gamma_2} u_1 - C_0(\kappa_1)\ell^e f_1 \\ T_{1,\Gamma_2} u_2 + C_1(\kappa_2)\ell^o g_3 \end{pmatrix} \tag{5.27}$$

is a solution of (5.22) *(see* (5.18), (5.10) *and* (5.12)*).*

Let $L_2^+(\mathbb{R})$ denote the subspace of $L_2(\mathbb{R})$ formed by all the functions supported in $\overline{\mathbb{R}}^+$. The system of equations (5.23) in $H^+_{1/2}(\mathbb{R}) \times H^+_{-1/2}(\mathbb{R})$ is reducible to an equivalent Riemann-Hilbert problem in $[L_2^+(\mathbb{R})]^2$ through the usual lifting procedure [10], [11]. In fact, let $k \in \mathbb{C}^{++}$ and denote by $t_\pm$ the square root functions

$$t_\pm(\xi) = (\xi \pm k)^{1/2} = |\xi \pm \kappa|^{1/2} e^{\frac{i}{2}\arg(\xi \pm \kappa)} \quad , \quad \xi \in \mathbb{R} \tag{5.28}$$

with branch cuts $\Gamma_\pm$ and $\arg(\xi-\kappa)\in\,]-\frac{3\pi}{2},\frac{\pi}{2}]$, $\arg(\xi+\kappa)\in\,]-\frac{\pi}{2},\frac{3\pi}{2}]$, such that $t=t_-t_+$ holds (see (2.19)). Now, let

$$\begin{pmatrix}\varphi^+\\ \\ \psi^+\end{pmatrix}=\mathcal{F}^{-1}\operatorname{diag}[\frac{1}{t_+},t_+]\,\mathcal{F}\begin{pmatrix}\varphi_0^+\\ \\ \psi_0^+\end{pmatrix}\tag{5.29}$$

where

$$\mathcal{F}^{-1}\operatorname{diag}[\frac{1}{t_+},t_+]\mathcal{F}\;:\;[L_2^+(\mathbb{R})]^2\to H^+_{1/2}(\mathbb{R})\times H^+_{-1/2}(\mathbb{R})\tag{5.30}$$

defines a Bessel potential operator with inverse given by $\mathcal{F}^{-1}\operatorname{diag}[t_+,\frac{1}{t_+}]\mathcal{F}$. Further, as $Jt_+=it_-J$ and $J\mathcal{F}=\mathcal{F}J$, we have

$$\begin{pmatrix}J\varphi^+\\ \\ J\psi^+\end{pmatrix}=\mathcal{F}^{-1}\operatorname{diag}[\frac{1}{it_-},\,it_-]\,\mathcal{F}\begin{pmatrix}J\varphi_0^+\\ \\ J\psi_0^+\end{pmatrix}\tag{5.31}$$

Hence, a straightforward computation yields (see (5.23)), (5.24))

$$\mathcal{F}^{-1}G_0\mathcal{F}\begin{pmatrix}\varphi_0^+\\ \\ \psi_0^+\end{pmatrix}=\begin{pmatrix}f_0\\ \\ g_0\end{pmatrix}+\begin{pmatrix}J\varphi_0^+\\ \\ J\psi_0^+\end{pmatrix}\tag{5.32}$$

with

$$\begin{aligned}G_0&=\operatorname{diag}[it_-,\frac{1}{it_-}]\,G\operatorname{diag}[\frac{1}{t_+},t_+]=\\ &=\frac{it_2}{t_1/\rho_1+t_2/\rho_2}\begin{bmatrix}-\dfrac{t_-}{t_+}\dfrac{t_2/\rho_2-t_1/\rho_1}{t_2} & \dfrac{2}{\rho_2}\dfrac{t}{t_2}\\ -\dfrac{2}{\rho_1}\dfrac{t_1}{t} & -\dfrac{t_+}{t_-}\dfrac{t_2/\rho_2-t_1/\rho_1}{t_2}\end{bmatrix}\end{aligned}\tag{5.33}$$

and

$$\begin{pmatrix}f_0\\ \\ g_0\end{pmatrix}=\begin{pmatrix}\mathcal{F}^{-1}it_-\,\mathcal{F}f\\ \\ \mathcal{F}^{-1}\dfrac{1}{it_-}\mathcal{F}g\end{pmatrix}\in[L_2(\mathbb{R})]^2\tag{5.34}$$

The study of the solvability of the Riemann-Hilbert problem (5.32) in $[L_2^+(\mathbb{R})]^2$ can be done through the determination of a generalized factorization of its presymbol G_0 [1]. The existence of such factorization for G_0 is established in the following proposition, where we use the well known criteria for piecewise matrix-valued functions (see [1, Chapter 8], [12, Chapter 5]).

Proposition 5.2: *The matrix-valued function G_0 in (5.32) is L_2-non-singular if and only if*

$$\frac{\rho_1}{\rho_2} \in \mathbb{C} \backslash \overline{\mathbb{R}^-} \tag{5.35}$$

If this condition is fulfilled, then the factorization of G_0 has zero total index.

Proof: Using the criteria cited above, we associate with G_0 the symbol $\widetilde{G}_0 : \dot{\mathbb{R}} \times [0,1] \rightarrow \mathbb{C}^{2\times 2}$ defined by

$$\begin{aligned} \widetilde{G}(\xi,\mu) &= G_0(\xi+0)\mu + (1-\mu)G_0(\xi) \quad , \quad \xi \in \mathbb{R} \, , \, \mu \in [0,1] \\ \widetilde{G}(\infty,\mu) &= G_0(-\infty)\mu + (1-\mu)G_0(\infty) \quad , \quad \mu \in [0,1] \end{aligned}$$

The matrix-valued function G_0 is L_2-non-singular if and only if

$$\widetilde{g}(\xi,\mu) = \det \widetilde{G}(\xi,\mu) \neq 0 \quad , \quad (\xi,\mu) \in \dot{\mathbb{R}} \times [0,1] \tag{5.36}$$

Because infinity is the only discontinuity point of G_0 (as $\frac{t_-}{t_+}(\pm\infty) = \pm 1$, see (5.28)), we have (see (5.33))

$$\widetilde{g}(\xi,\mu) = \det G_0(\xi) = -1 \quad , \quad \xi \in \mathbb{R}$$

and

$$\widetilde{g}(\infty,\mu) = \frac{-1}{(\rho_1+\rho_2)^2}\left[(\rho_1-\rho_2)^2(2\mu-1)^2 + 4\rho_1\rho_2\right] \quad , \quad \mu \in [0,1]$$

Hence the condition (5.36) is equivalent to

$$(2\mu-1)^2 \neq \frac{-4\rho_1\rho_2}{(\rho_1-\rho_2)^2} \quad , \quad \mu \in [0,1] \tag{5.37}$$

from which (5.35) follows, after a simple computation. Furthermore, if (5.36) is satisfied, the curve $\widetilde{g}$ has a zero winding number with respect to the origin, which proves the last assertion of the proposition. ■

We note that condition (2.13) implies (5.35). Nevertheless, the above proposition only gives an existence result for a generalized factorization of G_0, without any information about the partial indices [1] of the factorization and the explicit representation of the factors. A more detailed discussion of the solvability properties for the Riemann-Hilbert problem (5.32) is dependent on the possibility of obtaining these results, which in general remain as open questions.

Remark 5.3: If we restrict ourselves to the case $\kappa_1 = \kappa_2$ then $t = t_1 = t_2$ and we get (see (5.33))

$$G_0 = -\frac{i}{\rho_1+\rho_2}\begin{bmatrix} \frac{t_-}{t_+}(\rho_1-\rho_2) & -2\rho_1 \\ 2\rho_2 & \frac{t_+}{t_-}(\rho_1-\rho_2) \end{bmatrix} \tag{5.38}$$

The generalized factorization for this class of matrix-valued functions was studied in [16], [17]. For all the values of ρ_1, ρ_2 which satisfy (5.35), it was shown that G_0 given by (5.38) possess a canonical generalized factorization, which was explicitly worked out. From a careful discussion of the factors $G_{0\pm}$ obtained therein, it turns out that

$$\begin{pmatrix} \varphi_0^+ \\ \psi_0^+ \end{pmatrix} = \mathcal{F}^{-1}\, G_{0+}^{-1}\, \mathcal{F}\, \mathcal{P}^+\, \mathcal{F}^{-1}\, G_{0-}^{-1}\, \mathcal{F} \begin{pmatrix} f_0 \\ g_0 \end{pmatrix} \tag{5.39}$$

is the (unique) solution of (5.32). Therefore, from Theorem 5.1 we conclude the unique solvability of Problem $\mathcal{P}_M$ for this particular case, and its solution is given by (5.26), with (5.29), (5.31) and (5.39).

If κ_1 and κ_2 are different wave numbers fulfilling (2.12) and ρ_1, ρ_2 are positive real numbers (which obviously satisfy (5.35)), the unique solvability of the boundary-transmission problem $\mathcal{P}_M$ is already known from a variational approach (see [14]). Then, in this case, Theorem 5.1 implies the unique solvability of (5.23), and consequently of the Riemann-Hilbert problem (5.32).

Although in this case the explicit canonical factorization of G_0 is not available, it is possible to obtain the asymptotic behaviour at the origin of the solution of (5.32) for sufficiently smooth functions in the right-hand side, by use of the method developed in [13].

It is clear that if $(\varphi_0^+, \psi_0^+)^T$ is a solution to (5.32), then it also satisfies the equation

$$\mathcal{P}^+\mathcal{F}^{-1}G_0\mathcal{F}\begin{pmatrix} \varphi_0^+ \\ \psi_0^+ \end{pmatrix} = \mathcal{P}^+\begin{pmatrix} f_0 \\ g_0 \end{pmatrix} \tag{5.40}$$

where $\mathcal{P}^+ : [L_2(\mathbb{R})]^2 \to [L_2^+(\mathbb{R})]^2$ is the usual projection operator (the multiplication by the characteristic function of $\mathbb{R}^+$), or yet, applying the Fourier transformation,

$$P^+G_0\begin{pmatrix} \hat{\varphi}_0^+ \\ \hat{\psi}_0^+ \end{pmatrix} = P^+\begin{pmatrix} \hat{f}_0 \\ \hat{g}_0 \end{pmatrix} \tag{5.41}$$

with $(\hat{\varphi}_0^+, \hat{\psi}_0^+)^T = \mathcal{F}(\varphi_0^+, \psi_0)^T$, $(\hat{f}_0, \hat{g}_0)^T = \mathcal{F}(f_0, g_0)^T$ and $\mathcal{P}^+ = \mathcal{F}P^+\mathcal{F}^{-1} : [L_2(\mathbb{R})]^2 \to [\hat{L}_2^+(\mathbb{R})]^2$.

The following theorem can be derived from Corollary 3.2 in [13, pp 221] by the use of a slight modification of the proof of Theorem 2.2 in [13, pp 217], accordingly to the regularity assumptions which will be imposed on $(\hat{f}_0, \hat{g}_0)$.

Theorem 5.4: *Let $\hat{f}_0, \hat{g}_0$ be continuously differentiable functions on $\mathbb{R}$, such that*

$$\hat{f}_0(\xi) = c_1 |\xi|^{-3/2} + c_2 |\xi|^{-2} + o(|\xi|^{-5/2}) \quad , \quad \xi \to \pm\infty \tag{5.42}$$

$$\hat{g}_0(\xi) = c_1' |\xi|^{-3/2} + c_2' |\xi|^{-2} + o(|\xi|^{-5/2}) \quad , \quad \xi \to \pm\infty \tag{5.43}$$

for some constants c_j, c_j' $(j = 1, 2)$.

Then the solution $(\hat{\varphi}_0^+, \hat{\psi}_0^+)^T$ of (5.40) has the asymptotic behaviour

$$\begin{pmatrix} \hat{\varphi}_0^+(\xi) \\ \hat{\psi}_0^+(\xi) \end{pmatrix} = C_\pm \begin{bmatrix} |\xi|^{-\eta_1-1} & 0 \\ 0 & |\xi|^{-\eta_2-1} \end{bmatrix} (\vec{c} + o \begin{pmatrix} |\xi|^{-\omega_1+\varepsilon} \\ |\xi|^{-\omega_2+\varepsilon} \end{pmatrix}) \text{ for } \xi \to \pm\infty \tag{5.44}$$

where $C_\pm$ are constant 2×2 matrices and $\vec{c}$ is a constant 2×1 vector, dependent on the data functions (5.41), (5.42), the numbers η_j and ω_j are given by

$$\eta_j = (-1)^{j+1} \frac{1}{\pi} \arg(1 + i \frac{\rho_1 - \rho_2}{2\sqrt{\rho_1\rho_2}}) \quad , \tag{5.45}$$

$$\omega_j = \frac{1}{2} - \eta_j \quad , \quad j = 1, 2 \tag{5.46}$$

and (5.43) holds for any $\varepsilon \in]0, \min(\omega_1, \omega_2)[$.

As an immediate consequence of Abel's theorem (cf. [6, §5]) we have the following result.

Corollary 5.5: With the same notation of the previous theorem, if the Fourier transforms of $(f_0, g_0)^T$ fulfil the conditions (5.41) and (5.42), *then the solution $(\varphi_0^+, \psi_0^+)^T$ of (5.31) has the asymptotic behaviour*

$$\begin{pmatrix} \varphi_0^+(x) \\ \psi_0^+(x) \end{pmatrix} = C \begin{bmatrix} x^{\eta_1} & 0 \\ 0 & x^{\eta_2} \end{bmatrix} (\vec{c} + o \begin{pmatrix} x^{\omega_1-\varepsilon-1} \\ x^{\omega_2-\varepsilon-1} \end{pmatrix}) \quad , \quad x \to 0+ \tag{5.47}$$

for a constant 2×2 matrix C (note that $\eta_j \in]-1/2, 1/2[$).

Remark 5.6: We point out that if ρ_1, ρ_2 are complex numbers which satisfy (5.35) and a solution (φ_0^+, ψ_0^+) to (5.32) does exist, then Theorem 5.4 and Corollary 5.5 still hold, if we take

$$\eta_j = \mathrm{Re}\left\{(-1)^{j+1}\frac{1}{2\pi i}\log\frac{1+i\,\dfrac{\rho_1-\rho_2}{2\sqrt{\rho_1\rho_2}}}{1-i\,\dfrac{\rho_1-\rho_2}{2\sqrt{\rho_1\rho_2}}}\right\} \quad , \quad j=1,2 \tag{5.48}$$

where log denotes the principal branch of the logarithm.

REFERENCES

[1] Clancey, K. and Gohberg, I.C., *Factorization of Matrix Functions and Singular Integral Operators,* Operator Theory: Advances and Applications, Vol. 3, Birkhäuser, Basel–Boston–Stuttgart, 1981

[2] Costabel, M., Stephan, E.P. and Wendland, W.L., "Boundary integral equations of the first kind for the bi-Laplacian in a polygonal plane domain", *Ann. Scuola Norm. Sup. Pisa, Ser. IV,* **10**, 197–241 (1983)

[3] Eskin, G.I., *Boundary Value Problems for Elliptic Pseudodifferential Equations,* Amer. Math. Soc., Providence R.I., 1981

[4] Grisvard, P., *Elliptic Problems in Nonsmooth Domains,* Pitman Publishing Inc., London, 1985

[5] Hurd, R.A., "Scattering by a ferrite quadrant", *J. Electromag. Waves Appl.,* **3**, no. 5, 463–479 (1989)

[6] Meister, E., "*Integraltransformationen mit Anwendungen auf Probleme der mathematischen Physik*", Lang, Frankfurt am Main, 1983

[7] Meister, E., "Einige gelöste und ungelöste kanonische Probleme der mathematischen Beugungstheorie", *Expo. Math.,* **5**, 193–237 (1987)

[8] Meister, E. and Penzel, R., "Einige Randwerttransmissionsprobleme der Beugungstheorie für Keile bei mehreren Medien", *Kleinheubacher Berichte,* Nr. 30, FTZ-Darmstadt, S. 429–438, SSN 0343–5729

[9] Meister, E. and Speck, F.-O., "A contribution to the quarter-plane problem in diffraction theory", *J. Math. Anal. Appl.,* **130**, 223–236 (1988)

[10] Meister, E. and Speck, F.-O., "Boundary integral equation methods for canonical problems in diffraction theory", in: *Boundary Elements IX* (eds. C.A. Brebbia, W.L. Wendland, G. Kuhn), Vol. I, Proc. Conf. Stuttgart 1987, Springer 1987, pp 59–77

[11] Meister, E., Speck, F.-O. and Teixeira, F.S., "Wiener-Hopf and Hankel operators for some wedge diffraction problems with mixed boundary conditions", *Preprint-Nr. 1355, Fachbereich Mathematik, T.H.D.,* Darmstadt 1991

[12] Mikhlin, S.G. and Prössdorf, S., *Singular Integral Operators,* Springer, Berlin, 1986

[13] Penzel, F., "On the asymptotics of the solution of systems of singular integral equations with piecewise Hoelder-continuous coefficients", *Asympt. Anal.,* **1**, 213–225 (1988)

[14] von Petersdorff, T., "Boundary integral equations for mixed Dirichlet, Neumann and transmission conditions", *Math. Meth. Appl. Sci.,* **11**, 185–213 (1989)

[15] Speck, F.-O., "Mixed boundary-value problems of the type of Sommerfeld's half-plane problem", *Proc. Roy. Soc. Edinburgh,* **104A**, 261–277 (1986)

[16] Speck, F.-O., "Sommerfeld diffraction problems with first and second kind boundary conditions", *Siam J. Math. Anal.,* **20**, no. 2, 396–407 (1989)

[17] Teixeira, F.S., "Generalized Factorization for a class of symbols in $[PC(\dot{\mathbb{R}})]^{2\times 2}$", *Appl. Anal.,* **36**, 95–117 (1990)

[18] Teixeira, F.S., "Diffraction by a rectangular wedge: Wiener-Hopf-Hankel formulation", *Integral Equations and Operator Theory,* 14, 436–454 (1991)

Fachbereich Mathematik
Technische Hochschule Darmstadt
Schlossgartenstrasse 7,
D-6100 Darmstadt
F.R. Germany

and

Departamento de Matemática
Instituto Superior Técnico
Av. Rovisco Pais
1096 Lisboa Codex
Portugal

Y XU

Scattering of acoustic waves by an obstacle in a stratified medium

1. Introduction

The purpose of this paper is to apply boundary integral equation methods to study scattering problems of harmonic acoustic waves in a stratified medium.

In a stratified medium, sound waves can be trapped by acoustic ducts and caused to propagate horizontally. (see [1], [11]). Therefore, the scattering of sound waves in stratified medium by a bounded obstacles is more complicated than that in homogeneous medium. Thanks to the well-known Sommerfeld radiation condition, scattering theories for compact object or compact inhomogeneity in homogeneous medium have been well developed. Integral equation methods play a very important role in the study of boundary-value problems associated with the direct and inverse scattering of acoustic waves by bounded obstacles (cf. [2] [10]). However, the Sommerfeld radiation condition cannot be applied to scattering in a stratified medium when guided waves exist. In fact, in an underwater sound channel (USC), the propagation of a sound field decays at the rate of $\frac{1}{\sqrt{r}}$ instead of $\frac{1}{r}$ which occurs in a homogeneous medium.

In section 2, we will first study the radiating behavior of sound waves from a point source. Based on this analysis, a new radiation condition is found for exterior boundary problems. In section 3 we will study uniqueness and existence of exterior boundary problems by the boundary integral method. This paper could be viewed as a continuation of our research in the direct and inverse scattering in shallow oceans. Anyone who is interested could refer to [4] [5] [13] [14].

2. Green's function and its radiating amplitude in a stratified medium

The propagation of monochromatic sound in a stratified medium is described by the non-homogeneous Helmholtz equation

$$\Delta u + k^2 n(z) u = 0 , \tag{2.1}$$

where $\Delta = \frac{\partial^2}{\partial x_1^2} + \frac{\partial^2}{\partial x_2^2} + \frac{\partial^2}{\partial z^2}$ is the Laplacian, u is the time harmonic acoustic velocity potential, $k > 0$ the wave number and $n(z)$ the refraction index.

We will use the following conventions in this paper. A point in $\mathbf{R}^3$ will be described as

$$(\underset{\sim}{x}, z) = (x_1, x_2, z) = (r, \theta, z) = (R, \theta, \varphi),$$
$$(\underset{\sim}{\xi}, \zeta) = (\xi_1, \xi_2, \zeta) = (r', \theta', \zeta) = (\rho, \theta', \varphi')$$

in different coordinates systems respectively. There are relations

$$R^2 = r^2 + z^2 = |\underset{\sim}{x}|^2 + z^2 ,$$
$$r = R\sin\varphi , \qquad z = R\cos\varphi ,$$
$$\rho^2 = r'^2 + \zeta^2 = |\underset{\sim}{\xi}|^2 + \zeta^2 ,$$
$$r' = \rho\sin\varphi' , \quad \zeta = \rho\cos\varphi' .$$

We assume that $n(z)$ satisfies the condition (A), i.e., $n(z)$ is a strictly positive function which is continuous except for a finite number of jump discontinuities. A discontinuity in $n(z)$ represents a change in medium. Moreover, we assume

(2.2) $$\lim_{z\to\infty} n(z) = n_+ < \infty ,$$

(2.3) $$\lim_{z\to-\infty} n(z) = n_- < \infty ,$$

and for $\alpha > -\infty$, $\beta < \infty$, $m > 0$,

(2.4) $$\int_{\alpha}^{\infty} (1 + |z|)^m |n(z) - n^+| dz < \infty ,$$

(2.5) $$\int_{-\infty}^{\beta} (1 + |z|)^m |n(z) - n^-| dz < \infty .$$

Denote by $p_1(z)$ and $p_2(z)$ the two Jost solutions of the equation

(2.6) $$p''(z) + k^2[n^2(z) - a^2]p = 0 , \quad -\infty < z < \infty ,$$

which have asymptotic representations

(2.7) $$p_1(z) = e^{-ik\sqrt{n_-^2 - a^2}\, z} + O\left(\frac{1}{|z|}\right) , \text{ as } z \to -\infty ,$$

and

(2.8) $$p_2(z) = e^{ik\sqrt{n_+^2 - a^2}\, z} + O\left(\frac{1}{|z|}\right) , \text{ as } z \to +\infty ,$$

where a is a constant. Note that (2.6) cannot have solutions in the classical sense unless $n(z)$ is continuous. A suitable class of solutions for our discussion is defined as following: A function $\phi : \mathbf{R} \to \mathbf{C}$ is said to be a solution of (2.6) in $\mathbf{R}$ if and only if

$$\phi \in AC(\mathbf{R}) , \quad \phi' \in AC(\mathbf{R})$$

and (2.6) holds for almost all $z \in \mathbf{R}$, where $AC(\mathbf{R})$ denotes the set of all absolutely continuous functions in $\mathbf{R}$. Using Hankel transformation, we can represent the outgoing Green's function of (2.1) as

$$G(r,z,z_0) = \int_0^\infty \frac{p_1(z_<)p_2(z_>)}{w(ka)} J_0(kar)kad(ka) \;, \tag{2.9}$$

where $z_< = \min\{z,z_0\}$, $z_> = \max\{z,z_0\}$, J_0 is the Bessel function of the first kind of order zero. $w(ka)$ is the Wronskian of p_1 and p_2.

The integral (2.9) is improper because the Wronskian $w(ka)$ may have a finite number of simple zeros for $a > n_0 = \max\{n_+, n_-\}$. (cf. [6] p123). We use contour integration in the complex plane $\tau = a + ib$ to evaluate the integral, as discussed in [3]. The meaning of (2.9) is the limit of the integral along path C_1 which is the positive real axis with small detour to the fourth quadrant at the zeros of the denominator. In view of $H_0^{(2)}(kar) = -H_0^{(1)}(e^{i\pi}kar)$ where $H_0^{(i)}$ is the Hankel function of the ith kind of order zero, and A is a branch point of the Riemann surface, we can represent $G(r,z,z_0)$ as

$$\begin{aligned} G(r,z,z_0) &= \int_0^\infty \frac{p_1(z_<)p_2(z_>)}{2w(ka)} [H_0^{(1)}(kar) + H_0^{(2)}(kar)]kad(ka) \\ &= \int_C \frac{p_1(z_<)p_2(z_>)}{2w(ka)} H_0^{(1)}(kar)kad(da) \\ &+ \int_{C'} \frac{p_1(z_<)p_2(z_>)}{2w(ka)} H_0^{(1)}(kar)kad(ka) \end{aligned}$$

where the contours $C = C_1 \cup C_R \cup C_3 \cup C_4 \cup C'_R \cup C_2$, $C' = -(C_4 \cup C_3)$ and the cuts AOD $A'O'E$ are shown in Fig 2.1.

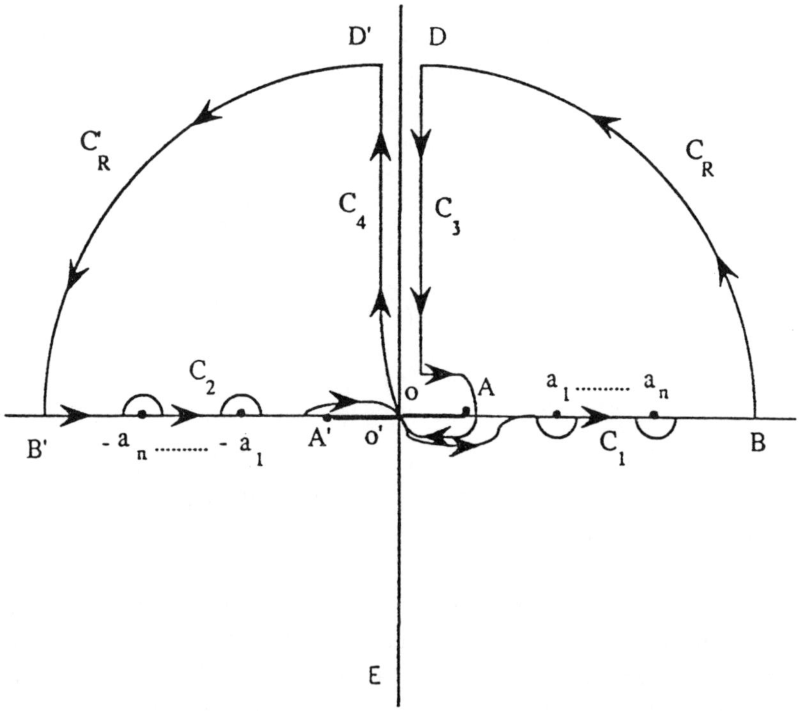

FIG. 2.1

Hence, if $\hat{n}$ is the branch point, then

$$G(r,z,z_0) = \int_{-\infty}^{\infty} \frac{p_1(z_<,\xi)p_2(z_>,\xi)}{2w(k\sqrt{\hat{n}^2-\xi^2})} H_0^{(1)}(k\sqrt{\hat{n}^2-\xi^2}\, r)k\xi d(k\xi)$$

$$+\sum_{n=1}^{N} c_n\phi_n(z)\phi_n(z_0)H_0^{(1)}(ka_n r)\ , \qquad (2.10)$$

where $c_n = \frac{ka_n}{2w'(ka_n)}$, a_n is the nth zero of $w(ka)$ and $\phi_n(z)$ the corresponding eigenfunction. Here to emphasize the dependence of p_1, p_2 on $\xi = \pm(\hat{n}^2 - a^2)^{1/2}$, we write $p_1(z_<), p_2(z_>)$ as $p_1(z_<,\xi)$, $p_2(z_>,\xi)$. As discussed in [11], a_n corresponds to the discrete spectrum of operator (2.6). N is a finite number if $n(z)$ satisfies (2.4) and (2.5). (cf. [6]). Physically, the first integral on the right hand side of (2.10) represents a free wave and the second represents guided waves. $\phi_n(z)$, $n = 1,2,\ldots,N$ are proper eigenfunctions of (2.6) and have asymptotic behaviors

$$\phi_n(z) \to e^{-k\sqrt{a_n^2-n_\pm^2}|z|}\ , \quad \text{as} \quad \to \pm\infty\ .$$

If there is no guided wave, then $N = 0$.

Now we consider the asymptotic behavior of $G(r, z, z_0)$ as r and $z \to \infty$. We write $G(r, z, z_0) = G_f(r, z, z_0) + G_g(r, z, z_0)$ where G_f, G_g represent the free wave and guided wave respectively.

In view of (2.7), (2.8) and the Hankel function's asymptotic behavior

$$H_0^{(1)}(kar) = \frac{1}{\sqrt{2\pi kar}} e^{ikar - i\frac{\pi}{4}} + O\left(\frac{1}{r^{3/2}}\right), \quad r \to \infty,$$

we have for $z = R\cos\phi$, $\quad r = R\sin\phi$, $\quad \phi < \frac{\pi}{2}$ and $R \to \infty$,

$$G_f(|\underset{\sim}{x} - \underset{\sim}{\xi}|,\ z, \zeta) = \int_{\hat{n}}^{\hat{n}} \frac{p_1(\zeta, k\sqrt{n_+^2 - a^2}) e^{ik\sqrt{n_+^2 - a^2} z}}{2w(ka)} \cdot \frac{e^{ika|\underset{\sim}{x} - \underset{\sim}{\xi}| - i\frac{\pi}{4}}}{\sqrt{2\pi kar}} \left(1 + O\left(\frac{1}{R}\right)\right) \cdot kad(ka).$$

Since

$$|\underset{\sim}{x} - \underset{\sim}{\xi}| = R\sin\phi\left(1 - \frac{r'\cos(\theta - \theta')}{R\sin\phi} + O\left(\frac{1}{R^2}\right)\right),$$

$$G_f(|\underset{\sim}{x} - \underset{\sim}{\xi}|, z, \zeta) = \int_{-n_+}^{n_+} \frac{p_1(\zeta, k\sqrt{n_+^2 - a^2})\sqrt{ka}}{2w(ka)} \cdot \frac{e^{iRk\sqrt{n_+^2 - a^2}\cos\phi + iRka\sin\phi}}{\sqrt{2\pi kR\sin\phi}} e^{-ikar'\cos(\theta - \theta') - i\frac{\pi}{4}} \cdot \left[1 + O\left(\frac{1}{R}\right)\right] d(ka).$$

By the method of stationary phase, we get

(2.11)

$$G_f(|\underset{\sim}{x} - \underset{\sim}{\xi}|, z, \zeta) = -\frac{e^{ikn_+[R - r'\sin\phi\cos(\theta - \theta')]}}{R} \cdot \frac{p_1(\zeta, kn_+\cos\phi) ikn_+\cos\phi}{2w(kn_+\cos\phi)} + O\left(\frac{1}{R^2}\right).$$

If $\phi > \frac{\pi}{2}$ and $R \to \infty$, then in the same way we have

$$G_f(|\underset{\sim}{x} - \underset{\sim}{\xi}|, z, \zeta) = -\frac{e^{ikn_-[R - r'\sin\phi\cos(\theta - \theta')]}}{R} \cdot \frac{p_2(\zeta, kn_-\cos\phi) ikn_-\cos\phi}{2w(kn_-\cos\phi)} + O\left(\frac{1}{R^2}\right). \tag{2.12}$$

On the other hand, for any finite z we have

$$G_g(|\underset{\sim}{x} - \underset{\sim}{\xi}|, z, \zeta) = \sum_{n=1}^{N} c_n \phi_n(\zeta)\phi_n(z) \frac{e^{ika_n r - ika_n r'\cos(\theta - \theta') - i\frac{\pi}{4}}}{\sqrt{2\pi ka_n r}} + O\left(\frac{1}{r^{3/2}}\right), \quad \text{as} \quad r \to \infty. \tag{2.13}$$

Define the function $F_f^0(\theta,\phi;r',\theta',z)$ by

$$F_f^0(\theta,\phi;r',\theta',\zeta)=\frac{ikn_+\cos\phi p_1(\zeta,kn_+\cos\phi)}{2w(kn_+\cos\phi)}\cdot e^{-ikn_+r'\sin\phi\cos(\theta-\theta')}$$

$$\text{if}\quad 0\le\phi<\frac{\pi}{2}\,,$$

$$F_f^0(\theta,\phi;r',\theta',\zeta)=\frac{ikn_-\cos\phi p_2(\zeta,kn_-\cos\phi)}{2w(kn_-\cos\phi)}e^{-ikn_-r'\sin\phi\cos(\theta-\theta')}$$

$$\text{if}\quad \frac{\pi}{2}<\phi\le\pi\,,$$

(2.14) $$\text{where}\, 0\le\theta,\theta'\le 2\pi\ ,\ 0\le r'<\infty,\ -\infty<\zeta<+\infty\,.$$

We call $F_f^0(\theta,\phi;r'\theta,z)$ the far-field pattern of point source free wave in stratified medium. Furthermore, we define a vector function

$$\mathsf{F}_g^0(z,\theta;r',\theta',\zeta)=\left(\sqrt{\frac{ka_1}{8\pi i}}\,\frac{\phi_1(z)\phi_1(z)}{w'(ka_1)}e^{-ika_1r'\cos(\theta-\theta')}\,,\right.$$

$$\left.\ldots,\sqrt{\frac{ka_N}{8\pi i}}\frac{\phi_N(\zeta)\phi_N(z)}{w'(ka_N)}e^{-ika_Nr'\cos(\theta-\theta')}\right)^T,$$

(2.15) $$-\infty<z,\zeta<\infty, 0\le\theta,\theta'\le 2\pi,\ 0\le r'<\infty\,,$$

as far-field pattern vector of point source guided wave in stratified medium. Then $G(|\underset{\sim}{x}-\underset{\sim}{\xi}|,z,\zeta)$ can be written as

$$G(|\underset{\sim}{x}-\underset{\sim}{\xi}|,z,\zeta)=\frac{e^{i\hat{k}R}}{R}F_f^0(\theta,\phi;r',\theta',\zeta)+\frac{1}{\sqrt{r}}A_g(r)\cdot\mathsf{F}_g^0(z,\theta;r',\theta',\zeta)$$

(2.16) $$+R_f^0(R,\theta,\phi)+R_g^0(r,z,\theta),$$

where

$$\hat{k}=\begin{cases} kn_+\,, & \text{for}\quad \phi\in\left[0,\frac{\pi}{2}\right)\,,\\ kn_-\,, & \text{for}\quad \phi\in\left(\frac{\pi}{2}\,,\pi\right]\,,\end{cases}$$

(2.17) $$A_g(r)=(e^{ika_1r},e^{ika_2r},\ldots,e^{ika_Nr}),$$

and R_f^0 , R_g^0 are differences of G_f and $\frac{e^{i\hat{k}R}}{R}F_f^0$, G_g and $\frac{1}{\sqrt{r}}A_g(r)\cdot\mathsf{F}_g^0$ respectively. Clearly

$$R_f^0(R,\theta,\phi)=O\left(\frac{1}{R^2}\right)\,,\quad\text{uniformly for } (\theta,\phi)\in[0,2\pi]\times[0,\pi]$$

$$\text{as } R\to\infty,$$

$$R_g^0(r,z,\theta)=O\left(\frac{1}{r^{3/2}}\right)\quad\text{uniformly for } (z,\theta)\in(-\infty,\infty)\times[0,2\pi]$$

$$\text{as}\quad r\to\infty\,.$$

The following Lemmas will be useful in the later discussion.

LEMMA 2.1. *For any given $(r',\theta',z) \in \mathbf{R}^3$, $F_f^0(\theta,\phi;r',\theta',\zeta)$ is bounded for $(\theta,\phi) \in B_1 = [0,2\pi] \times [0,\pi]$ and analytic in θ and ϕ for $\theta \in [0,2\pi]$ and $\phi \in [0,\pi/2) \cup (\pi/2,\pi]$. By analytic here we mean that the real and imaginary parts are real analytic functions of θ and ϕ.*

Proof. If $n(z)$ is continuous in $\mathbf{R}$, we can represent $p_1(z,k\nu_1), p_2(z,k\nu_2)$ where $\nu_1 = \sqrt{n_-^2 - a^2}$, $\nu_2 = \sqrt{n_+^2 - a^2}$ by transmutation

$$p_1(z,k\nu_1) = e^{-ik\nu_1 z} + \int_{-\infty}^{z} A^-(z,s)e^{-ik\nu_1 s}ds \ , \tag{2.18}$$

$$p_2(z,k\nu_2) = e^{ik\nu_2 z} + \int_{z}^{\infty} A^+(z,s)e^{ik\nu_2 s}ds \ , \tag{2.19}$$

where $A^-(z,s), A^+(z,s)$ are determined by a hyperbolic equation and proper boundary conditions (cf. [6]). They are independent of ν_1 and ν_2. Thus $p_1(\zeta, kn_- \cos\phi)$, $p_2(\zeta, kn_+ \cos\phi)$ and $W(kn_\pm \cos\phi)$ are analytic in ϕ. Since $G_f(|\underset{\sim}{x} - \underset{\sim}{\xi}|, z, \zeta)$ is continuous for $(\underset{\sim}{x}, z) \neq (\underset{\sim}{\xi}, \zeta)$ where $(\underset{\sim}{\xi}, \zeta)$ is given, $F_f^0(\theta,\phi;r',\theta',\zeta)$ is bounded at $\phi = \frac{\pi}{2}$. For this case the lemma follows.

If $n(z)$ has a finite number of jump discontinuities, we consider (2.6) as initial value problem in each layer where $n(z)$ is continuous. By above discussion we can assume $p_1(z,k\nu_1)$ is analytic for z in the first layer. $p_1(z,k\nu_1)$ satisfies (2.6) in the second layer and initial conditions which are analytic in a at the interface of two layers. Hence, a similar discussion follows $p_1(z,k\nu_1)$ is analytic in a for (r',θ',ζ) in the second layer. The same argument can apply to p_2 with the reverse order of the layers. It finishes the proof.

A straightforward calculation shows

LEMMA 2.2. *For given $(\theta,\phi) \in B_1$, $F_f^0(\theta,\phi;r',\theta',\zeta)$ is a solution of (2.1) in terms of $(r',\theta',\zeta) \in \mathbf{R}^3$.*

For given $(z,\theta) \in D_1 = (-\infty,\infty) \times [0,2\pi]$ each component of $\mathsf{F}_g(z,\theta;v',\theta',\zeta)$ is a solution of (2.1) in $\mathbf{R}^3$.

LEMMA 2.3. *If $n(z)$ satisfies the condition (B), i.e., $n(z)$ has Taylor expansion at infinity in the form*

$$n(z) = n_+ + \sum_{n=2}^{\infty} \frac{C_n^+}{z^n} \ , \quad z > \alpha \quad \text{for some } \alpha > 0, \tag{2.20}$$

and

$$n(z) = n_- + \sum_{n=2}^{\infty} \frac{C_n^-}{z^n} \ , \quad z < \beta \quad \text{for some } \beta > 0, \tag{2.21}$$

where $C_n^{\pm}$ are real constants, then

$$G(|\underset{\sim}{x}-\underset{\sim}{\xi}|,z,\zeta)=\frac{e^{ikR}}{R}\sum_{m=0}^{\infty}\frac{F_f^m(\theta,\phi;r',\theta',\zeta)}{R^m}$$

$$+\sum_{n=1}^{N}\sqrt{\frac{2}{k\pi\sigma_n r}}e^{ik\sigma_n r}\phi_n(z)\phi_n(\zeta)\sum_{m=0}^{\infty}\frac{F_g^{mn}(\theta,r',\theta')}{r^m}\,,$$

(2.22) $$\text{for } \phi\neq 0,\frac{\pi}{2}\,,\ \pi\,,\ R>R_0,\ \text{and } r>r',$$

where $R_0>0$ is some positive number, and

$$\sigma_n=(a_n^2-n_+^2)^{1/2}\qquad \textit{if}\qquad 0<\phi<\frac{\pi}{2}\,,$$
$$\sigma_n=-(a_n^2-n_-^2)^{1/2}\qquad \textit{if}\qquad \frac{\pi}{2}<\phi<\pi\,.$$

F_f^m, $m=0,1,\ldots,\infty$, are analytic in θ and ϕ. F_g^{mn}, $m=0,1,\ldots,\infty$, $n=1,2,\ldots,N$, are analytic in θ.

Proof. If $n(z)$ satisfies (2.20) and (2.21), then a solution of (2.6) can be expanded in the form

(2.23) $$p(z)=e^{ik\sigma_+ z}\sum_{n=1}^{\infty}\frac{p_n^+}{z^n}\,,\qquad z>\alpha\,,$$

and

(2.24) $$p(z)=e^{-ik\sigma_- z}\sum_{n=1}^{\infty}\frac{p_n^-}{z^n}\,,\qquad z<\beta\,,$$

where $p_n^{\pm}$ are constants and $\sigma_+=(n_+^2-a^2)^{1/2}$, $\sigma_-=(n_-^2-a^2)^{1/2}$. From (2.10) and by the method of stationary phase, we obtain the first summation of (2.22). The second part of (2.22) is implied immediately from the expansion of Hankel function

$$H_0^{(1)}(ka_n|\underset{\sim}{x}-\underset{\sim}{\xi}|)=\sum_{m=0}^{\infty}\varepsilon_m H_m^{(1)}(ka_n r)J_m(ka_n r')$$
$$\cdot[\cos m\theta\cos m\theta'+\sin m\theta\sin m\theta']\,,$$

where $\varepsilon_0=1,\varepsilon_m=2$ for $m\geq 1$.

3. Scattering problems for an bounded obstacle in stratified medium

In this section we study the direct scattering problem. Let $\Omega \subset \mathbf{R}^3$ be a bounded region with C^2 boundary such that the normal derivative on the boundary exists in the sense that the limit

$$\frac{\partial u}{\partial \nu}(\underset{\sim}{x}, z) = \lim_{h \to 0} (\nu(\underset{\sim}{x}, z), \ \mathrm{grad}\, u((\underset{\sim}{x}, z) - h\nu(\underset{\sim}{x}, z)), \quad (\underset{\sim}{x}, z) \in \partial\Omega \tag{3.1}$$

exists uniformly on $\partial\Omega$. We assume $n(z)$ satisfies the condition (A). $n(z)$ may have a finite number of jump discontinuities at points $-\infty < z_1 < \cdots < z_k < \infty$. The following notations will be used

$$I_l = \{z | z_l < z < z_{l+1}\} \ , \ \Omega_l = \mathbf{R}^2 \times I_l,$$
$$z_0 = -\infty \ , \ z_{k+1} = \infty \ , l = 0, 1, \ldots, k \ .$$

$$H(\mathbf{R}^3) = \left\{ u \in C^1(\mathbf{R}^3) \cap C^2(\bigcup_{l=0}^{k} \Omega_l) \middle| \ u \ \text{ satisfies } (2.1) \text{ in } \bigcup_{l=0}^{k} \Omega_l \right\} .$$

$$H(\mathbf{R}^3 \backslash \overline{\Omega}) = \left\{ u \in C^1(\mathbf{R}^3 \backslash \overline{\Omega}) \cap C^2 \left(\bigcup_{l=0}^{k} \Omega_l \backslash \overline{\Omega} \right) \middle| \ u \text{ satisfies } (2.1) \text{ in } \bigcup_{l=0}^{k} \Omega_l \backslash \overline{\Omega} \right\} .$$

We call u a solution of (2.1) in $\mathbf{R}^3$ (or in $\mathbf{R}^3 \backslash \overline{\Omega}$) if $u \in H(\mathbf{R}^3)$ (or $u \in H(\mathbf{R}^3 \backslash \overline{\Omega})$). If $u \in H(\mathbf{R}^3)$ or $u \in H(\mathbf{R}^3 \backslash \overline{\Omega})$, it follows that

$$u(\underset{\sim}{x}, z_l^+) = u(\underset{\sim}{x}, z_l^-) \ , \frac{\partial u}{\partial z} (\underset{\sim}{x}, z_l^+) = \frac{\partial u}{\partial z} (\underset{\sim}{x}, z_l^-) \ , \quad l = 1, 2, \ldots, k \ , \tag{3.4}$$

where z_l^+ (or z_l^-) means z approaching z_l from $z > z_l$ (or $z < z_l$).

Let $u^i(\underset{\sim}{x}, z)$ be a given incoming wave which is a solution of (2.1) in $\mathbb{R}^3$. The corresponding outgoing scattered wave $u^s(\underset{\sim}{x}, z)$ is defined by the decomposition of the total field $u = u^i + u^s$ which satisfies

$$\Delta_3 u + k^2 n(z) u = 0 \qquad \text{in} \qquad \mathbb{R}^3 \backslash \overline{\Omega} \,, \tag{3.2}$$

and where u^s has the asymptotic behavior

$$\begin{aligned} u^s(\underset{\sim}{x}, z) = \frac{e^{i\hat{k}R}}{R} F_f(\theta, \phi) + \frac{1}{\sqrt{r}} \mathsf{A}_g(r) \cdot \mathsf{F}_g(z, \theta) + R_f(R, \theta, \phi). \\ + R_g(r, z, \theta) \,, \end{aligned} \tag{3.3}$$

Here $F_f(\theta, \phi)$ is a function defined on B_1. $\mathsf{F}_g(z, \theta)$ is a function vector with components $\phi_n(z) f_n(\theta)$, $n = 1, 2, \ldots N$, where $f_n(\theta)$ is a function defined on $[0, 2\pi]$ and $\phi_n(z)$ is an eigenfunction. R_f and R_g have estimates

$$R_f(r, \theta, \phi) = O\left(\frac{1}{R^2}\right) , \qquad \text{as} \qquad R \to \infty \,,$$

$$R_g(r, z, \theta) = O\left(\frac{1}{r^{3/2}}\right) , \qquad \text{as} \qquad r \to \infty \,.$$

Defining

$$u_f(\underset{\sim}{x}, z) = u(\underset{\sim}{x}, z) - \sum_{n=1}^{N} \int_{-\infty}^{\infty} \phi_n(\zeta) u(\underset{\sim}{x}, \zeta) d\zeta \phi_n(z) \,, \quad |\underset{\sim}{x}| > r_0 \,,$$

$$u_n(\underset{\sim}{x}) = \int_{-\infty}^{\infty} \phi_n(\zeta) u(\underset{\sim}{x}, \zeta) d\zeta \,, \quad |\underset{\sim}{x}| > r_0 \,.$$

and u_f^s, u_n^s be corresponding scattered waves, we know any solution of (3.2) can be represented in the form

$$u(\underset{\sim}{x}, z) = u_f(\underset{\sim}{x}, z) + \sum_{n=1}^{N} \phi_n(z) u_n(\underset{\sim}{x})$$

for $R > R_0$, $r > r_0$ where R_0, r_0 are some positive numbers. That u^s has asymptotic behavior (3.3) is equivalent to the condition that u^s satisfies the following radiation conditions uniformly:

$$\frac{\partial u_f^s}{\partial R} - i\hat{k}u_f^s = O\left(\frac{1}{R^2}\right) , \quad \text{for} \quad |\phi - \frac{\pi}{2}| \neq 0 , \ R \to \infty ,$$

$$\frac{\partial u_n^s}{\partial r} - ika_n u_n^s = o\left(\frac{1}{\sqrt{r}}\right) , \quad r \to \infty ,$$

$$u_f^s = O\left(\frac{1}{R}\right) , \quad R \to \infty .$$

Depending on the physical property of the scatterer, the total field u satisfies different boundary conditions.

(1) Object with sound-soft boundary:

$$u = 0 , \qquad \text{on} \quad \partial\Omega . \tag{3.5}$$

(2) Object with sound-hard boundary:

$$\frac{\partial u}{\partial \nu} = 0 , \qquad \text{on} \quad \partial\Omega . \tag{3.6}$$

(3) Object with impedance boundary

$$\frac{\partial u}{\partial \nu} + \sigma u = 0 , \qquad \text{on} \quad \partial\Omega \tag{3.7}$$

where $\sigma \in C(\partial\Omega)$ and $\sigma > 0$.

We will call the problem (3.2), (3.3) and (3.5) the problem (D), call the problem (3.2), (3.3) and (3.6) the problem (N), and call the problem (3.2), (3.3) and (3.7) the problem (R).

3.1. Representation formula

Let $\Gamma_R = \Gamma_R^1 \cup \Gamma_R^2$ where $\Gamma_R^1 = \{(\underset{\sim}{x}, z) \in \mathbf{R}^3 | |\underset{\sim}{x}|^2 + z^2 = R^2, \ |\phi - \frac{\pi}{2}| \geq \frac{\pi}{4}\}$, $\Gamma_R^2 = \{(\underset{\sim}{x}, z) \in \mathbf{R}^3 | |\underset{\sim}{x}| = \frac{\sqrt{2}}{2}R , \ |z| \leq \frac{\sqrt{2}}{2}R\}$. By Green's second formula and

$$\Delta_3 G(r, z, \zeta) + k^2 n^2(z) G(r, z, \zeta) = \frac{-1}{4\pi R_\zeta^2}\delta(R_\zeta) ,$$

where $R_\zeta = [r^2 + (z - \zeta)^2]^{1/2}$, we can represent $u^s(\underset{\sim}{x}, z)$ as

$$u^s(\underset{\sim}{x},z) = \int_{\partial\Omega} \left[u^s(\underset{\sim}{\xi},\zeta) \frac{\partial G(|\underset{\sim}{x} - \underset{\sim}{\xi}|, z, \zeta)}{\partial \nu_\xi} - \frac{\partial u^s(\underset{\sim}{\xi},\zeta)}{\partial \nu_\xi} G(|\underset{\sim}{x} - \underset{\sim}{\xi}|, z, \zeta) \right] d\sigma_\xi$$

$$+ I_R(\underset{\sim}{x}, z)\ , \qquad \text{for} \quad (\underset{\sim}{x}, z) \in \Omega_R \backslash \Omega\ , \tag{3.8}$$

with

$$0 = \int_{\partial\Omega} \left[u^s(\underset{\sim}{\xi},\zeta) \frac{\partial G(|\underset{\sim}{x} - \underset{\sim}{\xi}|, z, \zeta)}{\partial \nu_\xi} - \frac{\partial u^s(\underset{\sim}{\xi}_1\zeta)}{\partial \nu_\xi} G(|\underset{\sim}{x} - \underset{\sim}{\xi}|, z, \zeta) \right] d\sigma_\xi$$

$$+ I_R(\underset{\sim}{x}, z)\ , \qquad \text{for} \quad (\underset{\sim}{x}, z) \in \Omega\ , \tag{3.9}$$

where

$$I_R(\underset{\sim}{x}, z) = -\int_{\Gamma_R} \left[u^s(\underset{\sim}{\xi},\zeta) \frac{\partial G(|\underset{\sim}{x} - \underset{\sim}{\xi}|, z, \zeta)}{\partial \nu_\xi} - \frac{\partial u^s(\underset{\sim}{\xi},\zeta)}{\partial \nu_\xi} G(|\underset{\sim}{x} - \underset{\sim}{\xi}|, z, \zeta) \right] d\sigma_\xi\ . \tag{3.10}$$

The subscript ξ is used to indicate the normal derivative and integration are taken with respect to $(\underset{\sim}{\xi}, \zeta)$. Ω_R is the interior region of Γ_R and $\Omega \subset \Omega_R$. We will prove $I_R(\underset{\sim}{x}, z) \to 0$ as $R \to \infty$ for any given $(\underset{\sim}{x}, z)$. Let $u^s(\underset{\sim}{x}, z) = u^s_f(\underset{\sim}{x}, z) + u^s_g(\underset{\sim}{x}, z)$, which has asymptotic behavior

$$u^s_f(\underset{\sim}{x}, z) = \frac{e^{i\hat{k}R}}{R} F_f(\theta, \phi) + O\left(\frac{1}{R^2}\right)\ , \quad \text{as} \quad R \to \infty\ , \tag{3.11}$$

$$u^s_g(\underset{\sim}{x}, z) = \frac{1}{\sqrt{r}} A_g(r) \cdot F_g(z, \theta) + O\left(\frac{1}{r^{3/2}}\right), \quad \text{as} \quad r \to \infty\ . \tag{3.12}$$

$$\begin{aligned} I_R(\underset{\sim}{x}, z) = &-\int_{\Gamma_R^1} \left(u^s \frac{\partial G}{\partial \nu_\xi} - \frac{\partial u^s}{\partial \nu_\xi} G \right) d\sigma_\xi \\ &- \int_{\Gamma_R^2} \left(u^s_f \frac{\partial G_f}{\partial \nu_\xi} - \frac{\partial u^s_f}{\partial \nu_\xi} G_f \right) d\sigma_\xi - \int_{\Gamma_R^2} \left(u^s_f \frac{\partial G_g}{\partial \nu_\xi} - \frac{\partial u^s_f}{\partial \nu_\xi} G_g \right) d\sigma_\xi \\ &- \int_{\Gamma_R^2} \left(u^s_g \frac{\partial G_f}{\partial \nu_\xi} - \frac{\partial u^s_g}{\partial \nu_\xi} G_f \right) d\sigma_\xi - \int_{\Gamma_R^2} \left(u^s_g \frac{\partial G_g}{\partial \nu_\xi} - \frac{\partial u^s_g}{\partial \nu_\xi} G_g \right) d\sigma_\xi \\ &=: I_R^1 + I_R^2 + I_R^3 + I_R^4 + I_R^5\ . \end{aligned} \tag{3.13}$$

We consider I_R^1 first. On Γ_R^1, $|\phi' - \frac{\pi}{2}| \geq \frac{\pi}{4}$ and $|\zeta| = R|\cos\phi'| \to \infty$ as $R \to \infty$. Hence

$$u^s = \frac{e^{i\hat{k}R}}{R} F_f(\theta', \phi') + O\left(\frac{1}{R^2}\right),$$

$$G|\underset{\sim}{x} - \underset{\sim}{\xi}|, z, \zeta) = \frac{e^{i\hat{k}R}}{R} F_f^0(\theta', \phi') + O\left(\frac{1}{R^2}\right)$$

due to $\phi_n(\zeta) \sim e^{-k|\sigma_n||\zeta|}$ when $|\zeta| \to \infty$. Rewriting the expression for I_R^1,

$$I_R^1 = -\int\limits_{\Gamma_R^1} \left[u^s\left(\frac{\partial G}{\partial R} - i\hat{k}G\right) - G\left(\frac{\partial u^s}{\partial R} - i\hat{k}u^s\right)\right] d\sigma,$$

we can conclude that $I_R^1 \to 0$ as $R \to \infty$ since as $R \to \infty$,

$$\frac{\partial G}{\partial R} - i\hat{k}G = O\left(\frac{1}{R^2}\right), \qquad G = O\left(\frac{1}{R}\right),$$

$$\frac{\partial u^s}{\partial R} - i\hat{k}u^s = O\left(\frac{1}{R^2}\right), \qquad u^s = O\left(\frac{1}{R}\right).$$

The same conclusion can be reached for I_R^2 in the same way if we first deform the integral surface from Γ_R^2 to $\left\{(\underset{\sim}{x}, z) \in \mathbf{R}^3 \big| |\underset{\sim}{x}|^2 + z^2 = R^2, \ |\phi - \frac{\pi}{2}| \leq \frac{\pi}{4}\right\}$.

I_R^3 and I_R^4 can be studied in a similar way. In view of $\frac{\sqrt{2}}{2} \leq \frac{r'}{\rho} \leq 1$ and

$$\text{(3.14)} \qquad \frac{\partial u_f^s}{\partial r'} - i\hat{k}\frac{r'}{\rho}u_f^s = \frac{r'}{\rho}\left(\frac{\partial u_f^s}{\partial \rho} - i\hat{k}u_f^s\right) + \frac{\partial u_f^s}{\partial \phi'}\frac{\zeta}{\rho^2} = O\left(\frac{1}{\rho^2}\right)$$

on Γ_R^2, we obtain

$$I_R^3 = \int\limits_{\Gamma_R^2} \left[G_g\left(\frac{\partial u_f^s}{\partial r'} - i\hat{k}\frac{r'}{\rho}u_f^s\right) - u_f^s\left(\frac{\partial G_g}{\partial r} - i\hat{k}\frac{r'}{\rho}G_g\right)\right] d\sigma$$

$$\leq \int\limits_{\Gamma_R^2} |G_g| \left|\frac{\partial u_f^s}{\partial r'} - i\hat{k}\frac{r'}{\rho}u_f^s\right| d\sigma + \int\limits_{\Gamma_R^2} |u_f^s| \sum_{n=1}^{N} |\phi_n(\zeta)||g_n^0(r', \theta')| d\sigma$$

where

$$g_n^0(r', \theta') = c_n\phi_n(z)\left[\left|\frac{\partial}{\partial r'}H_0^{(1)}(ka_n|\underset{\sim}{\xi} - \underset{\sim}{x}|)\right| + \left|\hat{k}H_0^{(1)}(ka_n|\underset{\sim}{\xi} - \underset{\sim}{x}|)\right|\right]$$

and c_n is the same as that in (2.10). Recall $G_g = O\left(\frac{1}{\sqrt{r'}}\right)$, $u_f^s = O\left(\frac{1}{\rho}\right)$ and $g_n^0(r',\theta') = O\left(\frac{1}{\sqrt{r'}}\right)$. Together with (3.14), we have

$$|I_R^3| \leq \mathbb{C} \int_{-R/\sqrt{2}}^{R/\sqrt{2}} \frac{1}{\sqrt{r'}} \cdot \frac{1}{R^2} \cdot R dz + \sum_{n=1}^{N} \int_0^{2\pi} \left[\int_{-R/\sqrt{2}}^{R/\sqrt{2}} |R u_f^s(\xi,\zeta)\phi_n(\zeta)| d\zeta \right] \frac{\mathbb{C}}{R\sqrt{r'}} r' d\theta' .$$

Here we use $\mathbb{C}$ to denote any positive constant. But

$$\int_{-R/\sqrt{2}}^{R/\sqrt{2}} |R u_f^s(\underset{\sim}{\xi},\zeta)\phi_n(\zeta)| d\zeta \leq \mathbb{C} \int_{-\infty}^{\infty} |\phi_n(\zeta)| d\zeta < \infty ,$$

from which it follows that $I_R^3 \to 0$ as $R \to \infty$. I_R^4 can be estimated in the same way. We have $I_R^4 \to 0$ as $R \to \infty$.

Now we estimate I_R^5. For r' sufficiently large, u_g^s has eigenfunction expansion

$$u_g^s(\xi,\zeta) = \sum_{n=1}^{N} \phi_n(\zeta) u_n^s(\underset{\sim}{\xi}) ,$$

hence,

$$(3.15) \quad \begin{aligned} I_R^5 &= \int_0^{2\pi} \int_{-R/\sqrt{2}}^{R/\sqrt{2}} \sum_{m,n=1}^{N} \phi_n(\zeta)\phi_m(\zeta) \left[\frac{\partial u_n^s(\underset{\sim}{\xi})}{\partial r'} g_m(\underset{\sim}{\xi}) - u_n^s(\underset{\sim}{\xi}) \frac{\partial g_m(\underset{\sim}{\xi})}{\partial r'} \right] \frac{R}{\sqrt{2}} d\zeta d\theta \\ &= \sum_{n=1}^{N} \int_{-\infty}^{\infty} \phi_n^2(\zeta) d\zeta \int_0^{2\pi} \left[\frac{\partial u_n^s(\underset{\sim}{\xi})}{\partial r'} g_n(\underset{\sim}{\xi}) - u_n^s(\underset{\sim}{\xi}) \frac{\partial g_n(\underset{\sim}{\xi})}{\partial r'} \right] r' d\theta + R_\phi , \end{aligned}$$

where

$$R_\phi = \sum_{m,n=1}^{N} \alpha_{mn} \int_{-R/\sqrt{2}}^{R/\sqrt{2}} \phi_n(\zeta)\phi_m(\zeta)d\zeta - \sum_{n=1}^{N} \alpha_{nn} \int_{-\infty}^{\infty} \phi_n^2(\zeta)d\zeta ,$$

$$\alpha_{mn} = \int_0^{2\pi} \left[\frac{\partial u_n^s(\underset{\sim}{\xi})}{\partial r'} g_m(\underset{\sim}{\xi}) - u_n^s(\underset{\sim}{\xi}) \frac{\partial g_m(\underset{\sim}{\xi})}{\partial r'} \right] r'd\theta ,$$

$$g_m(\underset{\sim}{\xi}) = c_m \phi_m(z) H_0^{(1)}(ka_m|\underset{\sim}{x} - \underset{\sim}{\xi}|),$$

and c_m is the same as that in (2.10). The orthogonality of the $\phi_n(\zeta)$ implies that $R_\phi \to 0$ as $R \to \infty$. Now from (3.15),

$$\begin{aligned} I_R^5 = \sum_{n=1}^{N} \|\phi_n\|^2 \int_0^{2\pi} \Bigg[& g_n(\underset{\sim}{\xi}) \left(\frac{\partial u_n^s(\underset{\sim}{\xi})}{\partial r'} - ika_n u_n^s(\underset{\sim}{\xi}) \right) \\ & - u_n^s(\underset{\sim}{\xi}) \left(\frac{\partial g_n(\underset{\sim}{\xi})}{\partial r'} - ika_n g_n(\underset{\sim}{\xi}) \right) \Bigg] r'd\theta + R_\phi \\ & \to 0 , \qquad \text{as} \quad R \to \infty . \end{aligned}$$

We have used the relations:

$$\begin{aligned} & g_n(\underset{\sim}{\xi}) = O\left(\frac{1}{\sqrt{r'}}\right) , \qquad u_n(\underset{\sim}{\xi}) = O\left(\frac{1}{\sqrt{r'}}\right) , \\ & \frac{\partial u_n^s(\underset{\sim}{\xi})}{\partial r'} - ika_n u_n^s(\underset{\sim}{\xi}) = O\left(\frac{1}{r'^{3/2}}\right) , \\ & \frac{\partial g_n(\underset{\sim}{\xi})}{\partial r'} - ika_n g_n(\underset{\sim}{\xi}) = O\left(\frac{1}{r'^{3/2}}\right) , \qquad \text{as} \quad r' \to \infty . \end{aligned}$$

The following theorem summarizes the above discussion:

THEOREM 3.1. *If $u^s \in H(\mathbf{R}^3\backslash\overline{\Omega}) \cap C(\mathbf{R}^3\backslash\Omega)$ satisfies (3.3) or the equivalent radiation condition, then*

$$u^s(\underset{\sim}{x}, z) = \int_{\partial\Omega} \left[u^s(\underset{\sim}{\xi}, \zeta) \frac{\partial G(|\underset{\sim}{x} - \underset{\sim}{\xi}|, \zeta, z)}{\partial \nu_\xi} - \frac{\partial u^s(\underset{\sim}{\xi}, \zeta)}{\partial \nu_\xi} G(|\underset{\sim}{x} - \underset{\sim}{\xi}|, \zeta, z) \right] d\sigma_\xi ,$$

(3.16)

for $(\underset{\sim}{x}, z) \in \mathbf{R}^3\backslash\overline{\Omega}$

and

$$0 = \int_{\partial\Omega} \left[u^s(\underset{\sim}{\xi}, \zeta) \frac{\partial G(|\underset{\sim}{x} - \xi|, \zeta, z)}{\partial \nu_\xi} - \frac{\partial u^s(\underset{\sim}{\xi}, \zeta)}{\partial \nu_\xi} G(|\underset{\sim}{x} - \underset{\sim}{\xi}|, \zeta, z) \right] d\sigma_\xi \,,$$

$$\text{for} \quad (\underset{\sim}{x}, z) \in \Omega \,. \tag{3.17}$$

3.2. Uniqueness theorem

LEMMA 3.1. *If $n(z)$ satisfies the condition (A) and (B), then any $u \in H(\mathbf{R}^3 \backslash \Omega)$ satisfying (3.3) can be expanded in the form*

$$u(\underset{\sim}{x}, z) = \frac{e^{ikR}}{R} \sum_{j=0}^{\infty} \frac{F_f^j(\theta, \phi)}{R^j} + \sum_{n=1}^{N} \frac{e^{ika_n r}}{\sqrt{r}} \phi_n(z) \sum_{j=0}^{\infty} \frac{F_g^{nj}(\theta)}{r^j} \,,$$

$$\text{for } |\phi - \frac{\pi}{2}| \geq \delta > 0 \,, \; R > 0 \text{ and } r > r_0 \,, \tag{3.18}$$

Moreover, $F_f^j(\theta, \phi)$ in the expansion (3.18) are recursively determined by the formula

$$2ikjF_f^j(\theta, \phi) = j(j-1)F_f^{j-1}(\theta, \phi) + \mathbf{B}F_f^{j-1}(\theta, \phi) + k^2 \sum_{l=0}^{j-1} \frac{C_{j-l+1}}{\cos^{j-l+1}\phi} F_f^l(\theta, \phi) \,,$$

$$j = 1, 2, 3, \ldots, \tag{3.19}$$

where

$$\mathbf{B} := \frac{1}{\sin\theta} \frac{\partial}{\partial\theta} \left(\sin\theta \frac{\partial}{\partial\theta} \right) + \frac{1}{\sin^2\theta} \frac{\partial^2}{\partial\phi^2} \tag{3.20}$$

is Beltrami's operator for the sphere.

F_g^{nj} are determined by

$$2i(j+1)ka_n F_g^{n(j+1)}(\theta) = \left(j + \frac{1}{2}\right)^2 \left[F_g^{nj}(\theta) - \frac{\partial^2}{\partial\theta^2} F_g^{nj}(\theta) \right] \,,$$

$$j = 0, 1, 2, \ldots, \quad n = 1, 2, \ldots N \,. \tag{3.21}$$

Proof. Since $n(z)$ satisfies the condition (A) and (B), $G(|\underset{\sim}{x} - \underset{\sim}{\xi}|z, \zeta)$ has expansion (2.22). (3.18) follows by straightforward calculation from (2.22) and (3.16). The expansion (3.18) must satisfy (3.2). In view of the asymptotic behavior of $\phi_n(z)$ as $z \to \infty$, we know that both the free wave and guided wave satisfy (3.2). The series (3.18) converges absolutely and uniformly in any closed domain where $R > R_0, r > r_0$ and $|\phi - \frac{\pi}{2}| \geq \delta > 0$. Differentiating (3.18) term by term, we obtain (3.19), (3.21) in the interior of the aforementioned domain. Hence, the Lemma is implied.

Now we prove a Rellich-type lemma [7], [12].

LEMMA 3.2. *Let k be positive and $n(z)$ satisfy the condition (A) and (B). If $u \in H(\mathbf{R}\backslash\Omega)$ satisfies (3.3) and*

$$\int_{\Gamma_R} |u|^2 d\sigma = o(1)\ , \quad \text{as} \quad R \to \infty\ , \tag{3.22}$$

where $\Gamma_R = \Gamma_R^1 \cup \Gamma_R^2$ defined as before, then $u = 0$ in $\mathbf{R}^3\backslash\overline{\Omega}$.

Proof. From (3.3),

$$\int_{\Gamma_R} |u|^2 d\sigma = \int_{\Gamma_R^1} |u|^2 d\sigma + \int_{\Gamma_R^2} |u|^2 d\sigma$$

$$= \int_{\Gamma_1^1} |F_f(\theta,\phi)|^2 d\sigma + \int_{\Gamma_1^2} |F_f(\theta,\phi)|^2 d\sigma + \int_0^{2\pi} \|\mathsf{F}_g(\theta)\|^2 d\phi + o(1)$$

$$\to 0\ , \qquad \text{as} \quad R \to \infty\ .$$

It implies

$$F_f(\theta,\phi) = 0 \quad \text{in } B_1 \text{ and } \mathsf{F}_g(\theta) = 0 \text{ in } [0,2\pi]. \tag{3.23}$$

By Lemma 2.1 and noticing (3.23) means $F_f^0 = 0\ , F_g^{n0} = 0$, $n = 1,2,\ldots,N$ in (3.19) and (3.21), we conclude that $u = 0$ in a region where $|\phi - \frac{\pi}{2}| \geq \delta, R > R_0$ and $r > r_0$. However, unique continuation holds for $u \in H(\mathbf{R}^3\backslash\overline{\Omega})$ when $n(z)$ satisfies the condition (A) and (B). (cf. for example, [8]). Therefore, $u = 0$ in $\mathbf{R}^3\backslash\overline{\Omega}$.

LEMMA 3.3. *If $u \in H(\mathbf{R}^3\backslash\overline{\Omega}) \cap C(\mathbf{R}^3\Omega)$ satisfies (3.3) and homogeneous Dirichlet boundary condition (3.5), then*

$$\int_{\Gamma_R} |u|^2 d\sigma = o(1)\ , \quad R \to \infty\ .$$

Proof. By Green's second formula,

$$(3.24) \qquad \int_{\Gamma_R} \left(u\frac{\partial \overline{u}}{\partial \nu} - \overline{u}\frac{\partial u}{\partial \nu} \right) d\sigma = \int_{\partial\Omega} \left(u\frac{\partial \overline{u}}{\partial \nu} - \overline{u}\frac{\partial u}{\partial \nu} \right) d\sigma = 0$$

for any R such that Ω is contained in the interior of Γ_R. Denoting $u = u_f + u_g$, the sum of free wave and guided wave, and noticing u_g has eigenfunction expansion $u_g(\underset{\sim}{\xi}, \zeta) = \sum_{n=1}^{N} \phi_n(\zeta) u_n(\underset{\sim}{\xi})$ for r' large, we have

$$\int_{\Gamma_R} \left(u\frac{\partial \overline{u}}{\partial \nu} - \overline{u}\frac{\partial u}{\partial \nu} \right) d\sigma = \int_{\Gamma_R^1} \left[u\left(\frac{\partial \overline{u}}{\partial R} + i\hat{k}\overline{u} \right) - \overline{u}\left(\frac{\partial u}{\partial R} - i\hat{k}u \right) \right] d\sigma$$

$$- i2\hat{k}\int_{\Gamma_R^1} |u|^2 d\sigma + \int_{\Gamma_R^2} \left[u_f\left(\frac{\partial \overline{u}_f}{\partial \rho} + i\hat{k}\frac{r'}{\rho}\overline{u}_f \right) - \overline{u}_f\left(\frac{\partial u_f}{\partial \rho} - i\hat{k}\frac{r'}{\rho}u_f \right) \right] d\sigma$$

$$- i2\hat{k}\int_{\Gamma_R^2} \frac{r'}{\rho}|u_f|^2 d\sigma + \sum_{n=1}^{N} \int_{-\infty}^{\infty} \phi_n^2(\zeta) d\zeta \int_0^{2\pi} \left[u_n\left(\frac{\partial \overline{u}_n}{\partial r'} + ika_n\overline{u}_n \right) \right.$$

$$\left. - \overline{u}_n\left(\frac{\partial u_n}{\partial r'} - ika_n u_n \right) \right] r'd\theta - i2\hat{k}\sum_{n=1}^{N} a_n \int_{-\infty}^{\infty} \phi_n^2(\zeta) d\zeta \int_0^{2\pi} |u_n|^2 r'd\theta$$

$$+ \int_{\Gamma_R^2} \left(u_f\frac{\partial \overline{u}_g}{\partial \nu} - \overline{u}_g\frac{\partial u_f}{\partial \nu} \right) d\sigma + \int_{\Gamma_R^2} \left(u_g\frac{\partial \overline{u}_f}{\partial \nu} - \frac{\partial u_g}{\partial \nu}\overline{u}_f \right) d\sigma \ .$$

An estimate for each integral in the above equation can be obtained in the similar way as that in section 3.1; from which it follows that

$$- 2i\left(\hat{k}\int_{\Gamma_R^1} |u|^2 d\sigma + \hat{k}\int_{\Gamma_R^2} \frac{r'}{\rho}|u_f|^2 d\sigma + \sum_{n=1}^{N} ka_n\|\phi_n\|^2 \int_0^{2\pi} |u_n|^2 r'd\theta \right)$$
$$\to 0 \ , \qquad \text{as} \quad R \to \infty \ .$$

Hence

$$\int_{\Gamma_R} |u|^2 d\sigma \to 0 \quad , \quad \text{as} \quad R \to \infty \ .$$

Lemma 3.3 is proved.

Since (3.24) is also true for u satisfying homogeneous Neumann or impedance boundary condition, we have

LEMMA 3.4. *If* $u \in H(\mathbf{R}^3\backslash\overline{\Omega}) \cap C^1(\mathbf{R}^3\backslash\Omega)$ *satisfies (3.3) and (3.6) (or (3.7)), then*

$$\int_{\Gamma_R} |u|^2 d\sigma = o(1) \quad , \quad R \to \infty \, .$$

From Lemma 3.2, 3.3, 3.4, we can conclude

THEOREM 3.2. *The problem (D) (or (N), (R)) has at most one solution in* $H(\mathbf{R}^3 \backslash \overline{\Omega}) \cap C(\mathbf{R}^3 \backslash \Omega)$ *(or in* $H(\mathbf{R}^3 \backslash \overline{\Omega}) \cap C^1(\mathbf{R}^3 \backslash \Omega)$*).*

3.3. Existence theorem

Define operators $\mathsf{S}, \mathsf{K}, \mathsf{K}', \mathsf{T}$ from $C(\partial\Omega)$ to $C(\partial\Omega)$ by

$$\mathsf{S}\phi := 2 \int_{\partial\Omega} G(|\underset{\sim}{x} - \underset{\sim}{\xi}|, z, \zeta)\phi(\underset{\sim}{\xi}, \zeta) d\sigma_\xi \ , \ (\underset{\sim}{x}, z) \in \partial\Omega \tag{3.2}$$

$$\mathsf{K}\phi := 2 \int_{\partial\Omega} \frac{G(|\underset{\sim}{x} - \underset{\sim}{\xi}|, z, \zeta)}{\partial \nu_\xi} \phi(\underset{\sim}{\xi}, \zeta) d\sigma_\xi \ , \ (\underset{\sim}{x}, z) \in \partial\Omega \ , \tag{3.26}$$

$$\mathsf{K}'\phi := 2 \int_{\partial\Omega} \frac{2G(|\underset{\sim}{x} - \underset{\sim}{\xi}|, z, \zeta)}{\partial \nu_x} \phi(\underset{\sim}{\xi}, \zeta) d\sigma_\xi \ , \ (\underset{\sim}{x}, z) \in \partial\Omega \ , \tag{3.27}$$

$$\mathsf{T}\phi := 2 \int_{\partial\Omega} \frac{2G(|\underset{\sim}{x} - \underset{\sim}{\xi}|, z, \zeta)}{\partial \nu_x \partial \nu_\xi} \phi(\underset{\sim}{\xi}, \zeta) d\sigma_\xi \ , \ (\underset{\sim}{x}, z) \in \partial\Omega \ , \tag{3.28}$$

Recall that any Green's function is a Levi function whose potential integral satisfies jump conditions on the boundary (see, for example, [9]), we can conclude from the representation formula (3.16):

LEMMA 3.5. *(1) If* $u = u^i + u^s$, $u^i \in H(\mathbf{R}^3)$, $u^s \in H(\mathbf{R}^3 \backslash \overline{\Omega}) \cap C(\mathbf{R}^3 \backslash \Omega)$ *is a solution of the problem (D), then*

$$\frac{\partial u^s}{\partial \nu} + \mathsf{K}' \left(\frac{\partial u^s}{\partial \nu} \right) = -\mathsf{T} u^i \qquad \text{on} \quad \partial\Omega \, . \tag{3.29}$$

(2) If $u = u^i + u^s$, $u^i \in H(\mathbf{R}^3)$, $u^s \in H(\mathbf{R}^3 \backslash \overline{\Omega}) \cap C^1(\mathbf{R}^3 \backslash \Omega)$ *is a solution of the problem (N), then*

$$u^s - \mathsf{K} u^s = \mathsf{S} \left(\frac{\partial u^i}{\partial \nu} \right) \qquad \text{on} \quad \partial\Omega \, . \tag{3.30}$$

(3) If $u = u^i + u^s$, $u^i \in H(\mathbf{R}^3)$, $u^s \in H(\mathbf{R}^3 \backslash \overline{\Omega}) \cap C^1(\mathbf{R}^3 \backslash \Omega)$ *is a solution of the problem (R), then*

$$(3.31) \qquad u^s - \mathsf{K}u^s - \mathsf{S}(\sigma u^s) = \mathsf{S}\left(\frac{\partial u^i}{\partial \nu} + \sigma u^i\right) \quad \text{on} \quad \partial\Omega .$$

The solutions to equations (3.29), (3.30) and (3.31) are not necessary unique. To avoid this inconvenience we use a modified integral equation used in [2]. Represent a function in $H(\mathbf{R}^3\backslash\Omega)$ as

$$(3.32) \qquad u^s(\underset{\sim}{x}, z) = \int_{\partial\Omega} \left(\frac{\partial G}{\partial \nu_\xi} + i\lambda G\right) \varphi(\underset{\sim}{\xi}, \zeta) d\sigma_\xi, \quad (\underset{\sim}{x}, z) \notin \partial\Omega .$$

If u is a solution of the problem (D), where $\varphi \in C(\partial\Omega)$ and $\lambda > 0$ is any real number. Then (3.5) implies

$$(3.33) \qquad \varphi + (\mathsf{K} + i\lambda\mathsf{S})\varphi = -2u^i \quad \text{on} \quad \partial\Omega .$$

THEOREM 3.3. *(1) (3.33) has a unique solution $\varphi \in C(\partial\Omega)$. (2) Problem (D) has a unique solution represented by (3.32) where φ is determined uniquely by (3.33).*

Proof. (1) Since $\mathsf{K} + i\lambda\mathsf{S}$ is compact on $C(\partial\Omega)$, we need only to prove $\varphi + (\mathsf{K} + i\lambda\mathsf{S})\varphi = 0$ implies $\varphi = 0$. Let u^s be defined by (3.32) with φ satisfying the homogeneous equation. Then $u^s = 0$ on $\partial\Omega$. Uniqueness theorem 3.2 implies $u^s \equiv 0$ in $\mathbf{R}^3\backslash\Omega$. Notice (3.32) defines a solution of (2.1) in Ω when $(\underset{\sim}{x}, z) \in \Omega$. Let u^s_+, $(\frac{\partial u^s}{\partial \nu})_+$ be the limits of u^s , $\frac{\partial u^s}{\partial \nu}$ as $\Omega \ni (\underset{\sim}{x}, z) \to (\underset{\sim}{x}_0, z_0) \in \partial\Omega$, we obtain from the jump relation on $\partial\Omega$ that

$$u^s_+ = -\varphi ,$$
$$\left(\frac{\partial u^s}{\partial \nu}\right)_+ = i\lambda\varphi .$$

Hence, $(\frac{\partial u^s}{\partial \nu})_+ + \lambda u^s_+ = 0$ on $\partial\Omega$. The interior Robin problem for (2.1) has only trivial solution for $\lambda > 0$. It means $u^s = 0$ in Ω and $\varphi = -u^s_+ = 0$ on $\partial\Omega$.

(2) u^s defined by (3.32) satisfies (3.3) and belong to $H(\mathbf{R}^3\backslash\overline{\Omega})$. It is a solution of the problem (D) if and only if φ satisfies (3.33). So (2) is implied directly by (1).

To show the existence of the solution to problem (N), we represent u^s in the form

$$(3.34) \qquad u^s(\underset{\sim}{x}, z) = \int_{\partial\Omega} \left(G + i\lambda\frac{\partial G}{\partial \nu}\right) \varphi d\sigma , \quad (\underset{\sim}{x}, z) \notin \partial\Omega .$$

If u^s is a solution of the problem (N), then

$$(3.35) \qquad \phi - (\mathsf{K}' + i\lambda\mathsf{T})\phi = 2u^i \quad \text{on} \quad \partial\Omega .$$

Similarly to the proof of Theorem 3.3, we can prove

THEOREM 3.4. *(1) (3.35) has a unique solution in* $C(\partial\Omega)$.
(2) The problem (N) has a unique solution represented by (3.34) where φ *is determined uniquely by (3.35).*

If u^s represented by (3.34) is a solution of the problem (R), then

$$(1 - i\sigma\lambda)\phi - (\mathsf{K}' + i\lambda\mathsf{T} + i\lambda\sigma\mathsf{K} + \sigma\mathsf{S})\phi = 2u^i\ , \qquad \text{on} \quad \partial\Omega\ . \tag{3.36}$$

THEOREM 3.5. *(1) (3.36) has a unique solution.*
(2) Problem (R) has a unique solution which is given by (3.34) where $\phi \in C(\partial\Omega)$ *satisfies (3.36).*

Aknowledgement

This research was supported in part by the Institute for Mathematics and its Applications with funds provided by the National Science Fundation, the Minnesota Supercomputer Institute and the Alliant Techsystem Co.

References

[1] BREKHOUVSKIKH, L.M., *Waves in Layered Media*, Academic Press, (1960).

[2] COLTON, D. AND KRESS, R., *Integral Equation Methods in Scattering Theory*, John Wiley, New York (1983).

[3] EWING, W., JARDETZKY, W AND PRESS, F., *Elastic Waves in Layered Media,*, McGraw–Hill, New York (1957).

[4] GILBERT, R.P. AND XU, Y., *Starting fields and far fields in ocean acoustics*, Wave Motion, 11 (1989), 507–524.

[5] GILBERT, R.P. AND XU, Y, *Dense sets and the projection theorem for acoustic harmonic waves in homogeneous finite depth ocean*, Mathematical Methods in the Applied Sciences, 12 (1989), 69–76.

[6] LEVITAN, B.M., *Inverse Sturm–Liouville Problems*, VNU Sciences Press, Utrecht, the Netherlands (1987).

[7] HARTMAN, P. AND WILCOX, C., *On solution of the Helmholtz equation in exterior domains*, Math. Zeitschr. (1961), 75, 228–255.

[8] HORMANDER, L, *Linear Partial Differential Operators*, Springer–Verlag, Berlin (1963).

[9] MIRANDA, C., *Partial Differential Equations of Elliptic Type*, Springer–Verlag, New New (1970).

[10] RAMM, A.G., *Scattering by Obstacles*, D. Reidel Publishing Company, Dordrecht (1986).

[11] WILCOX, C.H., *Sound Propagation in Stratified Fluids*, Springer–Verlag, New York (1984).

[12] WILCOX, C.H., *Scattering Theory for the d'Alembert Equation in Exterior Domains*, Springer Lecture Notes in Mathematics, V. 442. Springer–Verlag, New York (1975).

[13] XU, Y., *The propagating solution and far-field patterns for acoustic harmonic waves in a finite depth ocean*, Applicable Analysis, 35 (1990), 129–151.

[14] XU, Y., *An injective far-field pattern operator and inverse scattering problem in a finite depth ocean*, Proc. Edinburgh Math. Soc. (1991) 34, 295-311..

L R BRAGG AND P SHI

Some non-classical heat problems associated with thermoelasticity

1. Introduction

In this paper, we will be concerned with a partial differential-integral equation of the form

$$c\Theta_t(r,t) - \Delta_\mu \Theta(r,t) = \epsilon \frac{\partial}{\partial t} \int_0^1 \sigma^\alpha \Theta(\sigma, t) d\sigma \tag{1.1}$$

in which $\Delta_\mu = D_r^2 + \mu r^{-1} D_r$ is the radial Laplacian operator with $\mu \geqslant 1$, c is a non-negative constant and ϵ is a small positive parameter. In general, we will select $\alpha = \mu$ but there are some exceptions. Our interest in this paper is on the construction of classical solutions to (1.1), of both the initial value problem and the initial boundary value problem. When $\epsilon = 0$, (1.1) reduces to the classical radial heat equation and homogeneous boundary value problems associated with this equation can be solved by the usual separation of variables along with the attendant Sturm-Liouville theory. When $\epsilon \neq 0$, the separation of variables technique leads to completeness questions that cannot be handled by standard theory. The equation (1.1) arises in connection with problems in thermoelasticity in which initial and boundary conditions are specified. In the case $\mu = 0$, initial-boundary value problems for (1.1) have been treated thoroughly by Day [6].

The rest of the paper is organized as follows. In section 2 we discuss the initial value problem for (1.1) and obtain a representation formula for its solution. The inappropriateness of the Sturm-Liouville theory will become evident in section 3, where we attempt a heuristic approach to the mixed initial-boundary value problem associated with (1.1) by separation of variables. The eigenfunctions for such problems can be expressed in terms of Bessel functions. On the other hand, the eigenvalues can only be obtained approximately. An algorithm that makes use of Newton's method for estimating these is given in section 4. Orthogonality questions that lead to a non-standard inner product are considered in section 5 and these, in turn, give rise to the completeness question of these eigenfunctions relative to that inner product, whose answer requires an interesting argument from the aspect of functional analysis. This latter material and the justification of the solution derived in section 3 are covered in section 6, where we treat an abstract evolution equation in Hilbert spaces that comprises (1.1) as a special case. Finally, in section 7, we briefly illustrate how the field equations for thermoelasticity in the radially symmetric case can be combined into a single equation having the form (1.1).

2. Solution to the Initial Value Problem

The initial value problem associated with (1.1) when $\epsilon = 0$ has been treated by Bragg [4] and Haimo [7]. An extension of those approaches provides some interesting insights on the solution structure of (1.1) when $\epsilon \neq 0$. We now construct a solution of (1.1) corresponding to the pair of conditions

$$\Theta_r(0,t) = 0, \quad t > 0; \tag{2.1a}$$

$$\Theta(r,0) = \varphi(r) \equiv \sum_{n=0}^{\infty} a_n\, r^{2n}, \quad r > 0, \tag{2.1b}$$

in which φ is entire in r^2 of growth $\big(1, 1/(4\tau)\big)$ (see Boas [3]). Now the solution $U(r,t)$ of (1.1) with $\epsilon = 0$ corresponding to the conditions (2.1) can be expressed as ([4], [7])

$$U(r,t) = \sum_{n=0}^{\infty} a_n\, R_n^{\mu+1}(r, t/c) \tag{2.2}$$

in which the $R_n^{\mu+1}(r,t)$ are radial heat polynomials (expressible in terms of the generalized Laguerre polynomials). The series (2.2) converges in the time strip $|t| < \tau$. We assume that a solution of (1.1) with $\epsilon \neq 0$ but small has the form

$$\Theta(r,t) = U(r,t) + F(t). \tag{2.3}$$

Introducing this into (1.1), we find that the function $F(\xi)$ satisfies the conditions

$$\left(c - \frac{\epsilon}{\alpha+1}\right) F'(\xi) = \epsilon \int_0^1 \sigma^\alpha \left[\frac{\partial}{\partial \xi} U(\sigma, \xi)\right] dr, \quad F(0) = 0. \tag{2.4}$$

Upon integrating this from $\xi = 0$ to $\xi = t$, we deduce

$$\left(c - \frac{\epsilon}{\alpha+1}\right)F(t) = \epsilon \int_0^1 \sigma^\alpha \{U(\sigma,t) - U(\sigma,0)\} d\sigma$$
$$= \epsilon \int_0^1 \sigma^\alpha U(\sigma,t) d\sigma - \epsilon \int_0^1 \sigma^\alpha \varphi(r) d\sigma. \tag{2.5}$$

Inserting this $F(t)$ into (2.3), we finally obtain

$$\Theta(r,t) = U(r,t) + \frac{\epsilon}{c - \frac{\epsilon}{\alpha+1}} \int_0^1 \sigma^\alpha U(\sigma,t) d\sigma - \frac{\epsilon}{c - \frac{\epsilon}{\alpha+1}} \int_0^1 \sigma^\alpha \varphi(\sigma) d\sigma \tag{2.6}$$

It is readily checked that this $\Theta(r,t)$ satisfies the conditions (2.1). One can similarly deduce the form (2.6) when $\varphi(r) \in L^2(0,\infty)$. We refer the reader to [7] for relavent discussions.

3. A Heuristic Approach to the Initial-Boundary Value Problem

We now consider the equation (1.1) with $\alpha = \mu$ along with the initial-boundary conditions

$$\Theta_r(0,t) = 0 \qquad t > 0, \tag{3.1a}$$
$$\Theta(1,t) = 0 \qquad t > 0, \tag{3.1b}$$
$$\Theta(r,0) = \varphi(r) \qquad 0 < r < 1. \tag{3.1c}$$

The situation is much more difficult for the construction of solutions than the initial value problem considered in section 2. We invoke the standard assumption of separation of variables for (1.1), namely $\Theta(r,t) = R(r)T(t)$. It then follows that

$$\Delta_\mu R(r) + \lambda^2 c R(r) = \epsilon \lambda^2 \int_0^1 \sigma^\mu R(\sigma) d\sigma \tag{3.2}$$

and

$$T'(t) + \lambda^2 T(t) = 0, \tag{3.3}$$

in which $\lambda > 0$ denotes eigenvalues to be determined. Now the solution of (3.2) corresponding to $\epsilon = 0$ is easily shown to be

$$\widetilde{R}(r) = \Gamma\left(\frac{\mu+1}{2}\right)\left(\frac{2}{\lambda\sqrt{c}r}\right)^{\frac{\mu-1}{2}} J_{\frac{\mu-1}{2}}(\lambda\sqrt{c}r), \tag{3.4}$$

where $J_\nu(x)$ denotes the standard Bessel function of index ν. Since the right hand side of (3.2) is a constant, all solutions to (3.2) must be of the form

$$R(r) = \widetilde{R}(r) + k\int_0^1 \sigma^\mu \widetilde{R}(\sigma)d\sigma \tag{3.5}$$

where k is a constant to be determined. Inserting this choice for $R(r)$ into (3.2) and noting the special properties of $\widetilde{R}(r)$, it is not difficult to show that $k = \epsilon/\left(c - \frac{\epsilon}{\mu+1}\right)$. Hence

$$R(r) = \widetilde{R}(\sigma) + \frac{\epsilon}{c - \frac{\epsilon}{\mu+1}}\int_0^1 \sigma^\mu \widetilde{R}(\sigma)d\sigma. \tag{3.6}$$

If we perform the integration in this and simplify, we can finally show that (see e.g. [9])

$$R(r) = \Gamma\left(\frac{\mu+1}{2}\right)\left(\frac{2}{\lambda\sqrt{c}}\right)^{\frac{\mu-1}{2}}\cdot$$
$$\left[\sigma^{\frac{1-\mu}{2}} J_{\frac{\mu-1}{2}}(\lambda\sqrt{c}r) + \frac{\epsilon}{c - \frac{\epsilon}{\mu+1}}\frac{J_{\mu+1}2(\lambda\sqrt{c})}{\lambda\sqrt{c}}\right]. \tag{3.7}$$

When $\mu = 1$ (corresponding to 2-space dimensions), this formula simplifies to

$$R(r) = J_0(\lambda\sqrt{c}r) + \frac{\epsilon}{c - \epsilon/2}\frac{J_1(\lambda\sqrt{c})}{\lambda\sqrt{c}}. \tag{3.8}$$

We also note that if $\mu = 2m$, then the right member of (3.7) can be expressed in terms of sine and cosine functions (see [6] for the case $\mu = 0$).

From these solution functions, we immediately see that $R'(0) = 0$. The function $R(r)$ satisfies the condition $R(1) = 0$ if

$$J_{\frac{\mu-1}{2}}(\lambda\sqrt{c}) + \frac{\epsilon}{c - \frac{\epsilon}{\mu+1}}\frac{J_{\frac{\mu+1}{2}}(\lambda\sqrt{c})}{\lambda\sqrt{c}} = 0. \tag{3.9}$$

As we will see in the next section, the values of λ that satisfy this are close to the zeros of $J_{\frac{\mu-1}{2}}(\lambda\sqrt{c})$. They form a discrete increasing set of real numbers that tends to $+\infty$. We denote the set by $\{\lambda_n;\ n = 1, 2, \ldots\}$.

Let $R_n(r)$ denote the function (3.7) corresponding to λ_n. Then the associated $T_n(t)$ is given by $\exp\,(-\lambda_n^2 t)$. Hence, a formal solution of (1.1) is defined by a series of the form

$$\Theta(r,t) = \sum_{n=1}^{\infty} A_n\, R_n(r)\, T_n(t) \tag{3.10}$$

Requiring that $\Theta(r,0) = \varphi(r)$ leads to the relation

$$\sum_{n=1}^{\infty} A_n\, R_n(r) = \varphi(r) \tag{3.11}$$

for determining the coefficients A_n in the representation (3.10). To obtain these A_n and to justify that (3.10)-(3.11) indeed solves (1.1) and (3.1) in an appropriate sense, we need to answer some deeper questions.

(1) Under what inner product would $\{R_n\}$ be orthogonal?

(2) Under the above orthogonality, what is the function space spanned by $\{R_n\}$ so that (3.11) holds for appropriate $\{A_n\}$?

Anewers to these questions will be given in section 5 and section 6.

4. The Approximation of Non-zero Eigenvalues

Because of the properties of Bessel functions, the classical Newton's method for the approximation of $\{\lambda_k\}$ turns out to be surprisingly simple in form. In this section we briefly demonstrate this and show some numerical examples. For this purpose, let

$$\nu = (\mu - 1)/2, \quad x = \lambda\sqrt{c}, \quad \text{and} \quad \delta = \epsilon/(c - \epsilon/\mu + 1).$$

Then solving for λ from (3.9) is equivalent to solving the equation

$$f(x) \equiv J_\nu(x) + \delta J_{\nu+1}(x)/x = 0.$$

Upon using the derivative relations [9]

$$J'_\nu(x) = \{\nu\, J_\nu(x) - x\, J_{\nu+1}(x)\}/x, \tag{4.1}$$

$$J'_{\nu+1}(x) = \{x\, J_\nu(x) - (\nu+1)\, J_{\nu+1}(x)\}/x, \tag{4.2}$$

we can show that the Newton approximation scheme for the kth non-trivial zero $x^{(k)}$ of $f(x)$ becomes

$$x_{n+1}^{(k)} = x_n^{(k)} - \frac{\left\{J_\nu(x_n^{(k)}) + \delta J_{\nu+1}(x_n^{(k)})\cdot(x_n^{(k)})^{-1}\right\}}{\left\{(\nu+\delta)J_\nu(x_n^{(k)})(x_n^{(k)})^{-1} - (1+(\nu+2)\delta(x_n^{(k)})^{-2})J_{\nu+1}(x_n^{(k)})\right\}} \tag{4.3}$$

Now the zeros of $f(x)$ separate the zeros of $J_\nu(x)$ and $J_{\nu+1}(x)$. We shall see that they are slightly larger than the zeros of $J_\nu(x)$. This useful observation allows us to initiate the iteration by zeros of J_ν without skipping possible values of $\{\lambda_k\}$. To approximate the kth root x^k of $f(x)$, we select $x_0^{(k)}$ to be the kth non-trivial zero of $J_\nu(x)$. Then the formula (4.3) for $x_1^{(k)}$ simplifies to

$$x_1^{(k)} = x_0^k + \frac{\delta x_0^{(k)}}{(x_0^{(k)})^2 + (\nu+2)\delta} \approx x_0^{(k)} + \delta/x_0^{(k)} \tag{4.4}$$

Clearly, $x_1^{(k)}$ is to the right of $x_0^{(k)}$. As k gets larger and larger, we see that $x_1^{(k)}$ gets closer and closer to $x_0^{(k)}$. Since δ is small, the scheme (4.3) will converge rapidly. In fact, for large k or small δ a single iteration is already within the range of small errors.

We conclude this section by the following table based on the algrithm (4.4). Only one iteration is performed. Our calculation is assisted by a pocket calculator and a table for zeros of Bessel functions (5 decimal places). We have taken $\mu = 1$, $c = 4$ and $\epsilon = 0.1$.

number of root	root of J_ν	root of f	difference
1	2.40483	2.41527	0.01044
2	5.52008	5.52465	0.00457
3	8.65373	8.65665	0.00290

The first three eigenvalues of (3.9) can thus be approximated by dividing the entries in the third column above by $\sqrt{c} = 2$.

5. Orthogonality Considerations

We now wish to investigate the orthogonality relations for $\{R_n(r);\ n = 1, 2\cdots\}$ obtained in section 3. Since Sturm-Liouville theory does not fit in with (3.2), we need to introduce a new inner product. We remark that the form of the orthogonality can be recovered by the abstract theory in section 6. However, the concrete inner product formula obtained below will provide a clear model on which our abstract theory is developed.

Theorem 5.1. *Let* $\langle R_m(r), R_n(r)\rangle$ *denote the inner product*

$$\langle R_m(r), R_n(r)\rangle = c\int_0^1 \sigma^\mu r R_m(\sigma)\, R_n(\sigma)\, d\sigma - \epsilon\Big(\int_0^1 \sigma^\mu R_m(\sigma) d\sigma\Big)\Big(\int_0^1 \sigma^\mu R_n(\sigma) d\sigma\Big) \tag{5.1}$$

Then

$$\langle R_m(r), R_n(r)\rangle = \begin{cases} 0 & \textit{if } m \neq n, \\ \Lambda_m & \textit{if } m = n, \end{cases}$$

where

$$\begin{aligned}\Lambda_m &= \lambda_n^{-2}\int_0^1 \sigma^\mu \big(R_n'(\sigma)\big)^2 d\sigma \\ &= \Big[\Gamma\big(\frac{\mu+1}{2}\big)\Big]^2 2^{\mu-2} c^{\frac{1-\mu}{2}} \lambda_n^{-1-\mu} \times \\ &\quad \Big[\lambda_n^2 c\big(J'_{\frac{\mu+1}{2}}(\lambda_n\sqrt{c})\big)^2 + \big(\lambda_n^2 c - (\frac{\mu+1}{2})^2\big)\, J^2_{\frac{\mu+1}{2}}(\lambda\sqrt{c})\Big]\end{aligned} \tag{5.2}$$

Proof. Case 1, $m \neq n$. In the usual way, we multiply the equation (3.2) corresponding to the eigenvalue λ_m by $r^\mu R_n(r)$ and subtract this from the equation obtained by multiplying (3.2) corresponding to the eigenvalue λ_n by $r^\mu R_m(r)$. We obtain

$$\begin{aligned}0 &= \frac{\partial}{\partial r}\{r^\mu(R_m'(r)R_n(r) - R_m(r)R_n'(r))\} + c(\lambda_m^2 - \lambda_n^2) r^\mu R_m(r) R_n(r) \\ &\quad - \epsilon\Big\{\lambda_m^2 r^\mu R_n(r)\int_0^1 \sigma^\mu R_m(r) d\sigma - \lambda_n^2 r^\mu R_n(r)\int_0^1 \sigma^\mu R_n(\sigma) d\sigma\Big\}\end{aligned}$$

Upon integrating with respect to r from 0 to 1 and using the facts that $R_n(1) = R'_n(0)$, we find

$$c(\lambda_m^2 - \lambda_n^2)\int_0^1 \sigma^\mu R_m(\sigma)R_n(\sigma)d\sigma$$
$$= \epsilon\left\{\left(\int_0^1 \sigma^\mu R_n(\sigma)d\sigma\right)\left(\int_0^1 \sigma^\mu R_m(\sigma)d\sigma\right)\right\}(\lambda_m^2 - \lambda_n^2).$$

Since $m \neq n$ and consequently $\lambda_m \neq \lambda_n$, this reduces to

$$c\int_0^1 \sigma^\mu R_m(\sigma)R_n(\sigma)d\sigma - \epsilon\left(\int_0^1 \sigma^\mu R_m(\sigma)d\sigma\right)\left(\int_0^1 \sigma^\mu R_n(\sigma)d\sigma\right) = 0. \qquad (5.3)$$

Case 2, $m = n$. Now, multiply (3.2), corresponding to λ_n, by $r^\mu R_n(r)$ and integrate on r from $r = 0$ to $r = 1$. We have

$$\int_0^1 (r^\mu R''_n(r)R_n(r) + \mu r^{\mu-1}R'_n(r)R_n(r))dr$$
$$+ \lambda_n^2 c\int_0^1 r^\mu R_n^2(t)dr - \epsilon\lambda_n^2\left(\int_0^1 r^\mu R_n(r)dr\right)^2 = 0$$

Integrating the first term in this by parts and simplifying, we get

$$c\int_0^1 \sigma^\mu (R_n(\sigma))^2 d\sigma - \epsilon\left(\int_0^1 \sigma^\mu R_n(\sigma)d\sigma\right)^2 = \lambda_n^{-2}\int_0^1 \sigma^\mu (R'_n(\sigma))^2 d\sigma. \qquad (5.4)$$

The evaluation of the right member of (5.4) can be carried out by employing standard integration formulas for the Bessel functions. From (3.7) and the relation

$$\frac{d}{dz}(z^{\frac{1-\mu}{2}} J_{\frac{\mu-1}{2}}(z)) = -z^{\frac{1-\mu}{2}} J_{\frac{\mu+1}{2}}(z)$$

we can show that

$$R'_n(r) = -\Gamma(\frac{\mu+1}{2})2^{\frac{\mu-1}{2}}(\lambda_n\sqrt{c})^{\frac{3-\mu}{2}} r^{\frac{1-\mu}{2}} J_{\frac{\mu+1}{2}}(\lambda_n\sqrt{c}r). \qquad (5.5)$$

The evaluation of the right member of (5.5) thus reduces to the computaton of the following expression:

$$\left\{\Gamma(\frac{\mu+1}{2})\right\}^2 2^{\mu-1}c^{\frac{3-\mu}{2}}\lambda_n^{1-\mu}\int_0^1 \sigma\left[J_{\frac{\mu+1}{2}}(\lambda_n\sqrt{c}\sigma)\right]^2 d\sigma \qquad (5.6)$$

Using the well known formula

$$\int_0^\beta x J_\nu^2(x) dx = \frac{1}{2}\left[\beta^2 \left(J_\nu^1(\beta)\right)^2 + \left(\beta^2 - \nu^2\right) J_\nu^2(\beta) + \nu^2 J_\nu^2(0)\right]$$

and the change of variables $\lambda_n \sqrt{c}\,\sigma = x$ in (5.6), we can finally show that (5.2) holds. The proof of Theorem 5.1 is complete.

Now, suppose that the function $\varphi(r)$ in (3.1) has the formal representation

$$\varphi(r) = \sum_{n=1}^{\infty} A_n R_n(r), \quad \{A_n\} \subset \mathbb{R}.$$

Then using Theorem 5.1, we find that

$$\langle \varphi, R_m \rangle = A_m \Lambda_m, \text{ or } A_m = \Lambda_m^{-1} \langle \varphi, R_m \rangle.$$

Hence, we formally have the representation

$$\varphi(r) = \sum_{n=1}^{\infty} \Lambda_n^{-1} \langle \varphi, R_n \rangle R_n(r)$$

and in view of (3.10) the solution to (1.1) with the conditions in (3.1) becomes

$$\Theta(r,t) = \sum_{n=1}^{\infty} \Lambda_n^{-1} \langle \varphi, R_n \rangle R_n(r) \exp(-\lambda_n^2 t). \tag{5.7}$$

There remains the mathematical justification of the above formal procedures. We complete this in the next section by treating an abstract generalization of problems involving (1.1). This generalization is also of independent interest.

6. A Nonclassical Semigroup

In this section we will be concerned with an abstract non-classical evolution problem that is modeled on (1.1) with the conditions (1.3). The problem has an interesting nonlocal structure and does not fall within the standard theory of semigroups. It also serves as a mathematical justification to the formal procedure of separation of variables developed in previous sections.

Let H be a Hilbert space with the inner product $(\cdot,\cdot)$ and the induced norm $\|\cdot\|$. Let $A : D(A) \subset H \to H$ be a symmetric, maximal dissipative operator, where $D(A)$ denotes the domain of A. We assume that A^{-1} is compact in the norm topology of H. Let φ and x be given elements in H, $f \in C^{\alpha}[0,+\infty; H)$ for some $0 < \alpha \le 1$. Under the these assumptions, we consider the the abstract evolution problem

$$\begin{cases} \dfrac{du}{dt} - Au = \epsilon \dfrac{d}{dt}(x,u)x + f(t) & \text{for } t > 0, \\ u(0) = \varphi. \end{cases} \tag{6.1}$$

To ensure the classical solvability[1] of (6.1), we further assume

$$0 < \epsilon < 1/\|x\|^2, \quad \text{and} \quad (x, f(t)) = 0 \quad \forall\, t > 0. \tag{6.2}$$

These assumptions will be in force throughout this section without mention again from place to place. It is interesting to note that (6.2) turns out to be a natural assumption for our applications in section 7.

We now invoke the method of separation of variables by setting $u = S(t)v$, where $v \in D(A) \subset H$ and $S(t)$ is a scalar function. Substituting this info (6.1), we obtain the separated system

$$Av + \lambda v = \epsilon\lambda(x,v)x, \qquad v \in D(A), \tag{6.3}$$

$$S(t) = \beta \exp\{-\lambda t\}, \qquad t > 0. \tag{6.4}$$

where λ and β are to be determined. Note that (6.3) is a perturbed eigenvalue problem associated with the operator A, but it is not obvious that these eigenvalues form a discrete set and that the corresponding eigenvectors generate a complete orthogonal basis. To establish these facts, we act on both sides of (6.3) by A^{-1}. This defines an operator K by the relation

$$Kv := A^{-1}v - \epsilon(x,v)A^{-1}x = -\lambda^{-1}v, \tag{6.5}$$

which also states that $-\lambda^{-1}$ is in the spectrum of the operator K.

[1]We refer the reader to Pazy [8] for the definition of the classical solution

Lemma 6.1. *The operator $K : H \to H$ defined in (6.5) is compact and self-adjoint under the equivalent inner product defined by*

$$\langle u, v\rangle = (u, v) - \epsilon(x, u)(x, v), \quad u,\ v \in H.$$

Proof. From (6.2) it follows that the norm induced by the new inner product is equivalent to $\|\cdot\|$ (by Schwartz inequality). Hence, the compactness of K follows from (6.5) since that relation yields the compactness of K under the original norm $\|\cdot\|$. Next, we prove the self-adjointness of K under the new inner product using the self-adjointness of A^{-1} under the original inner product. This follows from the straight forward computation:

$$\begin{aligned}\langle Ku, v\rangle &= (A^{-1}u - \epsilon(x, u)A^{-1}x, v) - \epsilon(x, A^{-1}u - \epsilon(x, u)A^{-1}x)(x, v)\\ &= (A^{-1}u, v) - \epsilon(x, u)(A^{-1}x, v) - \epsilon(x, v)(A^{-1}x, u)\\ &+ \epsilon^2(x, u)(x, v)(x, A^{-1}x) = \langle u, Kv\rangle.\end{aligned}$$

For our purposes, we say that a real number λ is a quasi-eigenvalue of the operator A if the following holds:

(1) the equation (6.3) admits nontrivial solutions and
(2) the set of solutions of (6.3) corresponding to λ is a finite dimensional subspace of H.

If λ is a quasi-eigenvalue of A, we call a non-trivial solution of (6.3) corresponding to λ a quasi-eigenvector of λ. The multiplicity of λ is the dimension of the subspace spanned by the corresponding quasi-eigenvectors. Based on these notions, we establish

Lemma 6.2. *The quasi-eigenvalues of A form a discrete set and can be arranged in the nondecreasing order*

$$0 < \lambda_1 \leqslant \lambda_2 < \cdots \leqslant \lambda_n \cdots \ \text{and} \ \lim_{n\to\infty} \lambda_n = +\infty.$$

Moreover, every λ_n is of single multiplicity. The set of all normalized quasi-eigenvectors form a complete orthonormal basis $\{e_n\}$ of H under the inner product $\langle\ ,\ \rangle$.

Proof. We first notice that 0 is not a quasi-eigenvalue of A. Next, we show that A admits no negative quasi-eigenvalues. Since A is maximal dissipative, it is well known that (see Tanabe [10]) $A + \lambda I$ is invertible for all negative λ and

$$\|(A+\lambda I)^{-1}\| \leq \frac{1}{-\lambda}, \quad \forall\, \lambda < 0. \tag{6.6}$$

If $\lambda < 0$ in (6.3) for some $v \in D(A)$, then applying $(A+\lambda I)^{-1}$ to both sides of (6.3) and using (6.6), we conclude $\|v\| \leqslant \epsilon\|x\|^2\,\|v\|$. Hence $v = 0$ in view of (6.2). Therefore, all quasi-eigenvalues of A are positive. Based on lemma 6.1 and the positivity of the quasi-eigenvalues of A, we conclude from the spectral theory of compact operators that all of the eigenvalues of K can be arranged as follows:

$$\mu_1 \leqslant \mu_2 \leqslant \cdots \leqslant \mu_n \cdots < 0 \text{ and } \lim_{n\to\infty} \mu_n = 0.$$

Each μ_n is of single multiplicity and the corresponding eigenvectors form a complete and normalized orthonormal basis for H. We denote this basis by $\{e_n\}$. In view of (6.3) and (6.5), $\lambda_n = -1/\mu_n$, $n = 1, 2, \ldots$, are all of the quasi-eigenvalues of A and the $\{e_n\}$ are the quasi-eigenvectors of A. This completes the proof.

We now come to the main result for problem (6.1). In order to state it clearly, we introduce the semigroup on H,

$$\widetilde{T}(t)\varphi = \sum_{n=1}^{\infty} \exp(-\lambda_n t)\langle\varphi, e_n\rangle e_n, \quad t > 0. \tag{6.7}$$

It is straightforward to prove that $\widetilde{T}(t)$ is an analytic semigroup of contraction whose infinitesimal generator is given by

$$\tilde{A}\xi = -\sum_{n=1}^{\infty} \lambda_n\langle\xi, e_n\rangle e_n, \quad \xi \in D(\tilde{A}). \tag{6.8}$$

Here $D(\tilde{A})$ denotes the domain of $\tilde{A}$. It consists of all elements in H such that the infinite series on the right hand side of (6.8) converges in the norm of H. Using notations (6.7) and (6.8) we have the following result.

Theorem 6.3. *The unique solution to problem (6.1) is given by*

$$u(t) = \widetilde{T}(t)\varphi + \int_0^t \widetilde{T}(t-s)f(s)ds \tag{6.9}$$

Moreover, u is a solution to (6.1) if and only if it solves the standard evolution equaition (in classical sense)

$$\frac{du}{dt} + \tilde{A}u = f(t), \quad u(0) = \varphi. \tag{6.10}$$

Proof. From the standard theory of linear semigroups, (6.9) yields the unique classical solution to (6.10). However, it remains to show that u, given by (6.9), is also a classical soltuion to (6.1). This will complete the proof of the theorem since (6.1) can have at most one classical solution in light of (6.2). The key step towards this goal is to prove that $u \in D(A)$ and to interchange the action by A with the summation in $\widetilde{T}(t)\varphi$. To this end, we define

$$\widetilde{T}_n(t)\xi = \sum_{j=1}^{n} \exp(-\lambda_j t)\,\langle \xi, e_j\rangle\, e_j, \quad \xi \in H$$

and

$$u_n = \widetilde{T}_n(t)\varphi + \int_0^t \widetilde{T}_n(t-s)f(s)\,ds, \quad n = 1, 2\cdots$$

Recall that $\tilde{A}e_j = -\lambda_j e_j$. It is easy to see that

$$\frac{du_n}{dt} - \tilde{A}\,u_n = \widetilde{T}_n(0)f(t) = \sum_{j=1}^{n}\langle f(t), e_j\rangle e_j. \tag{6.11}$$

For each $\xi \in D(\tilde{A})$, it follows from (6.3) and the definition of $\langle\ ,\ \rangle$ that

$$\begin{aligned}\langle \tilde{A}\xi, e_n\rangle &= -\lambda_n\langle \xi, e_n\rangle \\ &= -\lambda[(\xi, e_n) - \epsilon(x,\xi)(x, e_n)] \\ &= (-\lambda_n\, e_n + \epsilon\lambda_n(x, e_n)x, \xi) \\ &= (A\, e_n, \xi), \quad n = 1, 2, \ldots\end{aligned} \tag{6.12}$$

In particular, we have

$$\langle \tilde{A}\, u_n, e_k\rangle = (A\, e_k, u_n) = (e_k, A\, u_n), \quad k,\ n \in N \tag{6.13}$$

since A is symmetric and $\{e_k;\ k \in N\} \subset D(A) \cap D(\tilde{A})$. We now take the inner product of (6.11) with each e_k using $\langle\ ,\ \rangle$. After making simplifications by means of (6.13), we obtain

$$\left(\frac{d\,u_n}{dt} - A\,u_n - \epsilon\frac{d}{dt}(x, u_n)x, e_k\right) = \left(f_n - \epsilon(x, f_n)x,\ e_k\right) \quad k \in N,$$

where

$$f_n(t) = \sum_{j=1}^{n} \langle f(t)\ ,\ e_j\rangle e_j.$$

Therefore, it follows from the completeness of $\{e_k;\ k \in N\}$ that

$$A\,u_n = \frac{d\,u_n}{dt} - \epsilon\frac{d}{dt}(x, u_n)x - f_n(t) + \epsilon(x, f_n)x, \quad n \in N,\ t > 0. \tag{6.14}$$

On the other hand, a straight forward computation shows that

$$\frac{d\,u_n}{dt} = -\sum_{j=1}^{n} \lambda_j \exp(-\lambda_j t)\,\langle\varphi, e_j\rangle$$
$$+ \widetilde{T}_n(t)\,f(0) + \int_0^t \widetilde{T}_n(t-s)\,f'(s)\,ds.$$

from which it follows that

$$\frac{d\,u_n}{dt} \to \frac{d}{dt}\widetilde{T}(t)\,\varphi + \widetilde{T}(t)\,f(0) + \int_0^t \widetilde{T}(t-s)\,f'(s)\,ds \tag{6.15}$$

strongly in H for each fixed t as n tends to ∞. Recall that $\widetilde{T}(t)$ is a semigroup of contraction. Based on standard calculations, the right side of (6.15) is equal to $u'(t)$. Therefore (6.15) can be written as

$$\frac{d\,u_n}{dt} \to \frac{du}{dt} \text{ strongly in } H \text{ for each fixed } t. \tag{6.16}$$

Combining (6.14)-(6.16) and using $(x, f) = 0$, we have

$$A\,u_n \to \frac{du}{dt} - \epsilon\frac{d}{dt}(x, u)\,x - f(t) \tag{6.17}$$

strongly in H for each fixed t. Since A is maximal dissipative and therefore closed, it follows from (6.17) that

$$A\,u = \frac{du}{dt} - \epsilon \frac{d}{dt}(x,u)\,x - f(t)$$

This shows that $u(t)$, defined by (6.9), is in $D(A)$ and is a solution of (6.1). Theorem 6.3 is proved.

We now return to problem (1.1) with the initial boundary conditons (3.1a)-(3.1c). To apply Theorem 6.3 to the problem, we let H be the weighted $L^2(0,1)$ space with the weight r^α. It consists of all functions u in $L^2(0,1)$ such that

$$\int_0^1 r^\alpha [u(r)]^2\,dr < +\infty$$

with the usual weighted inner product

$$(u,v) = \int_0^1 r^\alpha u(r) v(r) dr, \quad u,\ v \in H \tag{6.18}$$

and the induced norm. Let A be the radial Laplace operator $\Delta_\mu = D_r^2 + \mu r^{-1} D_r$, whose domain consists of all functions with generalized derivatives up to second order belonging to H, and moreover, satisfying conditions (3.1a)-(3.1b). In such a setting, (1.1) with (3.1) is precisely of the form (6.1) and all the assumptions required for the operator A in Theorem 6.3 are satisfied. We refer the reader to Yosida [11] and Avantaggiati [1] for detail.

7. An Application to Thermoelasticity.

In this final section, we derive an equation analogous to (1.1) from the governing equations of the quasi-static approximation of linearized thermoelasticity. Following Carlson [5], these linearized equations in the isotropic and homogeneous case can be written as

$$\lambda \frac{\partial}{\partial x_i}\frac{\partial u_k}{\partial x_k} + \mu \left(\frac{\partial}{\partial x_j}\frac{\partial u_i}{\partial x_j} + \frac{\partial}{\partial x_j}\frac{\partial u_j}{\partial x_i} \right) - \beta \frac{\partial \theta}{\partial x_i} = f_i, \quad i = 1, 2 \cdots n, \tag{7.1}$$

and

$$q\Delta\theta = \rho s \frac{\partial \theta}{\partial t} + \beta c \frac{\partial}{\partial t} \operatorname{div} \mathbf{u}, \tag{7.2}$$

in which $\mathbf{u} = (u_1, u_2, \cdots, u_n)$ denotes the displacement vector and θ is the temperature. In (7.1) and (7.2) λ and μ are Láme containts; β is the interaction constant; q, ρ, s and c denote, respectively, the heat conductivity, the density, the specific heat, and the absolute reference temperature. We assume that the deformation of the material and the temperature distribution in it are radial symmetric with respect to the origin. This leads to the additional constitutive assumptions

$$\theta(x_1, \cdots, x_n, t) = \Theta(r,t) \quad \text{and} \quad u_i(x_1, \cdots, x_n, t) = R(r,t)\frac{x_i}{r}, \quad i = 1, 2 \cdots n. \tag{7.3}$$

Upon substituting (7.3) into (7.1), after somewhat tedious simplifications we obtain the radially symmetric equation for (7.1),

$$\frac{\partial}{\partial r}\left[r^{1-n}\frac{\partial}{\partial r}\left(r^{n-1}R\right)\right] = \frac{\beta}{\lambda + 2\mu}\frac{\partial \Theta}{\partial r} + f(r,t). \tag{7.4}$$

On the other hand, the equation (7.2) reduces to

$$qr^{1-n}\frac{\partial}{\partial r}\left[r^{n-1}\frac{\partial}{\partial r}\Theta\right] = \rho s \frac{\partial \Theta}{\partial t} + \beta c \frac{\partial}{\partial t}\left[r^{1-n}\frac{\partial}{\partial r}\left(r^{n-1}R\right)\right]. \tag{7.5}$$

Note that the time t is only a parameter in (7.4), as the result of the quasistatic approximation. This allows us to decouple the system using appropriate boundary conditions. Because of the symmetry with respect to the origin, we assume the condition of symmetry

$$\frac{\partial R}{\partial r}(0,t) = 0 \quad \text{and} \quad \frac{\partial \Theta}{\partial r}(0,t) = 0 \quad t > 0. \tag{7.6}$$

For the sake of simplicity, we consider the homogeneous Dirichlet boundary condition at $r = 1$.

$$R(1,t) = 0 \quad \text{and} \quad \Theta(1,t) = 0, \quad t > 0. \tag{7.7}$$

Following Day [6], we integrate (7.4) from r to 1, to obtain

$$\frac{\partial R}{\partial r}(1,t) - r^{1-n}\frac{\partial}{\partial r}\left(r^{n-1}R\right) = -\frac{\beta}{\lambda + 2\mu}\Theta + \int_r^1 f(\tau,t)\tau, \tag{7.8}$$

where we used (7.6) and (7.7). Multiplying (7.8) by $-r^{n-1}$ and integrating the result from 0 to 1 with respect to r, yield that

$$\frac{\partial R}{\partial r}(1,t) = -\frac{n\beta}{\lambda+2\mu}\int_0^1 r^{n-1}\Theta dr + n\int_0^1\int_r^1 r^{n-1} f(s,t)dsdr.$$

This and (7.8) implies that

$$r^{1-n}\frac{\partial}{\partial r}\left(r^{n-1}R\right) = -\frac{n\beta}{\lambda+2\mu}\int_0^1 r^{n-1}\Theta dr + \frac{\beta}{\lambda+2\mu}\Theta + F^*(r,t), \tag{7.9}$$

where

$$F^*(r,t) = n\int_0^1\int_r^1 r^{n-1} f(s,t)dsdr - \int_r^1 f(s,t)ds. \tag{7.10}$$

Substituting (7.9) into (7.5), after an obvious simplification we obtain

$$qr^{1-n}\frac{\partial}{\partial r}\left[r^{n-1}\frac{\partial}{\partial r}\Theta\right] = \left(\rho s + \frac{\beta^2 c}{\lambda+2\mu}\right)\frac{\partial\Theta}{\partial t} - \frac{\beta^2 nc}{\lambda+2\mu}\frac{\partial}{\partial t}\int_0^1 r^{n-1}\Theta dr + F(r,t), \tag{7.11}$$

where

$$F(r,t) = \beta c\frac{\partial}{\partial t}F^*(r,t).$$

To simplify (7.11), we let

$$a = \frac{q(\lambda+2\mu)}{\rho s(\lambda+2\mu)+\beta^2 c} \quad \text{and} \quad b = \frac{\beta^2 nc}{\rho s(\lambda+2\mu)+\beta^2 c}.$$

Then that equation can be written as

$$\frac{\partial\Theta}{\partial t} - ar^{1-n}\frac{\partial}{\partial r}\left(r^{n-1}\frac{\partial}{\partial r}\Theta\right) = b\frac{\partial}{\partial t}\int_0^1 r^{n-1}\Theta dr + \widetilde{F}(r,t), \tag{7.12}$$

where we used $\widetilde{F}(r,t)$ to denote $F(r,t)/[\rho s(\lambda+2\mu)+\beta^2 c]$. Thus we have obtained the decoupled equation in terms of the single unknown Θ.

This last equation is of the form (1.1) with an inhomogeneous term $\widetilde{F}(r,t)$. It can be shown that $\widetilde{F}(r,t)$ satisfies the second condition in (6.2),

$$(1, \widetilde{F}(\cdot,t)) = 0 \quad t > 0,$$

where $(\cdot,\cdot)$ is given by (6.18). We remark that the coefficient b on the right hand side of (7.12) is very small in applications since it is proportional to the square of the interaction coefficient β, which is very small (see e.g. [6]). This means that the first condition in (6.1) is not restrictive for applications. Therefore Theorem 6.3 can be applied directly to (7.12) with the initial boundary conditions (3.1).

REFERENCES

1. A. Avantaggiati, *On compact embedding theorems in weighted Sobolev spaces*, Czech. Math. J. **29** (1979), 635-647.

2. M. A. Abramowitz and I. A. Stegun, *Handbook of mathematical functions with formulas, graphs, and mathematical tables*, Nat. Bur. of Stds. 55, Washington, D.C., 1964.

3. R. P. Boas, *Entire functions*, Academic Press, New York, 1954.

4. L. R. Bragg, *The radial heat polynomials and related functions*, Trans. Amer. Math. Society **119** (1965), 270–290.

5. D. E. Carlson, *Linear thermoelasticity*, in Handbuch der Physik vol. VI a/2, (S. Flugg, ed.), Springer, Berlin.

6. W. A. Day, *Heat conduction within linear thermoelasticity*, Springer-Verlag, New York, 1985.

7. D. T. Haimo, *Functions with the Huygen's property*, Bull. Amer. Math. Society **71** (1965), 528–532.

8. A. Pazy, *Semigroups of linear operators and applications to partial differential equations*, Springer-Verlag, New York, 1983.

9. W. Magnus, F. Oberhettinger, R. P. Soni, *Formulas and theorems for the special functions of mathematical physics*, Springer-Verlag, New York, 1966.

10. H. Tanabe, *Equations of evolution*, Pitman, 1979.

11. K. Yosida, *Lectures on differential and integral equations*, Inter Sciences, Inc., New York, 1960.

Department of Mathematical Sciences, Oakland University, Rochester, Michigan 48309-4401, U.S.A

C O HORGAN AND L E PAYNE

The influence of geometric perturbations on the decay of Saint-Venant end effects in linear isotropic elasticity

Abstract.

The influence of domain boundary perturbation on the decay of Saint-Venant end effects in linear isotropic elastostatics is considered. The problem investigated is that of a semi-infinite cylinder with lateral sides fixed, prescribed displacements at the near ends and displacements tending to zero as the axial variable tends to infinity. Methods involving L_2 estimates are used to assess the influence of perturbations in the cross-sectional domain geometry on the exponential decay of solutions.

1. INTRODUCTION.

In recent years numerous investigations have dealt with the question of clarifying and extending Saint-Venant's principle in elasticity (see [6], [7] for recent reviews). In general one attempts to show for long thin members, clamped or traction-free along the long sides, that, away from the loaded ends, the solution of the problem in question decays exponentially in some appropriate measure to the solution of a simpler (usually lower dimensional) problem. Most of the work in this area which has been carried out since the survey articles [6], [7] appeared has been concerned with decay behavior for solutions of related nonlinear problems (see e.g. [8-16] and the references cited therein).

In actual practice it may be difficult to characterize the geometry of the member precisely, or the actual geometry may in fact be a perturbation of some simple geometry which is relatively easy to analyze. In either case the true geometry may differ from the geometry assumed in modeling, and in this paper we investigate the effect of such geometric perturbations on the decay behavior when the long sides are held fixed. We know that the difference between the solution of the actual problem and that corresponding to the perturbed geometry will decay exponentially in any convenient measure since the solutions of the separate problems exhibit this decay. The purpose of our investigation is to show that the two solutions remain "close" to one another throughout the length of the body if the end data are "close" and the geometric perturbation is "small". This statement will be made precise in what follows.

We remark that if the long sides of the member were traction- free rather than fixed an additional difficulty would arise as a consequence of the fact that tractions involve the normal vector on the boundary. In this case if the solutions to the actual and the perturbed problems were to remain "close" to one another we would expect not only that the distance between related points on the two boundaries should be small, but also that the normal vectors to the two surfaces at these related points should be directed in approximately the same direction.

We are concerned in this paper with solutions of the equations of equilibrium of linear isotropic elasticity which we write in the form

$$\Delta u_i + \omega u_{j,ji} = 0, \quad i = 1,2,3. \qquad (1.1)$$

Here we have adopted the usual convention of summing over repeated indices, Latin indices indicating summation from 1 to 3 and Greek indices indicating summation from 1 to 2. A subscript preceded by a comma denotes partial differentiation with respect to the corresponding Cartesian coordinate. In the elasticity context the functions u_i are components of displacement and the positive constant ω is expressed in terms of the usual Poisson's ratio ν $(-1 < \nu < 1/2)$ as

$$\omega = [1-2\nu]^{-1}, \quad 1/3 < \omega < \infty. \qquad (1.2)$$

For the range of Poisson's ratio indicated above, the associated strain-energy density is positive definite. We are concerned specifically with the Dirichlet problem for (1.1) on a semi-infinite cylinder $\tilde{R} = \{(x_1,x_2,x_3)|(x_1,x_2) \in \tilde{D},\ x_3 > 0\}$ where $\tilde{D}$ is a bounded simply-connected domain in $\mathbb{R}^2$ with Lipschitz boundary $\partial\tilde{D}$. We assume the existence of a solution vector $\underset{\sim}{u} \in C^1(\bar{\tilde{R}})$ satisfying (1.1) in $\tilde{R}$ and subject to the conditions

$$u_i(x_1,x_2,x_3) = 0, \ (x_1,x_2) \in \partial\tilde{D}, \ x_3 > 0, \tag{1.3}$$

$$u_i(x_1,x_2,x_3) \to 0 \text{ (uniformly in } \tilde{D}) \text{ as } x_3 \to \infty, \tag{1.4}$$

$$u_i(x_1,x_2,0) = \tilde{f}_i(x_1,x_2), \ (x_1,x_2) \in \tilde{D} \tag{1.5}$$

where $i = 1,2,3$. In (1.5) we assume that $\tilde{f}_i$ is a piecewise C^2 vector function in $\tilde{D}$ which vanishes on $\partial\tilde{D}$. It is well known that conditions (1.3)-(1.5) imply that solutions of (1.1) actually decay exponentially in L_2 as $x_3 \to \infty$ (see [2, 4, 5]).

Our aim here is to compare the solution u_i of (1.1), (1.3)-(1.5) with the solution v_i of

$$\Delta v_i + \omega v_{j,ji} = 0 \text{ in } \hat{R} \tag{1.6}$$

$$v_i(x_1,x_2,x_3) = 0, \ (x_1,x_2) \in \partial\hat{D}, \ x_3 > 0 \tag{1.7}$$

$$v_i(x_1,x_2,x_3) \to 0 \text{ (uniformly in } \hat{D}) \text{ as } x_3 \to \infty \tag{1.8}$$

$$v_i(x_1,x_2,0) = \hat{f}_i(x_1,x_2), \ (x_1,x_2) \in \hat{D}. \tag{1.9}$$

Here $\hat{D}$ is a bounded simply-connected domain in $\mathbb{R}^2$ which may be regarded as a perturbation of $\tilde{D}$, and $\hat{R} = \{(x_1,x_2,x_3)|(x_1,x_2) \in \hat{D}, \ x_3 > 0\}$. We assume as before that $v_i \in C^1(\bar{\hat{R}})$ and that $\hat{f}_i(x_1,x_2) = 0$ on $\partial\hat{D}$.

Specifically we wish to compare u_i and v_i as the axial variable x_3 increases. If $\hat{D}$ were a simple perturbation of $\tilde{D}$ we might make the comparison by simply mapping the region $\hat{D}$ onto $\tilde{D}$. However, since we wish to consider general (not necessarily specified) perturbations we shall adopt the comparison technique of Crooke and Payne [3]. Thus we define

$$D = \tilde{D} \cap \hat{D} \tag{1.10}$$

and set

$$W_i + w_i = u_i - v_i, \tag{1.11}$$

where W_i satisfies

$$\begin{aligned} &\Delta W_i + \omega W_{j,ji} = 0 \quad \text{in} \quad R_z, \\ &W_i = 0 \quad \text{on} \quad \partial D \times [0,\infty), \\ &W_i \to 0 \quad \text{(uniformly in D) as} \quad x_3 \to \infty, \\ &W_i(x_1,x_2,0) = \tilde{f}_i(x_1,x_2) - \hat{f}_i(x_1,x_2), \ (x_1,x_2) \in D, \end{aligned} \tag{1.12}$$

and

$$\begin{aligned} &\Delta w_i + \omega w_{j,ji} = 0 \quad \text{in} \quad R_z, \\ &w_i = u_i - v_i \quad \text{on} \quad \partial D \times [0,\infty), \\ &w_i \to 0 \quad \text{(uniformly in D) as} \quad x_3 \to \infty, \\ &w_i(x_1,x_2,0) = 0, \ (x_1,x_2) \in D, \end{aligned} \tag{1.13}$$

where $R_z = \{(x_1, x_2, x_3) | (x_1, x_2) \in D,\ x_3 > z\}$. In order to study the behavior of $u_i - v_i$ in R_z as z increases we introduce the L_2 integral which, by virtue of (1.11), satisfies

$$\left\{\int_{R_z} (u_i - v_i)(u_i - v_i) dV\right\}^{1/2} \leq \left\{\int_{R_z} W_i W_i dV\right\}^{1/2} + \left\{\int_{R_z} w_i w_i dV\right\}^{1/2}$$

$$\equiv (\Phi_1)^{1/2} + (\Phi_2)^{1/2}. \qquad (1.14)$$

A bound for Φ_1 could be obtained from results of Biollay [2]. Here, however, a more direct method (see Appendix B) is used to obtain a bound of the form

$$\Phi_1(z) \leq K_1 e^{-k_1 z}, \qquad (1.15)$$

where K_1, k_1 are computable constants. In the main body of the paper we confine attention to investigating the decay of $\Phi_2(z)$ as z increases. Since we use it later we introduce the notation

$$D' = \tilde{D} \cup \hat{D} \qquad (1.16)$$

and assume for convenience that both $\tilde{D}$ and $\hat{D}$ (and hence D and D') are star-shaped with respect to a common interior point of D which we take as the origin in the (x_1, x_2) plane.

If we now define δ to be the largest distance along a ray (from the origin) between a point on $\partial\hat{D}$ and a point on $\partial\tilde{D}$ we shall prove that

$$\Phi_2(z) \leq K_2 \delta e^{-\gamma z}, \quad z \geq 0, \tag{1.17}$$

where K_2 and γ are computable constants.

On combining (1.15) and (1.17) in (1.14) we obtain our main result

$$\left[\int_{R_z} (u_i - v_i)(u_i - v_i)\, dV\right]^{1/2}$$

$$\leq \left[K_1 e^{-k_1 z}\right]^{1/2} + \left[K_2 \delta e^{-\gamma z}\right]^{1/2}. \tag{1.18}$$

It is shown in Appendix B that K_1 will be small if the data $\tilde{f}_i$, $\hat{f}_i$ are "close" to one another. On the other hand, the constant δ will be small if $\tilde{D}$ and $\hat{D}$ are "close" to one another. If $\tilde{D}$ and $\hat{D}$ <u>coincide</u>, then $\delta = 0$ and (1.18) reduces to the usual Saint-Venant decay estimate. On the other hand, if $\tilde{f}_i$ and $\hat{f}_i$ are equal in D, then $K_1 = 0$.

We remark that in general one could allow $\tilde{D}$ and $\hat{D}$ (and hence D and D') to depend on x_3 so that for instance, it would be possible to consider cross-sections that expand as x_3 increases, but this would add additional complications and the decay would not in general be exponential.

In the next section we will make use of λ, the first eigenvalue in the fixed membrane problem for D, i.e., λ is determined by

$$\begin{aligned} \Delta\varphi + \lambda\varphi &= 0 \quad \text{in} \quad D \\ \varphi &= 0 \quad \text{on} \quad \partial D \\ \varphi &> 0 \quad \text{in} \quad D. \end{aligned} \tag{1.19}$$

Actually it suffices to have a lower bound for λ, and such bounds may be found from standard isoperimetric inequalities, monotony principles, etc. (see e.g. Bandle [1]).

We also make use of the notation

$$p = \min_{\partial D} x_\beta n_\beta > 0 \tag{1.20}$$

where n_β is the β-component of the unit normal vector directed outward on ∂D. The positivity of p follows from the fact that D is star-shaped. We also define d as

$$d = \max_{D'}[x_1^2+x_2^2]^{1/2}. \tag{1.21}$$

The techniques of this paper could be modified to study the effects of geometric perturbation in the case of linear anisotropic elasticity where (1.1) is replaced by

$$[c_{ijk\ell}u_{k,\ell}]_{,j} = 0, \quad i = 1,2,3. \tag{1.22}$$

Here the constants $c_{ijk\ell}$ are the usual elastic moduli which are assumed to lead to an associated strain-energy density which is positive definite. The constants K_2 and γ of (1.17) would, however, be considerably more complicated in this case.

2. DERIVATION OF (1.17).

We recall that the function w_i is a solution of

$$\Delta w_i + \omega w_{j,ji} = 0 \quad \text{in} \quad R_z, \tag{2.1}$$

$$w_i = u_i - v_i \quad \text{on} \quad \partial D \times [0,\infty), \tag{2.2}$$

$$w_i \to 0 \quad \text{(uniformly in D) as} \quad x_3 \to \infty, \tag{2.3}$$

$$w_i(x_1,x_2,0) = 0, \ (x_1,x_2) \in D. \tag{2.4}$$

We seek a bound for $\Phi_2(z)$ defined by (1.14). The first step is to make use of the following inequality derived in Appendix A:

$$\Phi_2(z) \le k^{-1}\left\{\sigma \int_z^\infty \oint_{\partial D} w_i w_i ds\, dx_3 - \Phi_2'(z)\right\}. \tag{2.5}$$

where

$$k^{-1} = 2(1+\omega)(2\lambda^{-1})^{1/2} \tag{2.6}$$

and

$$\sigma = [d+\tau\lambda^{-1/2}]/p, \quad \tau = (3\sqrt{2}-4)/8, \tag{2.7}$$

with λ given by (1.19), p by (1.20) and d by (1.21). In (2.5), the prime denotes differentiation with respect to z. To solve the differential inequality (2.5), we shall need a bound for the integral

$$J(z) = \int_z^\infty \oint_{\partial D} w_i w_i ds\, dx_3. \tag{2.8}$$

To this end we define for arbitrary $x_3 > 0$

$$u_i^* = \begin{cases} u_i & \text{in } \tilde{D} \\ 0 & \text{in } \mathbb{R}^2/\tilde{D} \end{cases}$$

$$v_i^* = \begin{cases} v_i & \text{in } \hat{D} \\ 0 & \text{in } \mathbb{R}^2/\hat{D} \end{cases} \tag{2.9}$$

and set

$$w_i^* = u_i^* - v_i^*. \tag{2.10}$$

Applying the divergence theorem in the exterior domain $\mathbb{R}^2/D = D^c$ noting that $x_\beta n_\beta$ (and hence p) is positive on ∂D, and using the fact that $w_i^* \equiv 0$ in $\mathbb{R}^2/D'$ we obtain

$$J(z) \le \frac{1}{p} \int_z^\infty \oint_{\partial D} w_i^* w_i^* \, x_\beta n_\beta \, ds \, dx_3$$

$$= -\frac{2}{p} \int_z^\infty \int_{D^c} w_i^* w_i^* \, dA \, dx_3 - \frac{2}{p} \int_z^\infty \int_{D^c} w_i^* w_{i,\beta}^* \, x_\beta dA \, dx_3$$

$$\le \frac{2d}{p} \left\{ \int_z^\infty \int_{D^c} w_i^* w_i^* \, dA \, dx_3 \int_z^\infty \int_{D^c} w_{i,\beta}^* w_{i,\beta}^* \, dA \, dx_3 \right\}^{1/2}, \tag{2.11}$$

where p and d are given by (1.20) and (1.21) respectively. In the third step we have dropped a negative term on the right and used Schwarz's inequality.

We now require a bound for the L_2 integral in (2.11) in terms of the Dirichlet integral. This could be derived by making use of lower bounds for the first eigenvalue of the associated Rayleigh Quotient; however, we derive a somewhat cruder bound by a much simpler method.

In polar coordinates let

$$r = f(\theta) \tag{2.12}$$

be the equation for ∂D. Then

$$\int_z^\infty \int_{D^c} w_i^* w_i^* \, dA \, dx_3 = \int_z^\infty \int_{D^c} [r-f(\theta)]_{,r} \, w_i^* w_i^* r \, dr \, d\theta \, dx_3. \quad (2.13)$$

An application of the divergence theorem gives

$$\int_z^\infty \int_{D^c} w_i^* w_i^* \, dA \, dx_3 = -2 \int_z^\infty \int_{D^c} [r-f(\theta)] w_i^* w_{i,r}^* r \, dr \, d\theta \, dx_3$$

$$- \int_z^\infty \int_{D^c} [r-f(\theta)] w_i^* w_i^* \, dr \, d\theta \, dx_3$$

$$< 2\left\{ \int_z^\infty \int_{D^c} w_i^* w_i^* \, dA \, dx_3 \int_z^\infty \int_{D'/D} [r-f(\theta)]^2 w_{i,\beta}^* w_{i,\beta}^* \, dA \, dx_3 \right\}^{1/2}. \quad (2.14)$$

In the last step we have dropped a negative term on the right and used Schwarz's inequality. Now since

$$0 \le [r-f(\theta)] \le \delta \quad (2.15)$$

in D'/D we find

$$\int_z^\infty \int_{D^c} w_i^* w_i^* \, dA \, dx_3 \le 4\delta^2 \int_z^\infty \int_{D'/D} w_{i,\beta}^* w_{i,\beta}^* \, dA \, dx_3. \quad (2.16)$$

Inserting (2.16) back into (2.11) we find

$$J(z) \leq \frac{4\delta d}{p} \int_z^\infty \int_{D'/D} w^*_{i,\beta} w^*_{i,\beta} \, dA \, dx_3. \qquad (2.17)$$

It would be desirable to obtain a bound for the integral in (2.17) which reflects the fact that the integral is small when δ is small. This would, however, require more smoothness of $\partial\tilde{D}$ and $\partial\hat{D}$ than we have assumed, and since the boundary ∂D is anyway usually not C^1 we might find it convenient to define D in a different manner. We may, however, derive a crude bound for $\int_z^\infty \int_{D'/D} w^*_{i,\beta} w^*_{i,\beta} \, dA \, dx_3$ as follows: The arithmetic- geometric mean inequality yields

$$\int_z^\infty \int_{D'/D} w^*_{i,\beta} w^*_{i,\beta} \, dV \leq 2[\tilde{E}(z) + \hat{E}(z)], \qquad (2.18)$$

where

$$\tilde{E}(z) = \int_{\tilde{R}_z} [u_{i,j} u_{i,j} + \omega u^2_{j,j}] dV, \qquad (2.19)$$

$$\hat{E}(z) = \int_{\hat{R}_z} [v_{i,j} v_{i,j} + \omega v^2_{j,j}] dV. \qquad (2.20)$$

Bounds for $\tilde{E}(z)$ and $\hat{E}(z)$ are well known (see [2, 4]) but since these bounds vary depending upon how certain inequalities are used we rederive them in Appendix B obtaining

$$\tilde{E}(z) \le \tilde{E}(0)e^{-\gamma_1 z}, \quad z \ge 0,$$
$$\hat{E}(z) \le \hat{E}(0)e^{-\gamma_2 z}, \quad z \ge 0. \tag{2.21}$$

Here

$$\gamma_1 = 2\tilde{\lambda}^{1/2}[1+\omega]^{-1/2},$$
$$\gamma_2 = 2\hat{\lambda}^{1/2}[1+\omega]^{-1/2}, \tag{2.22}$$

and $\tilde{E}(0)$, $\hat{E}(0)$ are the total energies in $\tilde{R}$, $\hat{R}$ respectively. It is also shown in Appendix B how these quantities may be estimated in terms of the boundary data (1.5), (1.9). It follows from (2.17), (2.18), (2.21) that

$$J(z) \le \frac{8\delta d}{p}\left\{\tilde{E}(0)e^{-\gamma_1 z} + \hat{E}(0)e^{-\gamma_2 z}\right\}. \tag{2.23}$$

We now return to (2.5), insert the bound (2.23) for J and obtain after rearrangement

$$\Phi_2'(z) + k\Phi_2 \le \frac{8\sigma\delta d}{p}\left[\tilde{E}(0)e^{-\gamma_1 z} + \hat{E}(0)e^{-\gamma_2 z}\right]. \tag{2.24}$$

Integrating from 0 to z, we obtain

$$\Phi_2(z) \le \Phi_2(0)e^{-kz} + \frac{8\delta d\sigma}{p}\left\{\tilde{E}(0)\frac{[e^{-kz}-e^{-\gamma_1 z}]}{\gamma_1 - k}\right.$$

$$\left. + \hat{E}(0)\frac{[e^{-kz}-e^{-\gamma_2 z}]}{\gamma_2 - k}\right\}. \qquad (2.25)$$

We require now a bound for $\Phi_2(0)$. By virtue of the fact that $w_i = 0$ at $x_3 = 0$ (see (1.13)), we have $\Phi_2'(0) = 0$ and so (2.5) yields

$$\Phi_2(0) \le \sigma k^{-1} \int_0^\infty \oint_{\partial D} w_i^* w_i^* \, ds \, dx_3$$

$$\le \frac{4\delta d\sigma}{pk} \int_0^\infty \int_{D'/D} w_{i,\beta}^* w_{i,\beta}^* \, dA \, dx_3$$

$$\le \frac{8\delta d\sigma}{pk} [\tilde{E}(0)+\hat{E}(0)]. \qquad (2.26)$$

To obtain the second inequality in (2.26) we have used (2.17), at $z = 0$, and the final inequality makes use of (2.18). Thus we see from (2.25), (2.26) that $\Phi_2(z)$ may be bounded as follows

$$\Phi_2(z) \le K_2 \delta e^{-\gamma z} \qquad (2.27)$$

where K_2 and γ are computable from (2.25), (2.26) once bounds for $\tilde{E}(0)$ and $\hat{E}(0)$ are derived. But these follow easily from the Dirichlet Principle for elastostatics (see Appendix B). We note that if

$$k < \min[\gamma_1, \gamma_2] \tag{2.28}$$

it follows from (2.25), (2.26) that

$$\Phi_2(z) \le \frac{4\delta d\sigma}{pk}[(k^{-1}\gamma-1)^{-1}+2][\tilde{E}(0)+\hat{E}(0)]e^{-kz}. \tag{2.29}$$

Now (2.28) is equivalent, by (2.6) and (2.22), to

$$2(1+\omega)[2/\lambda]^{1/2} > 2^{-3/2}(1+\omega)^{1/2}\{2/[\min(\tilde{\lambda},\hat{\lambda})]\}^{1/2}. \tag{2.30}$$

Since λ, $\tilde{\lambda}$, and $\hat{\lambda}$ will be close in value, inequality (2.30) will hold in general. When (2.30) holds we therefore obtain a decay rate that is conservative since we know that the decay rate is not smaller than $\min[\gamma_1, \gamma_2]$. However, our principal concern in this paper is with the factor multiplying the decaying exponential which we see from (2.27) or (2.29) is at worst of order δ.

In the special case in which $\tilde{f}_i$ and $\hat{f}_i$ are equal and of compact support in D we may use a different comparison. By the standard uniqueness result for the Dirichlet problem of linear isotropic elastostatics, it follows from (1.12) that

$$W_i \equiv 0. \tag{2.31}$$

Defining w_i^* as in (2.9), (2.10), we set

$$\Phi_2^*(z) = \int_z^\infty \int_{D'} w_i^* w_i^* \, dA \, dx_3. \tag{2.32}$$

Using the notation

$$D_1 = \tilde{D}/D \tag{2.33}$$

$$D_2 = \hat{D}/D \tag{2.34}$$

we have

$$\Phi_2^*(z) = \Phi_2(z) + \int_z^\infty \int_{D_1} u_i u_i \, dA \, dx_3 + \int_z^\infty \int_{D_2} v_i v_i \, dA \, dx_3, \tag{2.35}$$

where $\Phi_2(z)$ is still defined as in (1.14). The bound for $\Phi_2(z)$ we have already obtained. To bound the second integral on the right we note that

$$\begin{aligned}\int_z^\infty \int_{D_1} u_i u_i \, dA \, dx_3 &= \int_z^\infty \int_{D_1} [r-f(\theta)]_{,r} \, u_i u_i \, dA \, dx_3 \\ &= -2 \int_z^\infty \int_{D_1} [r-f(\theta)]\left[u_i \frac{\partial u_i}{\partial r} + \frac{1}{2r} u_i u_i\right] dA \, dx_3 \\ &\leq -2 \int_z^\infty \int_{D_1} [r-f(\theta)] u_i \frac{\partial u_i}{\partial r} \, dA \, dx_3. \end{aligned} \tag{2.36}$$

Thus from Schwarz's inequality, the bound (2.15) and other simple inequalities we have

$$\int_z^\infty \int_{D_1} u_i u_i \, dA \, dx_3 \le 4\delta^2 \int_z^\infty \int_{D_1} u_{i,j} u_{i,j} \, dA \, dx_3$$

$$\le 4\delta^2 \tilde{E}(z). \qquad (2.37)$$

In a similar way we find

$$\int_z^\infty \int_{D_2} v_i v_i \, dA \, dx_3 \le 4\delta^2 \hat{E}(z). \qquad (2.38)$$

Thus using (1.17) and (2.21) it follows from (2.35) that

$$\Phi_2^*(z) \le K_2 \delta e^{-\gamma z} + 4\delta^2 \left\{ \tilde{E}(0) e^{-\gamma_1 z} + \hat{E}(0) e^{-\gamma_2 z} \right\}. \qquad (2.39)$$

APPENDIX A. Proof of (2.5).

In order to derive the bound (2.5) for Φ_2 defined by (1.14), we introduce the auxiliary function ψ_i which satisfies

$$\Delta\psi_i + \omega\psi_{j,ji} = w_i \quad \text{in} \quad D \times (z,\infty), \qquad (A.1)$$

$$\psi_i = 0 \quad \text{on} \quad \partial D \times [z,\infty), \qquad (A.2)$$

$$\psi_i = 0, \ (x_1,x_2) \in D, \ x_3 = z, \qquad (A.3)$$

$$\psi_i, \psi_{i,j} \to 0 \ (\text{uniformly in } D) \ \text{as} \ x_3 \to \infty. \qquad (A.4)$$

We observe that $\psi_i = \psi_i(x_1,x_2,x_3;z)$. Clearly

$$\Phi_2 = \int_{R_z} w_i[\Delta\psi_i+\omega\psi_{j,ji}]dV$$

$$= \int_z^\infty \oint_{\partial D} w_i[\psi_{i,\beta}n_\beta+\omega\psi_{j,j}\delta_{i\beta}n_\beta]ds\ dx_3$$

$$- \int_{D_z} w_i[\psi_{i,3}+\omega v_{j,j}\delta_{i3}]dA, \qquad (A.5)$$

where we use the notation

$$\int_{D_z} \psi\ dA = \int_D \psi(x_1,x_2,z)dA, \qquad (A.6)$$

and δ_{ij} is the Kronecker delta.

An application of a weighted Schwarz's inequality for vectors leads to

$$\Phi_2 \le \left\{\sigma_1 \int_z^\infty \oint_{\partial D} w_iw_i\ ds\ dx_3 + \sigma_2 \int_{D_z} w_iw_i\ dA\right\}^{1/2}$$

$$\left\{\frac{1}{\sigma_1} \int_z^\infty \oint_{\partial D}(\psi_{i,\beta}n_\beta+\omega\psi_{j,j}n_i)(\psi_{i,\alpha}n_\alpha+\omega\psi_{k,k}n_i)ds\ dx_3\right.$$

$$\left. + \frac{1}{\sigma_2} \int_{D_z}[\psi_{i,3}+\omega\psi_{j,j}\delta_{i3}][\psi_{i,3}+\omega\psi_{k,k}\delta_{i3}]dA\right\}^{1/2}$$

$$= \left\{\sigma_1 \int_z^\infty \oint_{\partial D} w_iw_i ds\ dx_3-\sigma_2\Phi_2'(z)\right\}^{1/2}\left\{\frac{1}{\sigma_1} J_1 + \frac{1}{\sigma_2} J_2\right\}^{1/2}. \qquad (A.7)$$

for arbitrary positive constants σ_1 and σ_2. Here

$$J_1 = \int_z^\infty \oint_{\partial D} (\psi_{i,\beta}n_\beta+\omega\psi_{\beta,\beta}n_i)(\psi_{i,\alpha}n_\alpha+\omega\psi_{\alpha,\alpha}n_i)ds\ dx_3,$$

$$J_2 = \int_{D_z}(\psi_{i,3}+\omega\psi_{3,3}\delta_{i3})(\psi_{i,3}+\omega\psi_{3,3}\delta_{i3})dA. \tag{A.8}$$

We now make use of Rellich identities to bound J_1 and J_2. We start with the identity

$$\int_{R_z} x_\beta\psi_{i,\beta}[\Delta\psi_i+\omega\psi_{j,ji}]dV = \int_{R_z} x_\beta\psi_{i,\beta}w_i dV. \tag{A.9}$$

An application of the divergence theorem, using the boundary conditions on ψ_i, leads to

$$\frac{1}{2}\int_z^\infty \oint_{\partial D} x_\gamma n_\gamma[\psi_{i,\alpha}n_\alpha\psi_{i,\beta}n_\beta+\omega\psi^2_{\beta,\beta}]ds\ dx_3$$

$$= -\int_{R_z}[\psi_{i,j}\psi_{i,j}-\psi_{i,\beta}\psi_{i,\beta}+\omega(\psi^2_{j,j}-\psi_{i,i}\psi_{\beta,\beta})-x_\beta\psi_{i,\beta}w_i]dV. \tag{A.10}$$

We now bound I, the right hand side of (A.10). Noting that $\psi_{i,j}\psi_{i,j}\geq \psi_{i,\beta}\psi_{i,\beta}$ we observe that

$$I \leq -\omega\int_{R_z}[\psi^2_{3,3}+\psi_{3,3}\psi_{\beta,\beta}]dV + \int_{R_z}x_\beta\psi_{i,\beta}w_i dV. \tag{A.11}$$

We next investigate the integrand of the first term on the right of (A.11). Thus for arbitrary positive constants δ_1 and δ_2 we have

$$-\psi_{3,3}^2-\psi_{3,3}\psi_{\beta,\beta} = -\psi_{3,3}^2-(1-\delta_1)\psi_{3,3}\psi_{\beta,\beta}-\delta_1\psi_{3,3}\psi_{\beta,\beta}$$

$$\leq -\psi_{3,3}^2-(1-\delta_1)\psi_{3,3}\psi_{\beta,\beta} + \frac{\delta_1\psi_{3,3}^2}{4\delta_2} + \frac{\delta_1}{2}\psi_{\beta,\beta}\psi_{3,3}$$

$$+ \frac{\delta_1\delta_2}{4}\psi_{\beta,\beta}^2, \tag{A.12}$$

where in the last step we have used the fact that for positive numbers a, b, δ_2

$$ab \leq \frac{a^2}{4\delta_2} + \frac{ab}{2} + \frac{b^2\delta_2}{4} . \tag{A.13}$$

Thus

$$-\psi_{3,3}^2-\psi_{3,3}\psi_{\beta,\beta} \leq \left[\frac{\delta_1}{4\delta_2} - 1\right]\psi_{3,3}^2 + \frac{(3\delta_1-2)}{2}\psi_{3,3}\psi_{\beta,\beta}$$

$$+ \frac{\delta_1\delta_2}{4}\psi_{\beta,\beta}^2. \tag{A.14}$$

In order to make the right hand side a perfect square we set

$$\frac{\delta_1-4\delta_2}{\delta_2} = 3\delta_1-2 = \delta_1\delta_2, \tag{A.15}$$

which leads to

$$\delta_1 = 2^{-1/2}, \quad \delta_2 = 3-2^{3/2}.$$

Thus (A.14) reads

$$-\psi_{3,3}^2-\psi_{3,3}\psi_{\beta,\beta} \leq \tau\psi_{j,j}^2 \tag{A.16}$$

where

$$\tau = [3\sqrt{2}-4]/8. \tag{A.17}$$

On substituting (A.16) into (A.11) we obtain

$$I \leq \omega\tau \int_{R_z} \psi_{j,j}^2 dV + \int_{R_z} x_\beta\psi_{i,\beta}w_i dV$$

$$\leq \tau \int_{R_z} [\psi_{i,j}\psi_{i,j}+\omega\psi_{j,j}^2]dV + \int_{R_z} x_\beta\psi_{i,\beta}w_i dV$$

$$= -\tau \int_{R_z} \psi_i w_i dV + \int_{R_z} x_\beta\psi_{i,\beta}w_i dV. \tag{A.18}$$

In the last step, we have used the fact that

$$\int_{R_z} [\psi_{i,j}\psi_{i,j}+\omega\psi_{j,j}^2]dV = -\int_{R_z} \psi_i w_i dV,$$

which follows from the divergence theorem and (A.1)-(A.4). Inserting (A.18) into (A.10) we obtain

$$\frac{1}{2}\int_z^\infty \oint_{\partial D} x_\gamma n_\gamma[\psi_{i,\alpha}n_\alpha\psi_{i,\beta}n_\beta+\omega\psi_{\beta,\beta}^2]ds\ dx_3$$

$$\leq \int_{R_z}[x_\beta\psi_{i,\beta}-\tau\psi_i]w_i dV. \tag{A.19}$$

Rewriting the expression for J_1 in (A.8) and using the fact that ψ_i vanishes on ∂R_z and hence that $\psi_{i,\beta}n_\beta\psi_{i,\alpha}n_\alpha = \psi_{i,\beta}\psi_{i,\beta}$ on ∂D we have

$$J_1 = \int_z^\infty \oint_{\partial D}[\psi_{i,\beta}\psi_{i,\beta}+(\omega^2+2\omega)\psi_{\beta,\beta}^2]ds\ dx_3$$

$$= (1+\omega)\int_z^\infty \oint_{\partial D}[\psi_{i,\beta}\psi_{i,\beta}+\omega\psi_{\beta,\beta}^2]ds\ dx_3$$

$$-\ \omega\int_z^\infty \oint_{\partial D}[\psi_{i,\beta}\psi_{i,\beta}-\psi_{\beta,\beta}^2]ds\ dx_3. \tag{A.20}$$

But

$$\int_z^\infty \oint_{\partial D}[\psi_{i,\beta}\psi_{i,\beta}-\psi_{\beta,\beta}^2]ds\ dx_3$$

$$= \int_z^\infty \oint_{\partial D}[\psi_{3,\beta}\psi_{3,\beta}+\psi_{\alpha,\beta}\psi_{\alpha,\beta}-\psi_{\beta,\beta}^2]ds\ dx_3$$

$$\geq \int_z^\infty \oint_{\partial D}\psi_{3,\beta}\psi_{3,\beta}ds\ dx_3 \geq 0, \tag{A.21}$$

where we have used the inequality

$$\oint_{\partial D} \psi_{\alpha,\beta}\psi_{\alpha,\beta}\, ds \geq \oint_{\partial D} \psi^2_{\beta,\beta}\, ds$$

which follows from the fact that ψ_i vanishes on ∂D. Thus (A.20) becomes

$$J_1 \leq (1+\omega) \int_z^\infty \oint_{\partial D} [\psi_{i,\alpha}n_\alpha\psi_{i,\beta}n_\beta+\omega\psi^2_{\beta,\beta}]ds\, dx_3 \tag{A.22}$$

and the use of (A.19) gives

$$J_1 \leq \frac{2(1+\omega)}{p} \int_{R_z} [x_i\psi_{i,\beta}-\tau\psi_i]w_i dV, \tag{A.23}$$

where (1.20) has also been employed. Using Schwarz's inequality we obtain

$$J_1 \leq \frac{2(1+\omega)}{p} \left\{d\left[\int_{R_z} \psi_{i,\beta}\psi_{i,\beta}dV\right]^{1/2} + \tau\left[\int_{R_z} \psi_i\psi_i dV\right]^{1/2}\right\}\Phi_2^{1/2}. \tag{A.24}$$

But

$$\int_{R_z} \psi_i\psi_i dV \leq \frac{1}{\lambda} \int_{R_z} \psi_{i,\beta}\psi_{i,\beta}dV, \tag{A.25}$$

where λ is the first eigenvalue in the fixed membrane problem (1.19). As mentioned earlier if λ is not known explicitly lower bounds for it are easily computed. Thus

$$J_1 \leq \frac{2(1+\omega)}{p} (d+\tau\lambda^{-1/2})\left\{\int_{R_z} \psi_{i,\beta}\psi_{i,\beta}dV\right\}^{1/2} \Phi_2^{1/2}. \tag{A.26}$$

Rather than completing the bound for J_1 at this point we derive a similar bound for J_2 and combine J_1 and J_2 to complete the bound for $\sigma_1^{-1}J_1 + \sigma_2^{-1}J_2$, as required on the right hand side of (A.7).

On using the divergence theorem it follows directly from

$$\int_{R_z} \psi_{i,3}[\psi_{i,jj}+\omega\psi_{j,ji}]dV = \int_{R_z} \psi_{i,3}w_i dV \tag{A.27}$$

that

$$\frac{1}{2}\int_{D_z} [\psi_{i,3}\psi_{i,3}+\omega\psi_{3,3}^2]dA = \int_{R_z} \psi_{i,3}w_i dV. \tag{A.28}$$

and hence as before that

$$J_2 \leq 2(1+\omega)\left\{\int_{R_z} \psi_{i,3}\psi_{i,3}dV\ \Phi_2\right\}^{1/2}. \tag{A.29}$$

Now (A.26) and (A.29) yield

$$\frac{1}{\sigma_1} J_1 + \frac{1}{\sigma_2} J_2 \le 2(1+\omega)\left\{\frac{(d+\tau\lambda^{-1/2})}{\sigma_1 p}\left[\int_{R_z} \psi_{i,\beta}\psi_{i,\beta}dV\right]^{1/2}\right.$$

$$\left. + \frac{1}{\sigma_2}\left[\int_{R_z} \psi_{i,3}\psi_{i,3}dV\right]^{1/2}\right\}\Phi_2^{1/2}. \qquad (A.30)$$

If we choose

$$\sigma_1 = \frac{[d+\tau\lambda^{-1/2}]}{p}\,\sigma_2 \qquad (A.31)$$

we find, using Schwarz's inequality for vectors, that

$$\frac{1}{\sigma_1} J_1 + \frac{1}{\sigma_2} J_2 \le \frac{2\sqrt{2}(1+\omega)}{\sigma_2}\left\{\int_{R_z} \psi_{i,j}\psi_{i,j}dV\right\}^{1/2}\Phi_2^{1/2}. \qquad (A.32)$$

But on using the identity following (A.18), Schwarz's inequality and (A.25) it follows that (since $\omega > 0$)

$$\int_{R_z} \psi_{i,j}\psi_{i,j}dV \le \lambda^{-1}\Phi_2, \qquad (A.33)$$

and so we have

$$\frac{1}{\sigma_1} J_1 + \frac{1}{\sigma_2} J_2 \le \frac{2(1+\omega)}{\sigma_2}\left[\frac{2}{\lambda}\right]^{1/2}\Phi_2(z). \qquad (A.34)$$

Returning now to (A.7) we use (A.31) to write (A.7) as

$$\Phi_2 \le \left\{\frac{[d+\tau\lambda^{-1/2}]}{p}\,\sigma_2 \int_z^\infty \oint_{\partial D} w_i w_i ds\, dx_3 - \sigma_2 \Phi_2'(z)\right\}^{1/2}$$

$$\left\{\frac{2(1+\omega)}{\sigma_2}[\frac{2}{\lambda}]^{1/2}\Phi_2(z)\right\}^{1/2}. \tag{A.35}$$

Squaring both sides and canceling we obtain finally

$$\Phi_2 \le 2(1+\omega)[\frac{2}{\lambda}]^{1/2}\left\{\frac{[d+\tau\lambda^{-1/2}]}{p} \int_z^\infty \oint_{\partial D} w_i w_i ds\, dx_3 - \Phi_2'(z)\right\} \tag{A.36}$$

which is (2.5) if we set

$$\sigma = \frac{(d+\tau\lambda^{-1/2})}{p}, \tag{A.37}$$

and

$$k^{-1} = 2(1+\omega)[2\lambda^{-1}]^{1/2}. \tag{A.38}$$

APPENDIX B. Derivation of (2.21) and (1.15).

We derive the first inequality in (2.21). The second is obtained in a similar manner. An integration and use of (1.3), (1.4) gives

$$\tilde{E}(z) = -\int_{\tilde{D}_z} [u_i u_{i,3} + \omega u_3 u_{j,j}]dA, \tag{B.1}$$

and an application of Schwarz's inequality for vectors leads to

$$\tilde{E}(z) \leq \left\{\int_{\tilde{D}_z} [u_i u_i + \omega u_3^2] dA \int_{\tilde{D}_z} (u_{i,3} u_{i,3} + \omega u_{j,j}^2) dA\right\}^{1/2}$$

$$\leq \tilde{\lambda}^{-1/2} (1+\omega)^{1/2} \left\{\int_{\tilde{D}_z} u_{i,\beta} u_{i,\beta} dA \int_{\tilde{D}_z} [u_{i,3} u_{i,3} + \omega u_{j,j}^2] dA\right\}^{1/2} \qquad \text{(B.2)}$$

Here $\tilde{\lambda}$ is given by (1.19) with D replaced by $\tilde{D}$. An application of the arithmetic-geometric mean inequality now leads to

$$\tilde{E}(z) \leq \frac{(1+\omega)^{1/2}}{2\tilde{\lambda}^{1/2}} \int_{\tilde{D}_z} [u_{i,j} u_{i,j} + \omega u_{j,j}^2] dA$$

$$= -\frac{(1+\omega)^{1/2}}{2\tilde{\lambda}^{1/2}} \tilde{E}'(z). \qquad \text{(B.3)}$$

This inequality may be integrated to give

$$\tilde{E}(z) \leq \tilde{E}(0) e^{-\frac{2\tilde{\lambda}^{1/2} z}{[1+\omega]^{1/2}}} \qquad \text{(B.4)}$$

which is the first of inequalities (2.21). The second is obtained similarly. Thus

$$\hat{E}(z) \le \hat{E}(0)e^{-\frac{2\hat{\lambda}^{1/2} z}{[1+\omega]^{1/2}}}, \tag{B.5}$$

and (2.21) is established with γ_1, γ_2 given by (2.22). If we define for positive constants $\tilde{\mu}$ and $\hat{\mu}$ (to be determined)

$$\begin{aligned} \tilde{g}_i(x_1,x_2,x_3) &= \tilde{f}_i(x_1,x_2)e^{-\tilde{\mu}z} \quad \text{in} \quad \tilde{R}_0 \\ \hat{g}_i(x_1,x_2,x_3) &= \hat{f}_i(x_1,x_2)e^{-\hat{\mu}z} \quad \text{in} \quad \hat{R}_0, \end{aligned} \tag{B.6}$$

then by the Dirichlet Principle for elastostatics

$$\begin{aligned} \tilde{E}(0) &\le \int_{\tilde{R}_0} \{\tilde{g}_{i,j}\tilde{g}_{i,j}+\omega\tilde{g}_{j,j}^2\}dV \\ \hat{E}(0) &\le \int_{\hat{R}_0} \{\hat{g}_{i,j}\hat{g}_{i,j}+\omega\hat{g}_{j,j}^2\}dV. \end{aligned} \tag{B.7}$$

We may now choose $\tilde{\mu}$ and $\hat{\mu}$ to minimize the bounds in (B.7).

It is worth remarking at this point that we have all of the ingredients here for obtaining a bound for the Φ_1 of (1.14), namely

$$\Phi_1(z) = \int_{R_z} W_i W_i dV. \tag{B.8}$$

Since

$$\Phi_1(z) \le \frac{1}{\lambda} \int_{R_z} w_{i,\beta} w_{i,\beta} dV$$

$$\le \frac{1}{\lambda} E(z)$$

$$\le \frac{1}{\lambda} E(0) e^{-\frac{2\lambda^{1/2}}{[1+\omega]^{1/2}} z}, \qquad (B.9)$$

we thus obtain an estimate of the form (1.15).

Again from the Dirichlet principle for elastostatics we have

$$E(0) \le \int_{R_0} \{g_{i,j} g_{i,j} + \omega g_{j,j}^2\} dV \qquad (B.10)$$

where for some $\mu > 0$

$$g_i(x_1, x_2, x_3) = [\tilde{f}_i(x_1, x_2) - \hat{f}_i(x_1, x_2)] e^{-\mu z}. \qquad (B.11)$$

Clearly this term will be small if $\tilde{f}_i$ and its tangential derivatives are close to $\hat{f}_i$ and its corresponding tangential derivatives on $x_3 = 0$.

ACKNOWLEDGEMENTS.

The work of C. O. Horgan was supported by the U. S. National Science Foundation (NSF) under Grant #MSM-89-04719, by the U. S. Air Force Office of Scientific Research under Grant #AFOSR-89-0470, and by the U. S. Army Research Office under Grant #DAAL 03-91-G-0022. The work of L. E. Payne was carried out while he held an appointment as Visiting Professor, Department of Applied Mathematics, University of Virginia, Spring 1991 and was partially supported by the NSF under Grant #DMS-89-22519.

C. O. Horgan
Department of Applied Mathematics
University of Virginia
Charlottesville, VA 22903

L. E. Payne
Department of Mathematics
Cornell University
Ithaca, N.Y 14853

REFERENCES

1. Bandle, C., Isoperimetric Inequalities and Their Applications, Pitman Publishers (1980).

2. Biollay, Y., First boundary value problem in elasticity: bounds for the displacements and Saint-Venant's principle, ZAMP, 31 (1980), 556-567.

3. Crooke, P. S. and L. E. Payne, Continuous dependence on geometry for the backward heat equation, Math. Methods in Appl. Sci., 6 (1984), 443-448.

4. Flavin, J. N., Another aspect of Saint-Venant's principle in elasticity, ZAMP, 29 (1978), 328-332.

5. Flavin, J. N., R. J. Knops and L. E. Payne, Decay estimates for the elastic cylinder of variable cross-section, Quart. Appl. Math., 47 (1989), 325-350.

6. Horgan, C. O., Recent developments concerning Saint-Venant's principle: an update, Applied Mech. Reviews, 42 (1989), 295-303.

7. Horgan, C. O. and J. K. Knowles, Recent developments concerning Saint-Venant's principle, Advances in Applied Mechanics, Academic Press, (J. W. Hutchinson ed.) 23 (1983), 179-269.

8. Horgan, C. O. and L. E. Payne, On the asymptotic behavior of solutions of inhomogeneous second-order quasilinear partial differential equations, Quart. Appl. Math., 47 (1989), 753-771.

9. Horgan, C. O. and L. E. Payne, On Saint-Venant's principle in finite anti-plane shear: an energy approach, Arch. Rat. Mech. Anal., 109 (1990), 107-137.

10. Horgan, C. O. and L. E. Payne, Exponential decay estimates for capillary surfaces and extensible films, Stability & Applied Anal. of Continuous Media, 1 (1991) (to appear).

11. Horgan, C. O. and L. E. Payne, A Saint-Venant principle for a theory of nonlinear plane elasticity, Quart. Appl. Math. (to appear).

12. Horgan, C. O. and L. E. Payne, The effect of constitutive law perturbations on finite anti-plane shear deformations of a semi-infinite strip (to appear).

13. Horgan, C. O., L. E. Payne and J. G. Simmonds, Existence, uniqueness and decay estimates for solutions in the nonlinear theory of elastic, edge-loaded, circular tubes, Quart. Appl. Math., 48 (1990), 341-359.

14. Horgan, C. O. and D. Siegel, On the asymptotic behavior of a minimal surface over a semi-infinite strip, J. Math. Anal. Appl., 153 (1990), 397-406.

15. Mielke, A., Normal hyperbolicity of center manifolds and Saint-Venant's principle, Arch. Rat. Mech. Anal., 110 (1990), 353-372.

16. Payne, L. E. and J. R. L. Webb, Spatial decay estimates for second order partial differential equations, Nonlinear Analysis (to appear).

H BEGEHR AND W LIN*

A mixed-contact boundary problem in orthotropic elasticity

In this paper a mixed-contact boundary value problem for orthotropic elasticity is considered. First we pose this problem and prove a uniqueness theorem. Then by means of the theory of bianalytic functions representations of solutions to the basic problems of orthotropic elasticity are obtained. Finally we reduce this mixed-contact problem to boundary value problems for analytic functions which will be solved by virtue of the theory of singular integral equations.

§1. Introduction

D.E. SHERMAN [1] and ZH.A. RUKHADGE [2] investigated some special contact problems of isotropic elasticity in the plane. Recently N.A. ZHURA solved a more general problem, a mixed-contact boundary value problem for isotropic elasticity and its corresponding boundary value problem for analytic functions in [3], [4]. It is natural to ask if their results may be generalized to the case of anisotropic elasticity, because this will be more interesting for applications.

The theory of so-called bianalytic functions and its applications to elasticity have been developed by HUA LOO-KENG and his pupils among them one of the present authors and R.P. GILBERT [5], [6]. In this paper we shall utilize the methods of bianalytic functions to solve the mixed-contact problem for orthotropic elasticity. First we pose the mixed contact problem for doubly connected domains and prove a uniqueness theorem of this problem. Then we introduce representations of the solutions to the basic problems for orthotropic elasticity by means of bianalytic functions. Finally we reduce the mixed-contact problem to boundary value problems for analytic functions. In another paper we will prove the solvability of this boundary value problem by virtue of the theory of singular integral equations.

*Project supported by the National Natural Science Foundation of China and the Foundation of Zhongshan University Advanced Research Centre.

§2. Mixed-contact problem

For the situation of plane strain of an orthotropic elastic body the equilibrium equations are

$$\begin{cases} \dfrac{\partial \sigma_x}{\partial x} + \dfrac{\partial \tau_{xy}}{\partial y} + F_1 = 0 \, , \\[2ex] \dfrac{\partial \tau_{xy}}{\partial x} + \dfrac{\partial \sigma_y}{\partial y} + F_2 = 0 \, , \end{cases} \tag{2.1}$$

where σ_x , σ_y , τ_{xy} are the components of stress, $F := (F_1, F_2)$ is the body force vector. In the case of small deformations, the generalized HOOKE's law will be given by the equations

$$\begin{cases} \sigma_x = \dfrac{1}{1-\nu_{12}\nu_{21}} \left(E_{11} \dfrac{\partial u}{\partial x} + \nu_{12} E_{22} \dfrac{\partial v}{\partial y} \right) \, , \\[2ex] \sigma_y = \dfrac{1}{1-\nu_{12}\nu_{21}} \left(\nu_{21} E_{11} \dfrac{\partial u}{\partial x} + E_{22} \dfrac{\partial v}{\partial y} \right) \, , \\[2ex] \tau_{xy} = G_{12} \left(\dfrac{\partial u}{\partial y} + \dfrac{\partial v}{\partial x} \right) \, , \end{cases} \tag{2.2}$$

where G_{12} is the modulus of motion, ν_{12} and ν_{21} are POISSON's ratios, E_{11} and E_{22} are YOUNG's moduli; they are related by the equation

$$\nu_{12} E_{22} = \nu_{21} E_{11} \, . \tag{2.3}$$

Substituting (2.2) into (2.1) we obtain

$$\left[A^{11} \frac{\partial^2}{\partial x^2} + A^{12} \frac{\partial^2}{\partial x \partial y} + A^{21} \frac{\partial^2}{\partial y \partial x} + A^{22} \frac{\partial^2}{\partial y^2} \right] \begin{pmatrix} u \\ v \end{pmatrix} = - \begin{pmatrix} F_1 \\ F_2 \end{pmatrix} , \tag{2.4}$$

where

$$A^{11} = \begin{pmatrix} \frac{E_{11}}{1-\nu_{12}\nu_{21}} & 0 \\ 0 & G_{12} \end{pmatrix} , \quad A^{12} = \begin{pmatrix} 0 & \frac{\nu_{12} E_{22}}{1-\nu_{12}\nu_{21}} \\ G_{12} & 0 \end{pmatrix} ,$$
$$A^{21} = \begin{pmatrix} 0 & G_{12} \\ \frac{\nu_{21} E_{11}}{1-\nu_{12}\nu_{21}} & 0 \end{pmatrix} , \quad A^{22} = \begin{pmatrix} G_{12} & 0 \\ 0 & \frac{E_{22}}{1-\nu_{12}\nu_{21}} \end{pmatrix} . \tag{2.5}$$

If u , v are continuously differentiable up to the second order, then system (2.4) may be written into

$$\mathbb{L} \begin{pmatrix} u \\ v \end{pmatrix} := \left[A^{11} \frac{\partial^2}{\partial x^2} + (A^{12} + A^{21}) \frac{\partial^2}{\partial x \partial y} + A^{22} \frac{\partial^2}{\partial y^2} \right] \begin{pmatrix} u \\ v \end{pmatrix} = - \begin{pmatrix} F_1 \\ F_2 \end{pmatrix} . \tag{2.6}$$

The displacement boundary value problem is to look for the solution u , v to system (2.4) in the domain Ω , which the elastic body occupies, if the displacement u , v are specified at all points of the boundary Γ of Ω . If the stresses are specified at all points of Γ , then this is the stress boundary value problem. Assume n to be the exterior unit normal to Γ , and X_n and Y_n to be the components of the external stress along the axes, then we may write the boundary conditions for this problem in the form

$$\begin{aligned}(\sigma_x \cos(n,x) + \tau_{xy}\cos(n,y))\,|_\Gamma &= X_n\ ,\\ (\tau_{xy}\cos(n,x) + \sigma_y\cos(n,y))\,|_\Gamma &= Y_n\ .\end{aligned} \tag{2.7}$$

Noticing (2.2) and using notations (2.5), we get the matrix form

$$\left[\cos(n,x)\left(A^{11}\frac{\partial}{\partial x} + A^{12}\frac{\partial}{\partial y}\right) + \cos(n,y)\left(A^{21}\frac{\partial}{\partial x} + A^{22}\frac{\partial}{\partial y}\right)\right]\begin{pmatrix}u\\ v\end{pmatrix}\bigg|_\Gamma = \begin{pmatrix}X_n\\ Y_n\end{pmatrix}. \tag{2.8}$$

We now start to formulate the problem which will be considered in this paper. Suppose the closed curves Γ_0 , Γ_1 , Γ_2 of class C^3 are given in the plane $I\!R^2$ such that Γ_0 is surrounded by Γ_1 and Γ_2 by Γ_0 . By Ω_1 denote the doubly-connected domain with boundary $\partial\Omega_1 = \Gamma_0 \bigcup \Gamma_1$, and by Ω_2 the domain with $\partial\Omega_2 = \Gamma_0 \bigcup \Gamma_2$. Suppose Ω_1 and Ω_2 are occupied by two distinct orthotropic elastic bodies, respectively. Obviously, they obey the system of equilibrium equations, respectively,

$$\begin{gathered}\left[\left(A_j^{11}\frac{\partial}{\partial x} + A_j^{12}\frac{\partial}{\partial y}\right)\frac{\partial}{\partial x} + \left(A_j^{21}\frac{\partial}{\partial x} + A_j^{22}\frac{\partial}{\partial y}\right)\frac{\partial}{\partial y}\right]\begin{pmatrix}u^j\\ v^j\end{pmatrix} + \begin{pmatrix}F_1^j\\ F_2^j\end{pmatrix} = 0\ ,\\ (x,y)\in\Omega_j\ ,\quad j = 1,2\ ,\end{gathered} \tag{2.9}$$

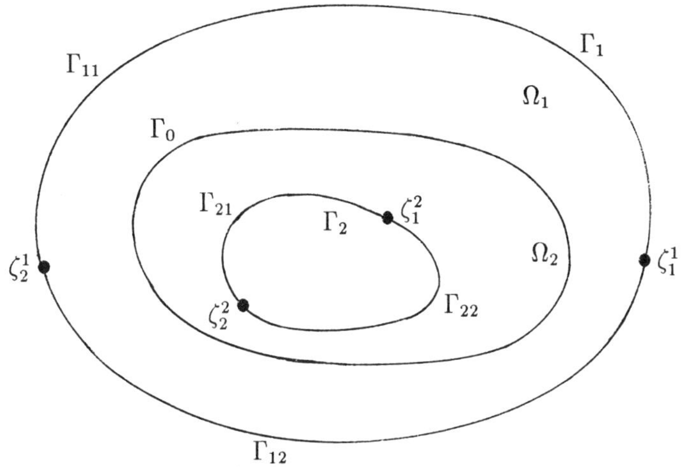

(Figure)

where (u^j, v^j) is the unknown displacement vector and the (2×2) -matrices A_j^{ik} take the forms

$$\left\{\begin{array}{ll} A_j^{11} = \begin{pmatrix} \frac{E_{11}^j}{1-\nu_{12}^j\nu_{21}^j} & 0 \\ 0 & G_{12}^j \end{pmatrix}, & A_j^{12} = \begin{pmatrix} 0 & \frac{\nu_{12}^j E_{22}^j}{1-\nu_{12}^j\nu_{21}^j} \\ G_{12}^j & 0 \end{pmatrix}, \\ A_j^{21} = \begin{pmatrix} 0 & G_{12}^j \\ \frac{\nu_{21}^j E_{11}^j}{1-\nu_{12}^j\nu_{21}^j} & 0 \end{pmatrix}, & A_j^{22} = \begin{pmatrix} G_{12}^j & 0 \\ 0 & \frac{E_{22}^j}{1-\nu_{12}^j\nu_{21}^j} \end{pmatrix}, \end{array}\right. \quad j = 1,2\,. \tag{2.10}$$

We divide the curves Γ_j into arcs Γ_{jr} by the points $\zeta_r^j \in \Gamma_j$, moreover, while surrounding Γ_j counterclockwise the beginning of the arc Γ_{jr} is the point ζ_r^j , $j, r = 1, 2$ (Figure). By n_j the exterior unit normal to Γ_j , $j = 0, 1, 2$, is denoted.

The mixed-contact boundary problem is to find the displacement vector

$$(u, v) = \begin{cases} (u^1(x,y)\,,\ \ v^1(x,y))\,, & (x,y) \in \Omega_1\,, \\ (u^2(x,y)\,,\ \ v^2(x,y))\,, & (x,y) \in \Omega_2\,, \end{cases}$$

such that $u^j, v^j \in C^2(\Omega_j)$ satisfy system (2.9) in the domains Ω_1 and Ω_2 , respectively, the boundary conditions

$$(u^j, v^j) = 0\,, \qquad (x,y) \in \Gamma_{j1}\,,$$

$$\left[\cos(n_j, x)\left(A_j^{11}\frac{\partial}{\partial x} + A_j^{12}\frac{\partial}{\partial y}\right) + \cos(n_j, y)\left(A_j^{21}\frac{\partial}{\partial x} + A_j^{22}\frac{\partial}{\partial y}\right)\right]\begin{pmatrix} u^j \\ v^j \end{pmatrix} = 0 \tag{2.11}$$

$$(x,y) \in \Gamma_{j2}\,, \qquad j = 1,2\,,$$

and the contact conditions

$$(u^1, v^1) = (u^2, v^2)\,,$$

$$\begin{aligned} &\left[\cos(n_0, x)\left(A_1^{11}\frac{\partial}{\partial x} + A_1^{12}\frac{\partial}{\partial y}\right) + \cos(n_0, y)\left(A_1^{21}\frac{\partial}{\partial x} + A_1^{22}\frac{\partial}{\partial y}\right)\right]\begin{pmatrix} u^1 \\ v^1 \end{pmatrix} \\ = &\left[\cos(n_0, x)\left(A_2^{11}\frac{\partial}{\partial x} + A_2^{12}\frac{\partial}{\partial y}\right) + \cos(n_0, y)\left(A_2^{21}\frac{\partial}{\partial x} + A_2^{22}\frac{\partial}{\partial y}\right)\right]\begin{pmatrix} u^2 \\ v^2 \end{pmatrix}, \end{aligned} \tag{2.12}$$

$$(x,y) \in \Gamma_0\,.$$

Before we start to discuss the uniqueness and the existence of the solution to this mixed-contact problem, we introduce some function classes. We refer a function W on Γ_j to belong to the class $H_\circ(\Gamma_j)$ if it is HÖLDER continuous on the closed arcs Γ_{jr} ,

$r = 1, 2$. A function W is said to belong to the class $H_*(\Gamma_j)$, if on each closed subarc of Γ_j not involving the points $\zeta_r^j \in \Gamma_j$ it is HÖLDER continuous, while near the points ζ_r^j it can be represented in the form $W(t) = \frac{W_*(t)}{|t - \zeta_r^j|^\alpha}$ $(0 \le \alpha < 1)$, where $W_* \in H_\circ(\Gamma_j)$. We also refer a function $W(z) = W(x, y)$, $z = x + iy$, defined in the domain Ω_j to belong to $H_*(\overline{\Omega}_j)$ if it is HÖLDER continuous in Ω_j up to the boundary $\partial\Omega_j$ everywhere except possibly at the points ζ_r^j near which the inequality $|W(z)| \le \frac{M}{|z - \zeta_r^j|^\alpha}$ holds, where $M > 0$ is constant and $0 \le \alpha < 1$. Moreover, its limit value $W(t)$ is assumed to exist on $\partial\Omega_j$ everywhere with the possible exception of the points ζ_r^j and belongs to the class $H_*(\partial\Omega_j)$. Similarly, $H_*^1(\overline{\Omega}_j)$ denotes the class of functions the first order partial derivatives of which belong to $H_*(\overline{\Omega}_j)$.

§3. Uniqueness theorem

We now prove the uniqueness of the solution to the mixed-contact problem for orthotropic elasticity. In the following T denotes matrix transposition.

Lemma 3.1 *If $w = (u, v)^T \in C^2(\Omega) \bigcap H_*^1(\overline{\Omega})$, and Ω^+ is a subdomain of Ω , then*

$$\int_{\partial\Omega^+} (u, v) \left\{ \left[\left(A^{11} \frac{\partial}{\partial x} + A^{12} \frac{\partial}{\partial y} \right) \begin{pmatrix} u \\ v \end{pmatrix} \right] \cos(n, x) \right.$$

$$\left. + \left[\left(A^{21} \frac{\partial}{\partial x} + A^{22} \frac{\partial}{\partial y} \right) \begin{pmatrix} u \\ v \end{pmatrix} \right] \cos(n, y) \right\} ds \qquad (3.1)$$

$$= \iint_{\Omega^+} [(u, v) \mathbb{L}(w) + Q(w)] \, dx \, dy \, ,$$

holds, where A^{ij} and $\mathbb{L}(w)$ are defined in (2.5) and (2.6), respectively, and $Q(w)$ is the quadratic form

$$Q(w) = \frac{\nu_{12} E_{22}}{(1 - \nu_{12}\nu_{21})\nu_{21}} \left[\left(\frac{\partial u}{\partial x} \right)^2 + 2\nu_{21} \frac{\partial u}{\partial x} \frac{\partial v}{\partial y} + \frac{\nu_{21}}{\nu_{12}} \left(\frac{\partial v}{\partial y} \right)^2 \right]$$

$$+ G_{12} \left(\frac{\partial v}{\partial x} + \frac{\partial u}{\partial y} \right)^2 \qquad (3.2)$$

Proof It is easy to verify the identity

$$\frac{\partial}{\partial x}\left[(u,v)\left(\begin{pmatrix}\frac{E_{11}}{1-\nu_{12}\nu_{21}} & 0\\ 0 & G_{12}\end{pmatrix}\frac{\partial}{\partial x}+\begin{pmatrix}0 & \frac{\nu_{12}E_{22}}{1-\nu_{12}\nu_{21}}\\ G_{12} & 0\end{pmatrix}\frac{\partial}{\partial y}\right)\binom{u}{v}\right]$$

$$+\frac{\partial}{\partial y}\left[(u,v)\left(\begin{pmatrix}0 & G_{12}\\ \frac{\nu_{21}E_{11}}{1-\nu_{12}\nu_{21}} & 0\end{pmatrix}\frac{\partial}{\partial x}+\begin{pmatrix}G_{12} & 0\\ 0 & \frac{E_{22}}{1-\nu_{12}\nu_{21}}\end{pmatrix}\frac{\partial}{\partial y}\right)\binom{u}{v}\right] \tag{3.3}$$

$$= (u,v)\mathbb{L}\binom{u}{v}+Q(w)\,.$$

Applying the GREEN formula, from (3.3) we obtain the desired formula (3.1).

Theorem 3.1 *The mixed-contact problem (2.4), (2.11) and (2.12) has at most one solution in the function class* $C^2(\Omega_j)\bigcap H^1_*(\overline{\Omega_j})$, $j=1,2$.

Proof Suppose there exist two solutions $w^*=(u^*,v^*)^T$ and $w^{**}=(u^{**},v^{**})^T$ to this problem. Then

$$w(x,y) = (u,v)^T := w^{**}(x,y)-w^*(x,y)$$

fulfils the homogeneous equation

$$\mathbb{L}(w) = 0\,. \tag{3.4}$$

Moreover, from conditions (2.11) and (2.12) we have

$$\begin{aligned} w^j(x,y) &= (u^j,v^j)\\ &:= w^{**j}(x,y)-w^{*j}(x,y) = 0\,, \qquad (x,y)\in\Gamma_{j1}\,,\ j=1,2\,; \end{aligned} \tag{3.5}$$

$$\left[\cos(n_j,x)\left(A_j^{11}\frac{\partial}{\partial x}+A_j^{12}\frac{\partial}{\partial y}\right)+\cos(n_j,y)\left(A_j^{21}\frac{\partial}{\partial x}+A_j^{22}\frac{\partial}{\partial y}\right)\right]w^j = 0\,, \quad (x,y)\in\Gamma_{j2}\,,\quad j=1,2\,; \tag{3.6}$$

$$w^1(x,y) = w^2(x,y)\,, \qquad (x,y)\in\Gamma_0\,; \tag{3.7}$$

$$\begin{aligned} &\left[\cos(n_0,x)\left(A_1^{11}\frac{\partial}{\partial x}+A_1^{12}\frac{\partial}{\partial y}\right)+\cos(n_0,y)\left(A_1^{21}\frac{\partial}{\partial x}+A_1^{22}\frac{\partial}{\partial y}\right)\right]w^1(x,y)\\ &= \left[\cos(n_0,x)\left(A_2^{11}\frac{\partial}{\partial x}+A_2^{12}\frac{\partial}{\partial y}\right)+\cos(n_0,y)\left(A_2^{21}\frac{\partial}{\partial x}+A_2^{22}\frac{\partial}{\partial y}\right)\right]w^2(x,y)\,,\\ &\qquad (x,y)\in\Gamma_0\,. \end{aligned} \tag{3.8}$$

From the domain Ω_j we remove the set of points z, belonging to the closed discs $| z-\zeta_r^j | \leq \varepsilon$, $j, r = 1,2$, with a small enough radius ε, and by $\Omega_{j,\varepsilon}$ we denote the residual domain. It is easy to see that $w^j = (u^j, v^j)^T$ in $\Omega_{j,\varepsilon}$ $(j = 1,2)$ satisfy the conditions of Lemma 3.1. So we get

$$\int\limits_{\partial\Omega_{j,\varepsilon}} (w^j)^T \left\{ \left[\left(A_j^{11} \frac{\partial}{\partial x} + A_j^{12} \frac{\partial}{\partial y} \right) w^j \right] \cos(n_j, x) \right.$$

$$\left. + \left[\left(A_j^{21} \frac{\partial}{\partial x} + A_j^{22} \frac{\partial}{\partial y} \right) w^j \right] \cos(n_j, y) \right\} ds$$

$$= \iint\limits_{\Omega_{j,\varepsilon}} [(w^j)^T I\!\!L(w^j) + Q_j(w^j)] \, dx \, dy \,, \qquad j = 1,2 \,.$$

We set $\Gamma_{j,\varepsilon} = \partial\Omega_{j,\varepsilon} \setminus \Gamma_0$ i.e., by $\Gamma_{j,\varepsilon}$ we denote the curve obtained from Γ_j by removing its part, lying in the disc $| z - \zeta_r^j | \leq \varepsilon$, and adding the circular arc $| z - \zeta_r^j | = \varepsilon$ $(z \in \Omega_j)$. So, noticing (3.4), we obtain

$$\begin{aligned} \int\limits_{\Omega_{1,\varepsilon}} Q_1(w^1) \, dx \, dy \; = \; & \int\limits_{\Gamma_{1,\varepsilon}} (w^1)^T \left\{ \left[\left(A_1^{11} \frac{\partial}{\partial x} + A_1^{12} \frac{\partial}{\partial y} \right) w^1 \right] \cos(n_1, x) \right. \\ & \left. + \left[\left(A_1^{21} \frac{\partial}{\partial x} + A_1^{22} \frac{\partial}{\partial y} \right) w^1 \right] \cos(n_1, y) \right\} ds \\ & - \int\limits_{\Gamma_0} (w^1)^T \left\{ \left[\left(A_1^{11} \frac{\partial}{\partial x} + A_1^{12} \frac{\partial}{\partial y} \right) w^1 \right] \cos(n_0, x) \right. \\ & \left. + \left[\left(A_1^{21} \frac{\partial}{\partial x} + A_1^{22} \frac{\partial}{\partial y} \right) w^1 \right] \cos(n_0, y) \right\} ds \,, \end{aligned} \tag{3.9}$$

$$\begin{aligned} \int\limits_{\Omega_{2,\varepsilon}} Q_2(w^2) \, dx \, dy \; = \; & \int\limits_{\Gamma_0} (w^2)^T \left\{ \left[\left(A_2^{11} \frac{\partial}{\partial x} + A_2^{12} \frac{\partial}{\partial y} \right) w^2 \right] \cos(n_0, x) \right. \\ & \left. + \left[\left(A_2^{21} \frac{\partial}{\partial x} + A_2^{22} \frac{\partial}{\partial y} \right) w^2 \right] \cos(n_0, y) \right\} ds \\ & - \int\limits_{\Gamma_{2,\varepsilon}} (w^2)^T \left\{ \left[\left(A_2^{11} \frac{\partial}{\partial x} + A_2^{12} \frac{\partial}{\partial y} \right) w^2 \right] \cos(n_2, x) \right. \\ & \left. + \left[\left(A_2^{21} \frac{\partial}{\partial x} + A_2^{22} \frac{\partial}{\partial y} \right) w^2 \right] \cos(n_2, y) \right\} ds \,. \end{aligned} \tag{3.10}$$

By adding (3.9) and (3.10) and combining with (3.5) up to (3.8) we arrive at

$$\begin{aligned}\sum_{j=1}^{2}\int_{\Omega_{j,\varepsilon}} Q_j(w^j)\,dx\,dy \;=\; & \int_{\gamma_{1,\varepsilon}} (w^1)^T\Big\{\Big[\Big(A_1^{11}\frac{\partial}{\partial x}+A_1^{12}\frac{\partial}{\partial y}\Big)\,w^1\Big]\cos(n_1,x)\\
&+\Big[\Big(A_1^{21}\frac{\partial}{\partial x}+A_1^{22}\frac{\partial}{\partial y}\Big)\,w^1\Big]\cos(n_1,y)\Big\}\,ds\\
&-\int_{\gamma_{2,\varepsilon}} (w^2)^T\Big\{\Big[\Big(A_2^{11}\frac{\partial}{\partial x}+A_2^{12}\frac{\partial}{\partial y}\Big)\,w^2\Big]\cos(n_2,x)\\
&+\Big[\Big(A_2^{21}\frac{\partial}{\partial x}+A_2^{22}\frac{\partial}{\partial y}\Big)\,w^2\Big]\cos(n_2,y)\Big\}\,ds\,,\end{aligned} \tag{3.11}$$

where $\gamma_{j,\varepsilon}=\bigcup_{r=1}^{2}\Big\{z\in\Omega_j\Big|\,|\,z-\zeta_r^j\,|=\varepsilon\Big\}$, $j=1,2$. By means of $w^j\in H^1_*(\overline{\Omega}_j)$, it is easy to show that on the right hand side of equation (3.11) the integrals tend to zero together with ε . Hence, finally we get

$$\sum_{j=1}^{2}\int_{\Omega_j} Q_j(w^j)\,dx\,dy \;=\; 0\;.$$

As from (3.2),

$$\begin{aligned}Q_j(w^j) \;=\; & \frac{\nu_{12}^j E_{22}^j}{(1-\nu_{12}^j\nu_{21}^j)\nu_{21}^j}\left[\left(\frac{\partial u^j}{\partial x}\right)^2+2\nu_{21}^j\frac{\partial u^j}{\partial x}\frac{\partial v^j}{\partial y}+\frac{\nu_{21}^j}{\nu_{12}^j}\left(\frac{\partial v^j}{\partial y}\right)^2\right]\\
&+G_{12}^j\left(\frac{\partial v^j}{\partial x}+\frac{\partial u^j}{\partial y}\right)^2\;,\qquad j=1,2\;,\end{aligned}$$

and from the theory of orthotropic elasticity

$$0<E_{22}^j\;,\qquad 0<G_{12}^j\;,\qquad 0<\nu_{12}^j\;,\;\nu_{21}^j<\frac{1}{2}\;,\qquad j=1,2\;,$$

it is easy to show that $Q_j(w^j)$ $(j=1,2)$ is a positive semidefinite quadratic form with respect to $\frac{\partial u^j}{\partial x}$, $\frac{\partial u^j}{\partial y}$, $\frac{\partial v^j}{\partial x}$, $\frac{\partial v^j}{\partial y}$. From $Q_j(w^j)=0$ we have in Ω_j $\frac{\partial u^j}{\partial x}=\frac{\partial v^j}{\partial y}=0$, $\frac{\partial u^j}{\partial y}+\frac{\partial v^j}{\partial x}=0$. Hence, from $\mathbb{L}\binom{u^j}{v^j}=0$ also $u^j_{yy}=v^j_{xx}=0$ in Ω_j follows. Taking $u^j=v^j=0$ on Γ_0 into account we conclude

$$u^j \;=\; v^j \;=\; 0\;,\qquad j=1,2$$

i.e.

$$w^1 \equiv 0 , \qquad w^2 \equiv 0 .$$

This completes the proof of the uniqueness theorem.

§4. Representation of the general solution and the boundary conditions

1. First we consider the case without body forces in a simply-connected domain, i.e. we first seek out the general solution to the homogeneous system

$$\frac{1}{1-\nu_{12}\nu_{21}}\left[\begin{pmatrix} E_{11} & 0 \\ 0 & G_{12}(1-\nu_{12}\nu_{21}) \end{pmatrix}\frac{\partial^2}{\partial x^2}\right.$$
$$+(\nu_{12}E_{22}+G_{12}(1-\nu_{12}\nu_{21})\begin{pmatrix} 0 & 1 \\ 1 & 0 \end{pmatrix}\frac{\partial^2}{\partial x\partial y} \tag{4.1}$$
$$\left.+\begin{pmatrix} G_{12}(1-\nu_{12}\nu_{21}) & 0 \\ 0 & E_{22} \end{pmatrix}\frac{\partial^2}{\partial y^2}\right]\begin{pmatrix} u \\ v \end{pmatrix} = 0 .$$

As was done in [6], we introduce new parameters ν , E , δ defined by

$$\nu^2 := \nu_{12}\nu_{21} , \qquad E^2 := E_{11}E_{22} , \qquad \delta^2 := \sqrt{\frac{E_{11}}{E_{22}}} . \tag{4.2}$$

Using the transformations

$$x = \delta x' , \quad y = y' ; \quad u = u' , \quad v = \delta v' , \tag{4.3}$$

and setting

$$k_1 = \frac{E}{2G_{12}} - \nu \tag{4.4}$$

we reduce system (4.1) into

$$\left[\begin{pmatrix} 1 & 0 \\ 0 & \frac{1-\nu^2}{2(k_1+\nu)} \end{pmatrix}\frac{\partial^2}{\partial x'^2} + \frac{1+2k_1\nu+\nu^2}{2(k_1+\nu)}\begin{pmatrix} 0 & 1 \\ 1 & 0 \end{pmatrix}\frac{\partial^2}{\partial x'\partial y'}\right.$$
$$\left.+\begin{pmatrix} \frac{1-\nu^2}{2(k_1+\nu)} & 0 \\ 0 & 1 \end{pmatrix}\frac{\partial^2}{\partial y'^2}\right]\begin{pmatrix} u' \\ v' \end{pmatrix} = 0 . \tag{4.5}$$

Its corresponding characteristic equation will be

$$\frac{1-\nu^2}{2(k_1+\nu)}(\xi^4+2k_1\xi^2\eta^2+\eta^4) = 0 .$$

When $k_1 = 1$, this will correspond to the case of homogeneous isotropic elasticity. So we only consider the case when $k_1 > 1$. Introducing as in [5] the new parameters

$$k = k_1 - \sqrt{k_1^2-1} , \qquad \lambda = \frac{1-\nu^2}{2(k_1+\nu)}k , \tag{4.6}$$

and the transformations

$$x' = x_1 , \qquad y' = \frac{1}{\sqrt{k}}y_1 ; \qquad u' = \frac{1+k\nu}{(k+\nu)\sqrt{k}}u_1 , \qquad v' = -v_1 , \tag{4.7}$$

(4.5) is reduced to

$$\begin{aligned}&\left[\begin{pmatrix} k & 0 \\ 0 & -\lambda \end{pmatrix}\frac{\partial}{\partial x_1} + \begin{pmatrix} 0 & \lambda \\ k & 0 \end{pmatrix}\frac{\partial}{\partial y_1}\right] \\ &\cdot\left[\begin{pmatrix} \frac{1}{k} & 0 \\ 0 & \frac{1}{k} \end{pmatrix}\frac{\partial}{\partial x_1} + \begin{pmatrix} 0 & -1 \\ 1 & 0 \end{pmatrix}\frac{\partial}{\partial y_1}\right]\begin{pmatrix} u_1 \\ v_1 \end{pmatrix} = 0 .\end{aligned} \tag{4.8}$$

If we set

$$\frac{1}{k}\frac{\partial u_1}{\partial x_1} - \frac{\partial v_1}{\partial y_1} = \theta , \qquad \frac{\partial u_1}{\partial y_1} + \frac{1}{k}\frac{\partial v_1}{\partial x_1} = \omega , \tag{4.9}$$

then from (4.8) we obtain

$$k\frac{\partial\theta}{\partial x_1} + \lambda\frac{\partial\omega}{\partial y_1} = 0 , \qquad k\frac{\partial\theta}{\partial y_1} - \lambda\frac{\partial\omega}{\partial x_1} = 0 ,$$

which implies that $\varphi = k\theta - i\lambda\omega$ is an analytic function of the variable $z_1 = x_1 + iy_1$. Putting $f = u_1 + iv_1$, from (4.9) we get

$$\frac{k+1}{2}\frac{\partial f}{\partial \overline{z_1}} - \frac{k-1}{2}\frac{\partial f}{\partial z_1} = \frac{\lambda-k}{4\lambda}\varphi(z_1) + \frac{\lambda+k}{4\lambda}\overline{\varphi(z_1)} . \tag{4.8'}$$

It is not difficult to show that the general solution to equation (4.8′), i.e. the solution to system (4.8) is representable as

$$f(z_1) = \alpha\Phi(z_1) + \beta\overline{\Phi(z_1)} + \Psi\left(\frac{k+1}{2k}z_1 - \frac{k-1}{2k}\overline{z_1}\right) , \tag{4.10}$$

$$\alpha = \frac{\lambda-k}{2(1-k)\lambda} , \qquad \beta = \frac{\lambda+k}{2(1+k)\lambda} , \tag{4.11}$$

where $\Phi(z_1) = \int\limits_{z_0}^{z_1} \varphi(\zeta)\, d\zeta$, and $\Psi(z_2)$ is an arbitrary analytic function of the variable

$$z_2 \ := \ \frac{k+1}{2k}\, z_1 + \frac{k-1}{2k}\, \overline{z_1} \ = \ x_1 + i\frac{y_1}{k} \ .$$

From (4.3) and (4.7) we have

$$x \ = \ \delta x_1 \ , \qquad y \ = \ \frac{1}{\sqrt{k}}\, y_1 \ ; \qquad u \ = \ \frac{1+k\nu}{(k+\nu)\sqrt{k}}\, u_1 \ , \qquad v \ = \ -\delta v_1 \ . \tag{4.12}$$

Applying (4.10) and (4.12) we have

$$\begin{aligned} u \ &= \ \frac{1+k\nu}{(k+\nu)\sqrt{k}}\, u_1 \\ &= \ \frac{1+k\nu}{(k+\nu)\sqrt{k}}\, (\alpha+\beta)\, \mathrm{Re}\,[\Phi(z_1)] + \frac{1+k\nu}{(k+\nu)\sqrt{k}}\, \mathrm{Re}\,[\Psi(z_2)] \ , \end{aligned} \tag{4.13}$$

$$v \ = \ -\delta v_1 \ = \ \mathrm{Re}\,[i\delta(\alpha-\beta)\Phi(z_1)] + \mathrm{Re}\,[i\delta\Psi(z_2)] \ . \tag{4.14}$$

After some computations we can verify

$$\begin{aligned} \frac{1+k\nu}{(k+\nu)\sqrt{k}}\, (\alpha+\beta) \ &= \ \frac{1+k\nu}{(k+\nu)\sqrt{k}}\, \frac{\lambda-k^2}{(1-k^2)\lambda} \ = \ -\frac{k+\nu}{E\sqrt{k}}\, \alpha_3 \ , \\ \frac{1+k\nu}{(k+\nu)\sqrt{k}} \ &= \ -\frac{1+k\nu}{kE\sqrt{k}}\, (k\beta_3) \ , \\ \delta(\alpha-\beta) \ &= \ \delta\frac{\lambda-1}{\lambda}\, \frac{k}{1-k^2} \ = \ -\frac{1}{E_{22}}\, \frac{1+\nu k}{\delta k}\, \alpha_3 \ , \\ \delta \ &= \ -\frac{\nu+k}{E_{22}k\delta}\, (k\beta_3) \ , \end{aligned} \tag{4.15}$$

where

$$\alpha_3 \ = \ \frac{1+k\nu}{1-k^2}\, \frac{E}{1-v^2} \ , \qquad \beta_3 \ = \ -\frac{E}{k+\nu} \ . \tag{4.16}$$

Substituting (4.15) into (4.13) and (4.14), we obtain

$$\begin{aligned} u \ &= \ -\frac{k+\nu}{E\sqrt{k}}\, \mathrm{Re}\,[\alpha_3\Phi(z_1)] - \frac{1+k\nu}{kE\sqrt{k}}\, \mathrm{Re}\,[k\beta_3\Psi(z_2)] \ , \\ v \ &= \ -\frac{1+\nu k}{E_{22}\delta k}\, \mathrm{Re}\,[i\alpha_3\Phi(z_1)] - \frac{\nu+k}{E_{22}\delta k}\, \mathrm{Re}\,[ik\beta_3\Psi(z_2)] \ . \end{aligned} \tag{4.17}$$

Setting

$$p_1 = -\frac{k+\nu}{2E\sqrt{k}} , \qquad p_2 = -\frac{1+k\nu}{2kE\sqrt{k}} ,$$

$$q_1 = -\frac{1+\nu k}{2E_{22}\delta k} i , \qquad q_2 = -\frac{\nu+k}{2E_{22}\delta k} i , \tag{4.18}$$

we finally obtain the general solution

$$u = 2\,\mathrm{Re}\,[p_1\alpha_3\Phi(z_1) + p_2 k\beta_3\Psi(z_2)] ,$$

$$v = 2\,\mathrm{Re}\,[q_1\alpha_3\Phi(z_1) + q_2 k\beta_3\Psi(z_2)] , \tag{4.19}$$

and consequently the boundary conditions for the displacement problem will be given in the form

$$2\,\mathrm{Re}\,[p_1\alpha_3\Phi(t_1) + p_2 k\beta_3\Psi(t_2)] = U(t) ,$$

$$2\,\mathrm{Re}\,[q_1\alpha_3\Phi(t_1) + q_2 k\beta_3\Psi(t_2)] = V(t) , \tag{4.20}$$

where

$$t_1 := \frac{1}{2}\left(\frac{1}{\delta} + \sqrt{k}\right) t + \frac{1}{2}\left(\frac{1}{\delta} - \sqrt{k}\right)\bar{t} , \qquad t_2 := \frac{k+1}{2k} t_1 + \frac{k-1}{2k}\overline{t_1}$$

and $U(t)$ and $V(t)$ are given functions.

We now express the boundary conditions for the stress problem

$$\sigma_x \cos(n,x) + \tau_{xy}\cos(n,y) = X_n ,$$

$$\tau_{xy}\cos(n,x) + \sigma_y \cos(n,y) = Y_n \tag{2.7}$$

noindent by the general solution (4.10). Substituting ν , E , δ , k_1 from (4.2) and (4.4) into (2.2) by replacing ν_{12} , ν_{21} , E_{11} , E_{22} , G_{12} , we get

$$\sigma_x = \frac{E}{1-\nu^2}\left(\delta^2\frac{\partial u}{\partial x} + \nu\frac{\partial v}{\partial y}\right) ,$$

$$\sigma_y = \frac{E}{1-\nu^2}\left(\nu\frac{\partial u}{\partial x} + \delta^{-2}\frac{\partial v}{\partial y}\right) , \tag{4.21}$$

$$\tau_{xy} = \frac{E}{2(k_1+\nu)}\left(\frac{\partial u}{\partial y} + \frac{\partial v}{\partial x}\right) .$$

From (4.12) we have

$$\frac{\partial u}{\partial x} = \frac{1+k\nu}{(k+\nu)\sqrt{k}\,\delta}\frac{\partial u_1}{\partial x_1} , \qquad \frac{\partial v}{\partial x} = -\frac{\partial v_1}{\partial x_1} ,$$

$$\frac{\partial u}{\partial y} = \frac{1+k\nu}{k+\nu}\frac{\partial u_1}{\partial y_1} , \qquad \frac{\partial v}{\partial y} = -\sigma\sqrt{k}\frac{\partial v_1}{\partial y_1} . \tag{4.22}$$

Because

$$z_1 = x_1 + iy_1 = \frac{x}{\delta} + i\sqrt{k}\,y\,, \qquad z_2 = x_1 + \frac{i}{k}y_1 = \frac{x}{\delta} + \frac{i}{\sqrt{k}}y\,, \tag{4.23}$$

it is easy to verify

$$\begin{aligned} \frac{\partial}{\partial x_1} &= \frac{\partial}{\partial z_1} + \frac{\partial}{\partial \overline{z_1}} = \frac{\partial}{\partial z_2} + \frac{\partial}{\partial \overline{z_2}}\,, \\ \frac{\partial}{\partial y_1} &= i\left(\frac{\partial}{\partial z_1} - \frac{\partial}{\partial \overline{z_1}}\right) = \frac{i}{k}\left(\frac{\partial}{\partial z_2} - \frac{\partial}{\partial \overline{z_2}}\right). \end{aligned} \tag{4.24}$$

Introducing $\psi(z_2) := \Psi'(z_2)$ and noticing $\varphi(z_1) = \Phi'(z_1)$, we obtain from (4.10)

$$\begin{aligned} \frac{\partial f}{\partial x_1} &= \alpha\varphi(z_1) + \beta\overline{\varphi(z_1)} + \psi(z_2)\,, \\ \frac{\partial f}{\partial y_1} &= i\big(\alpha\varphi(z_1) - \beta\overline{\varphi(z_1)}\big) + \frac{i}{k}\psi(z_2)\,, \end{aligned} \tag{4.25}$$

and consequently

$$\begin{aligned} \frac{\partial u_1}{\partial x_1} &= (\alpha+\beta)\operatorname{Re}\varphi(z_1) + \operatorname{Re}\psi(z_2)\,, \\ \frac{\partial u_1}{\partial y_1} &= -(\alpha+\beta)\operatorname{Im}\varphi(z_1) - \frac{1}{k}\operatorname{Im}\psi(z_2)\,, \\ \frac{\partial v_1}{\partial x_1} &= (\alpha-\beta)\operatorname{Im}\varphi(z_1) + \operatorname{Im}\psi(z_2)\,, \\ \frac{\partial v_1}{\partial y_1} &= (\alpha-\beta)\operatorname{Re}\varphi(z_1) + \frac{1}{k}\operatorname{Re}\psi(z_2)\,. \end{aligned} \tag{4.26}$$

Substituting (4.22) and (4.26) into (4.21) the representation of the components of stress by means of $\varphi(z_1)$ and $\psi(z_2)$

$$\begin{aligned} \sigma_x &= \alpha_1\operatorname{Re}\varphi(z_1) + \beta_1\operatorname{Re}\psi(z_2)\,, \\ \sigma_y &= \alpha_2\operatorname{Re}\varphi(z_1) + \beta_2\operatorname{Re}\psi(z_2)\,, \\ \tau_{xy} &= \alpha_3\operatorname{Im}\varphi(z_1) + \beta_3\operatorname{Im}\psi(z_2)\,, \end{aligned} \tag{4.27}$$

is obtained, where

$$\alpha_1 = -\delta\sqrt{k}\,\alpha_3\,, \qquad \alpha_2 = \frac{1}{\delta\sqrt{k}}\alpha_3\,, \qquad \beta_1 = -\frac{\delta}{\sqrt{k}}\beta_3\,, \qquad \beta_2 = \frac{\sqrt{k}}{\delta}\beta_3\,. \tag{4.28}$$

Now we simplify the boundary conditions (2.7). Obviously,

$$\cos(n,x) = \frac{dy}{ds}, \qquad \cos(n,y) = -\frac{dx}{ds}.$$

From (4.12) we have

$$\cos(n,x) = \frac{dy}{dy_1}\frac{dy_1}{ds} = \frac{1}{\sqrt{k}}\frac{dy_1}{ds}, \qquad \cos(n,y) = -\frac{dx}{dx_1}\frac{dx_1}{ds} = -\delta\frac{dx_1}{ds}. \tag{4.29}$$

Inserting (4.27) and (4.29) into (2.7), we obtain

$$\begin{aligned} X_n &= [\alpha_1 \operatorname{Re}\varphi(z) + \beta_1 \operatorname{Re}\psi(z_1)]\left(\frac{1}{\sqrt{k}}\frac{dy_1}{ds}\right) \\ &\quad + [\alpha_3 \operatorname{Im}\varphi(z) + \beta_3 \operatorname{Im}\psi(z_1)]\left(-\delta\frac{dx_1}{ds}\right) \\ &= -\left[\delta\sqrt{k}\,\alpha_3 \operatorname{Re}\Phi'(x_1+iy_1) + \frac{\delta}{\sqrt{k}}\beta_3 \operatorname{Re}\Psi'\left(x_1+\frac{iy_1}{k}\right)\right]\frac{1}{\sqrt{k}}\frac{dy_1}{ds} \\ &\quad - \operatorname{Re}\left[i\alpha_3\Phi'(x_1+iy_1) + i\beta_3\Psi'\left(x_1+\frac{iy_1}{k}\right)\right]\left(-\delta\frac{dx_1}{ds}\right), \end{aligned} \tag{4.30}$$

so that

$$\begin{aligned} X_n\,ds &= \operatorname{Re}[i\delta\alpha_3\Phi'(x_1+iy_1)\,dx_1 - \delta\alpha_3\Phi'(x_1+iy_1)\,dy_1] \\ &\quad + \operatorname{Re}\left[i\delta\beta_3\Psi'\left(x_1+\frac{iy_1}{k}\right)dx_1 - \frac{\delta\beta_3}{k}\Psi'\left(x_1+\frac{iy_1}{k}\right)dy_1\right] \\ &= d[\operatorname{Re}(i\delta\alpha_3\Phi(x_1+iy_1)] + d\left[\operatorname{Re}\left(i\delta\beta_3\Psi\left(x_1+\frac{i}{k}y_1\right)\right)\right]. \end{aligned}$$

Integrating along the boundary Γ with respect to the arc length s from some starting point on Γ to a variable point, we get

$$\operatorname{Re}\left[i\alpha_3\Phi(x_1+iy_1) + i\beta_3\Psi\left(x_1+\frac{iy_1}{k}\right)\right] = \int_0^s \frac{X_n\,d\tilde{s}}{\delta} + c_2. \tag{4.31}$$

Similarly, it is not difficult to obtain

$$Y_n\,ds = -d\left[\operatorname{Re}\left(\frac{\alpha_3}{\sqrt{k}}\Phi(x_1+iy_1)\right)\right] - d\left[\operatorname{Re}\left(\sqrt{k}\,\beta_3\Psi\left(x_1+\frac{iy_1}{k}\right)\right)\right],$$

$$\operatorname{Re}\left[\alpha_3\Phi(x_1+iy_1) + k\beta_3\psi\left(x_1+\frac{iy_1}{k}\right)\right] = -\int_0^s \sqrt{k}\,Y_n\,d\tilde{s} + c_1. \tag{4.32}$$

2. Since

$$\Phi(z_1) = \int_0^{z_1} \varphi(\zeta_1)\,d\zeta_1 \,, \qquad \Psi(z_2) = \int_0^{z_2} \psi(\zeta_2)\,d\zeta_2 \,,$$

in the above discussion the functions $\Phi(z_1)$ and $\Psi(z_2)$ in general are not single valued in the multiply-connected domain. This coincides with the mechanical interpretation, too. In fact, assume C is a curve in the domain Ω . If C revolves the boundary curve Γ_k , then the principal vector of the exterior forces along C is

$$(X+iY)\mid_C \;=\; (X+iY)\mid_{\Gamma_k} \;=\; (X_k+iY_k) \,.$$

Thus, when z moves along C once, the functions $\Phi(z_1)$ and $\Psi(z_2)$ shall have increments which can be represented by the logarithm function. For our doubly-connected domain we have

$$\alpha_3\Phi(z_1) = \Psi_1(z_1) + A\log(z_1 - z_{10}) \,,$$

$$k\beta_3\Psi(z_2) = \Psi_2(z_2) + \beta\log(z_2 - z_{20}) \,,$$

where Ψ_1 and Ψ_2 are single-valued analytic functions, and A , B are constants which may be determined by the principal vector of the exterior forces along the boundary. Without loss of generality, we may set $z_{10} = 0$ and $z_{20} = 0$, so that

$$\begin{aligned} \alpha_3\Phi(z_1) &= \Psi_1(z_1) + A\log z_1 \,, \\ k\beta_3\Psi(z_2) &= \Psi_2(z_2) + B\log z_2 \,, \end{aligned} \qquad (4.33)$$

where $z_1 = \dfrac{x}{\delta} + i\sqrt{k}\,y$, $z_2 = \dfrac{x}{\delta} + \dfrac{i}{\sqrt{k}}y$. Substituting (4.33) into (4.31) and (4.32) we obtain the representations of the boundary conditions for the stress problem

$$\begin{aligned} \mathrm{Re}\,[\Psi_1(t_1)+\Psi_2(t_2)] + \mathrm{Re}\,[A\log t_1 + B\log t_2] &= -\int_0^s \sqrt{k}\,Y_n\,d\tilde{s} + c_1 \,, \\ \mathrm{Re}\left[i\Psi_1(t_1) + \frac{i}{k}\Psi_2(t_2)\right] + \mathrm{Re}\left[iA\log t_1 + \frac{i}{k}B\log t_2\right] &= \int_0^s \frac{X_n\,d\tilde{s}}{\delta} + c_2 \,, \end{aligned} \qquad (4.34)$$

and for the displacement problem

$$\begin{aligned} 2\mathrm{Re}\,[p_1\Psi_1(t_1)+p_2\Psi_2(t_2)] + 2\mathrm{Re}\,[p_1A\log t_1 + p_2B\log t_2] &= U(t) \,, \\ 2\mathrm{Re}\,[q_1\Psi_1(t_1)+q_2\Psi_2(t_2)] + 2\mathrm{Re}\,[q_1A\log t_1 + q_2B\log t_2] &= V(t) \,. \end{aligned} \qquad (4.35)$$

3. We now consider the general case when body forces exist. From (2.4) we have

$$\frac{1}{1-\nu_{12}\nu_{21}}\left[\begin{pmatrix} E_{11} & 0 \\ 0 & G_{12}(1-\nu_{12}\nu_{21}) \end{pmatrix}\frac{\partial^2}{\partial x^2}\right.$$

$$+(\nu_{12}E_{22}+G_{12}(1-\nu_{12}\nu_{21}))\begin{pmatrix} 0 & 1 \\ 1 & 0 \end{pmatrix}\frac{\partial^2}{\partial x\partial y} \tag{4.36}$$

$$\left.+\begin{pmatrix} G_{12}(1-\nu_{12}\nu_{21}) & 0 \\ 0 & E_{22} \end{pmatrix}\frac{\partial^2}{\partial y^2}\right]\begin{pmatrix} u \\ v \end{pmatrix} = -\begin{pmatrix} F_1 \\ F_2 \end{pmatrix}.$$

Applying the notations (4.2) and (4.4) and using the transformations (4.3) we get

$$\left[\begin{pmatrix} 1 & 0 \\ 0 & \frac{1-\nu^2}{2(k_1+\nu)} \end{pmatrix}\frac{\partial^2}{\partial x'^2}+\frac{1+2k_1\nu+\nu^2}{2(k_1+\nu)}\begin{pmatrix} 0 & 1 \\ 1 & 0 \end{pmatrix}\frac{\partial^2}{\partial x'\partial y'}\right.$$

$$\left.+\begin{pmatrix} \frac{1-\nu^2}{2(k_1+\nu)} & 0 \\ 0 & 1 \end{pmatrix}\frac{\partial^2}{\partial y'^2}\right]\begin{pmatrix} u' \\ v' \end{pmatrix} = -\frac{1-\nu^2}{E}\begin{pmatrix} F_1 \\ F_2 \end{pmatrix}.$$

Applying (4.6) and another change of coordinates, $x' = x_1$, $y' = \frac{1}{\sqrt{k}}y_1$, we get

$$\left[\begin{pmatrix} 1 & 0 \\ 0 & \frac{1-\nu^2}{2(k_1+\nu)} \end{pmatrix}\frac{\partial^2}{\partial x_1^2}+\frac{1+2k_1\nu+\nu^2}{2(k_1+\nu)}\begin{pmatrix} 0 & \sqrt{k} \\ \sqrt{k} & 0 \end{pmatrix}\frac{\partial^2}{\partial x_1\partial y_1}\right.$$

$$\left.+\begin{pmatrix} \frac{1-\nu^2}{2(k_1+\nu)}k & 0 \\ 0 & k \end{pmatrix}\frac{\partial^2}{\partial y_1^2}\right]\begin{pmatrix} u' \\ v' \end{pmatrix} = -\frac{1-\nu^2}{E}\begin{pmatrix} F_1 \\ F_2 \end{pmatrix}. \tag{4.37}$$

Performing the transformation

$$\begin{pmatrix} u' \\ v' \end{pmatrix} = \begin{pmatrix} \frac{1+k\nu}{(k+\nu)\sqrt{k}} & 0 \\ 0 & -1 \end{pmatrix}\begin{pmatrix} u_1 \\ v_1 \end{pmatrix}$$

and noticing

$$\lambda = \frac{1-\nu^2}{2(k_1+\nu)}k,$$

we have

$$\left[\begin{pmatrix} 1 & 0 \\ 0 & -\frac{\lambda}{k} \end{pmatrix}\frac{\partial^2}{\partial x_1^2}+\begin{pmatrix} 0 & \frac{\lambda}{k}-k \\ -(\lambda-1) & 0 \end{pmatrix}\frac{\partial^2}{\partial x_1\partial y_1}+\begin{pmatrix} \lambda & 0 \\ 0 & -k \end{pmatrix}\frac{\partial^2}{\partial y_1^2}\right]\begin{pmatrix} u_1 \\ v_1 \end{pmatrix}$$

$$= -\frac{1-\nu^2}{E}\begin{pmatrix} \frac{(k+\nu)\sqrt{k}}{1+k\nu}F_1 \\ F_2 \end{pmatrix}, \tag{4.38}$$

i.e.

$$\left[\begin{pmatrix} k & 0 \\ 0 & -\lambda \end{pmatrix}\frac{\partial}{\partial x_1} + \begin{pmatrix} 0 & \lambda \\ k & 0 \end{pmatrix}\frac{\partial}{\partial y_1}\right]\left[\begin{pmatrix} \frac{1}{k} & 0 \\ 0 & \frac{1}{k} \end{pmatrix}\frac{\partial}{\partial x_1} + \begin{pmatrix} 0 & -1 \\ 1 & 0 \end{pmatrix}\frac{\partial}{\partial y_1}\right]\begin{pmatrix} u_1 \\ v_1 \end{pmatrix}$$

$$= -\frac{1-\nu^2}{E}\begin{pmatrix} \frac{(k+\nu)\sqrt{k}}{1+k\nu} F_1 \\ F_2 \end{pmatrix} .$$

Under the notation (4.9) we have

$$\frac{1}{k}\frac{\partial u_1}{\partial x_1} - \frac{\partial v_1}{\partial y_1} = \theta , \qquad \frac{\partial u_1}{\partial y_1} + \frac{1}{k}\frac{\partial v_1}{\partial x_1} = \omega ,$$

$$k\frac{\partial \theta}{\partial x_1} + \lambda\frac{\partial \omega}{\partial y_1} = -\frac{(1-\nu^2)}{E}\frac{(k+\nu)\sqrt{k}}{1+k\nu} F_1 , \qquad k\frac{\partial \theta}{\partial y_1} - \lambda\frac{\partial \omega}{\partial x_1} = -\frac{1-\nu^2}{E} F_2 , \tag{4.39}$$

which may be rewritten in complex form as

$$\left(\frac{\partial}{\partial x_1} + ik\frac{\partial}{\partial y_1}\right)(u_1 + iv_1) = k\theta + ik\omega ,$$

$$\frac{1}{2}\left(\frac{\partial}{\partial x_1} + i\frac{\partial}{\partial y_1}\right)(k\theta - i\lambda\omega) = F , \tag{4.40}$$

where

$$F := -\frac{1-\nu^2}{2E}\left(\frac{k+\nu}{1+k\nu}\sqrt{k}\, F_1 + iF_2\right) . \tag{4.41}$$

Further, noticing $\varphi = k\theta - i\lambda\omega$ and

$$\frac{1}{2}\left(\frac{\partial}{\partial x_1} + ik\frac{\partial}{\partial y_1}\right) = \frac{k+1}{2}\frac{\partial}{\partial \overline{z_1}} - \frac{k-1}{2}\frac{\partial}{\partial z_1} = \frac{\partial}{\partial \overline{z_2}} ,$$

we get

$$\frac{\partial f}{\partial \overline{z_2}} = \frac{\lambda - k}{4\lambda}\varphi(z_1) + \frac{\lambda + k}{4\lambda}\overline{\varphi(z_1)} , \tag{4.42}$$

$$\frac{\partial \varphi}{\partial \overline{z_1}} = F . \tag{4.43}$$

It is clear that (4.43) has the particular solution

$$\varphi_0(z_1) = -\frac{1}{\pi}\int_{G_1}\frac{F(\zeta_1)}{\zeta_1 - z_1}\, d\xi_1\, d\eta_1 , \qquad \zeta_1 = \xi_1 + i\eta_1 , \tag{4.44}$$

where G_1 is the image of Ω under the mapping
$z_1 = \frac{1}{2}\left(\frac{1}{\delta} + \sqrt{k}\right) z + \frac{1}{2}\left(\frac{1}{\delta} - \sqrt{k}\right) \overline{z} =: c(z)$. Since

$$z_2 = \frac{k+1}{2k} z_1 + \frac{k-1}{2k} \overline{z_1} =: b(z_1) ,$$

we have

$$z_1 = \frac{k+1}{2} z_2 + \frac{1-k}{2} \overline{z_2} =: a(z_2) ,$$

and hence

$$\frac{\partial f}{\partial \overline{z_2}} = \frac{\lambda - k}{4\lambda} \varphi(a(z_2)) + \frac{\lambda + k}{4\lambda} \overline{\varphi(a(z_2))} .$$

Thus (4.42) has the particular solution

$$\tilde{f}_0(z_2) = -\frac{1}{\pi} \int_{G_2} \left(\frac{\lambda - k}{4\lambda} \varphi_0(a(\zeta_2)) + \frac{\lambda + k}{4\lambda} \overline{\varphi_0(a(\zeta_2))} \right) \frac{d\xi_2 \, d\eta_2}{\zeta_2 - z_2} . \tag{4.45}$$

Consequently, system (4.36) has the particular solution

$$\begin{aligned} u_0(z) &= \frac{1 + k\nu}{(k+\nu)\sqrt{k}} \operatorname{Re}[f_0(z)] , \\ v_0(z) &= -\delta \operatorname{Im}[f_0(z)] , \\ f_0(z) &= \tilde{f}_0(b(c(z))) . \end{aligned} \tag{4.46}$$

Finally we have with $z_1 = c(z)$, $z_2 = b(z_1)$ the general solution

$$\begin{aligned} u(z) &= 2\mathrm{Re}\,[p_1 \Psi_1(z_1) + p_2 \Psi_2(z_2)] + 2\mathrm{Re}\,[p_1 A \log z_1 + p_2 B \log z_2] \\ &\quad + \frac{1 + k\nu}{(k+\nu)\sqrt{k}} \operatorname{Re}[f_0(z)] , \\ v(z) &= 2\mathrm{Re}\,[q_1 \Psi_1(z_1) + q_2 \Psi_2(z_2)] + 2\mathrm{Re}\,[q_1 A \log z_1 + q_2 B \log z_2] \\ &\quad - \delta \operatorname{Im}[f_0(z)] . \end{aligned} \tag{4.47}$$

§5. Reduction of the contact problem to a boundary value problem for analytic functions

On the base of the results from §4 we are able to reduce the contact problem to boundary value problems for the analytic functions $\Psi_1(z_1)$ and $\Psi_2(z_2)$. Let Γ'_j and Γ''_j be the images of Γ_j $(j = 0, 1, 2)$ under the affine transformations

$$z_1 = c(z) := x_1 + iy_1 = \frac{x}{\delta} + i\sqrt{k}\,y ,$$

$$z_2 = b(z_1) := x_1 + \frac{i}{k}\,y_1 = \frac{x}{\delta} + \frac{i}{\sqrt{k}}\,y ,$$

respectively. Substituting (4.47) into the first eqation of (2.11) we have

$$\begin{aligned} 2\mathrm{Re}\,[p_1^j\Psi_1^j(t_1) + p_2^j\Psi_2^j(t_2)] &= -2\mathrm{Re}\,[p_1^jA^j\log t_1 + p_2^jB^j\log t_2] \\ &\quad -\frac{1+k^j\nu^j}{(k^j+\nu^j)\sqrt{k^j}}\,\mathrm{Re}\,[f_0^j(t)] , \\ 2\mathrm{Re}\,[q_1^j\Psi_1^j(t_1) + q_2^j\Psi_2^j(t_2)] &= -2\mathrm{Re}\,[q_1^jA^j\log t_1 + q_2^jB^j\log t_2] + \delta^j\,\mathrm{Im}\,[f_0(t)] , \end{aligned} \tag{5.1}$$

$$t\in\Gamma_{j1} ,\ t_1\in\Gamma'_{j1} ,\ t_2\in\Gamma''_{j1} ,\ j = 1,2 .$$

Noticing from (2.8), (4.30) and (4.32) on Γ

$$\left\{\left[\cos(n,x)\left(A^{11}\frac{\partial}{\partial x} + A^{12}\frac{\partial}{\partial y}\right) + \cos(n,y)\left(A^{21}\frac{\partial}{\partial x} + A^{22}\frac{\partial}{\partial y}\right)\right]\begin{pmatrix} u-u_0 \\ v-v_0\end{pmatrix}\right\} ds$$

$$= \begin{pmatrix} X_n\,ds \\ Y_n\,ds\end{pmatrix}$$

$$= \begin{pmatrix} d\,\mathrm{Re}\left\{i\delta\alpha_3\Phi(x_1+iy_1) + i\delta\beta_3\Psi\left(x_1+\frac{i}{k}y_1\right)\right\} \\ d\,\mathrm{Re}\left\{-\frac{\alpha_3}{\sqrt{k}}\,\Phi(x_1+iy_1) - \sqrt{k}\,\beta_3\Psi\left(x_1+\frac{i}{k}y_1\right)\right\}\end{pmatrix} ,$$

we can reduce the second equation of (2.11) into the following condition

$$\begin{aligned} \mathrm{Re}\,[\Psi_1^j(t_1) + \Psi_2^j(t_2)] &= -\mathrm{Re}\,[A^j\log t_1 + B^j\log t_2] + c_1 - g_1^j(t) , \\ \mathrm{Re}\left[i\Psi_1^j(t_1) + \frac{i}{k}\Psi_2^j(t_2)\right] &= -\mathrm{Re}\left[iA^j\log t_1 + \frac{i}{k}B^j\log t_2\right] + c_2 - g_2^j(t) , \end{aligned} \tag{5.2}$$

$$t\in\Gamma_{j2} ,\ t_1\in\Gamma'_{j2} ,\ t_2\in\Gamma''_{j2} ,\ j = 1,2 ,$$

where g_1^j , g_2^j are functions completely determined by $u_0^j(t)$ and $v_0^j(t)$.

By means of the general solution (4.47) and from the first equation of the contact condition (2.12) we get

$$
\begin{aligned}
&2\mathrm{Re}\,[(p_1^1\Psi_1^1(t_1) - p_1^2\Psi_1^2(t_1)) + (p_2^1\Psi_2^1(t_2) - p_2^2\Psi_2^2(t_2))] \\
&= -2\mathrm{Re}\,[(p_1^1A^1 - p_1^2A^2)\log t_1 + (p_2^1B^1 - p_2^2B^2)\log t_2] \\
&\quad -\mathrm{Re}\left[\frac{1+k^1\nu^1}{(k^1+\nu^1)\sqrt{k^1}}\,f_0^1(t) - \frac{1+k^2\nu^2}{(k^2+\nu^2)\sqrt{k^2}}\,f_0^2(t)\right] , \\
&2\mathrm{Re}\,[(q_1^1\Psi_1^1(t_1) - q_1^2\Psi_1^2(t_1)) + (q_2^1\Psi_2^1(t_2) - q_2^2\Psi_2^2(t_2))] \\
&= -2\mathrm{Re}\,[(q_1^1A^1 - q_1^2A^2)\log t_1 + (q_2^1B^1 - q_2^2B^2)\log t_2] + \mathrm{Im}\,[\delta^1 f_0^1(t) - \delta^2 f_0^2(t)] , \\
&t\in\Gamma_0 \,,\ t_1\in\Gamma_0' \,,\ t_2\in\Gamma_0'' \,.
\end{aligned}
\tag{5.3}
$$

Noticing (4.34), from the second equation of (2.12) we obtain

$$
\begin{aligned}
&2\mathrm{Re}\left[\left(\frac{1}{\sqrt{k^1}}\,\Psi_1^1(t_1) - \frac{1}{\sqrt{k^2}}\,\Psi_1^2(t_1)\right) + \left(\frac{1}{\sqrt{k^1}}\,\Psi_2^1(t_2) - \frac{1}{\sqrt{k^2}}\,\Psi_2^2(t_2)\right)\right] \\
&= -2\mathrm{Re}\,[(A^1 - A^2)\log t_1 + (B^1 - B^2)\log t_2] + c_1^1 + h_1(t) , \\
&2\mathrm{Re}\left[i(\delta^1\Psi_1^1(t_1) - \delta^2\Psi_1^2(t_1)) + i\left(\frac{\delta^1}{k^1}\,\Psi_2^1(t_2) - \frac{\delta^2}{k^2}\,\Psi_2^2(t_2)\right)\right] \\
&= -2\mathrm{Re}\left[i(A^1 - A^2)\log t_1 + i\left(\frac{B^1}{k^1} - \frac{B^2}{k^2}\right)\log t_2\right] + c_2^1 + h_2(t) , \\
&t\in\Gamma_0 \,,\ t_1\in\Gamma_0' \,,\ t_2\in\Gamma_0'' \,,
\end{aligned}
\tag{5.4}
$$

where $h_1(t)$ and $h_2(t)$ are completely determined by the functions $f_0^1(t)$ and $f_0^2(t)$.

Summarizing the above discussion, we conclude that the mixed-contact problem may be reduced to a boundary value problem for the function $\Psi_1^j(z_1)$, analytic in the domain bounded by Γ_j' and Γ_0' , and the function $\Psi_2^j(z_2)$, analytic in the domain bounded by Γ_j'' and Γ_0'' .

References

[1] SHERMAN, D.I., Plane deformation in isotropic inhomogeneous media. Prikl. Mat. Mekh. 7 (1943), 301-309 (Russian).

[2] RUKHADGE, ZH. A., A mixed boundary-contact problem of the statics of the plane theory of elasticity. Tbiliss. Gos. Univ. Inst. Prikl. Mat. Trudy 10 (1981), 191-216 (Russian).

[3] ZHURA, N.A., On a mixed-contact elliptic boundary value problem. Dokl. Akad. Nauk SSSR 276 (1984), 22-26; English transl. in Soviet Math. Dokl. 29 (1984), 420-424.

[4] ZHURA, N.A., On a boundary value problem in function theory. Dokl. Akad. Nauk SSSR 276 (1984), 777-781; English transl. in Soviet Math. Dokl. 29 (1984), 578-582.

[5] HUA LOO-KENG, LIN WEI, WU CI-QUIAN, Second order systems of partial differential equations in the plane. Research Notes in Math. 128, Pitman Advanced Publishing Program, Boston etc. 1985, 291pp.

[6] GILBERT, R.P., LIN WEI, Function theoretic solutions to problems of orthotropic elasticity. J. Elasticity, 15 (1985), 143-154.

H. BEGEHR
I. Mathematisches Institut
Freie Universität Berlin
Arnimallee 3
1000 Berlin 33
Germany

LIN WEI
Department of Mathematics
Zhongshan University
Guangzhou
China